By Jon Gertner

The Ice at the End of the World

The Idea Factory

THE ICE AT THE END OF THE WORLD

THE ICE AT THE

END OF THE WORLD

An Epic Journey into Greenland's
Buried Past and Our Perilous Future

JON GERTNER

RANDOM HOUSE   NEW YORK

Published in the United States by Random House, an imprint and division of
Penguin Random House LLC, New York.

RANDOM HOUSE and the HOUSE colophon are registered trademarks of
Penguin Random House LLC.

Parts of this book appeared previously, sometimes in different form, in "The Secrets
of the Ice" and "What Could We Lose If a NASA Climate Mission Goes Dark?" in
The New York Times Magazine, and in "An Astronaut Finds Himself in Greenland,"
on newyorker.com.

Library of Congress Cataloging-in-Publication Data

Names: Gertner, Jon, author.
Title: The ice at the end of the world : an epic journey into Greenland's buried past
and our perilous future / by Jon Gertner.
Description: First edition. | New York: Random House, [2019] |
Includes bibliographical references and index.
Identifiers: LCCN 2018039395 | ISBN 9780812996623 (hbk.) |
ISBN 9780812996630 (ebk.)
Subjects: LCSH: Greenland—Discovery and exploration. |
Greenland—Environmental conditions. | Arctic regions—Research.
Classification: LCC G743 .G453 2019 | DDC 559.8/2—dc23
LC record available at lccn.loc.gov/2018039395

Printed in the United States of America on acid-free paper

randomhousebooks.com

9 8 7 6 5 4 3 2 1

First Edition

Title-page photograph: Sarah Das

Book design by Diane Hobbing

Once again, for Liz, Emmy, and Ben

Ice is time solidified.
—Gretel Ehrlich

Contents

Operation IceBridge flight over southeast Greenland (Joseph Macgregor/NASA)

INTRODUCTION

The View from Above

L ate one afternoon in April 2015, I found myself standing on the side of a desolate airport runway in Kangerlussuaq, Greenland, looking west toward an overcast sky. Next to me stood a group of NASA employees, all of them scanning the same gray clouds. There wasn't much small talk. As the temperatures dipped below freezing, we blew on our hands and kept our eyes fixed on the horizon. "We're running late," Luci Crittenden, a NASA flight operations engineer, finally declared. She looked down at her watch, then stomped her feet to keep off the chill.

Not long after, we heard a dull roar in the distance. And soon enough, we could see it coming—a stout-bellied U.S. Army C-130, trailing a plume of black exhaust. "Uh-oh, I think it's on fire," a woman standing next to me, Caitlin Barnes, remarked. I knew this was a joke—sort of. When the aircraft landed a few minutes later, the problem as far as I could tell wasn't an overheated engine, but old age. The plane taxied down the runaway, made a quick hairpin turn, and came to a vibrating halt in front of us, its rotating propellers making a thunderous racket. If a dozen rusted rivets had popped off the fuse-lage, or if a wheel from the landing gear had rolled off, I would not have been surprised. The machine looked positively ancient.

No one said much. But then Jhony Zavaleta, a NASA project manager, yelled over the din: "Nineteen sixty-five!"

Apparently, this was the year the aircraft had been built. I wasn't sure if Zavaleta was amazed by the fact or alarmed. And in some ways it didn't matter now. For the next few weeks, barring any mishap, the C-130 would fly this group around Greenland, six days a week, eight hours a day. It had spent months in the United States being custom-ized for the task. Under the wings, belly, and nose cone were the world's most sophisticated radar, laser, and optical photography instruments—the tools used for NASA "IceBridge" missions, like this one. The agency's strategy was to fly a specially equipped aircraft over the frozen landscapes of the Arctic so that a team of scientists could collect data on the ice below.

The IceBridge program had come into existence at a moment when the world's ice was melting at an astonishing rate. Our plane's instruments would measure how much the glaciers in Greenland had thinned from previous years, but the trend was already becom-ing obvious: On average, nearly 300 billion tons of ice and water were lost from Greenland every year, and the pace appeared to be accelerating. Yet it was also true that hundreds of billions of tons meant almost nothing in the vast expanse of the Arctic. The ice covering Greenland, known as the Greenland ice sheet, is about 1,500 miles long and almost 700 miles wide, comprising an area of 660,000 square miles; it is composed of nearly three quadrillion—that is, 3,000,000,000,000,000—tons of ice. In some places, it runs to a depth of two miles.[1] And so the larger concern, at least as I saw it, was not what was happening in Greenland in 2015, or even what might take place five or ten years hence. It was the idea that something had been set in motion, something immense and catastrophic that could not be easily stopped.

Ice sheet collapse was not a topic of everyday conversations in New York or London. Even if you happened to know some of the more unnerving details—about rapidly retreating glaciers, for instance, or about computer forecasts that suggested the Arctic's future could be calamitous—it was easier to think of the decline in ice as a faraway dripping sound, the white noise of a warming world. Still, by the time I signed on with the IceBridge team, the fate of the world's frozen regions seemed to me perhaps the most crucial scientific and economic question of the age: *The glaciers are going, but how fast?* The ice disappearing from Greenland, along with ice falling off distant glaciers in Antarctica, would inevitably raise sea levels and drown the great coastal cities that a global civilization—living amidst the assumption of steady climates and constant shorelines—had built over the course of centuries. But again, the pressing question: How soon would that world, our world, confront the floods?

Not long after the C-130 landed, we clambered up a staircase and stepped into the passenger cabin. We entered a long, cavernous room with a grime-streaked floor that smelled strongly of engine oil. Electrical cables snaked up the walls. Arctic survival packs swung from a ceiling net. I noticed a crude, freestanding lavatory toward the back of the room; an old drip coffeemaker and microwave oven were secured to a table along the far wall, with Domino Sugar sacks piled high on a pallet underneath. Positioned in the center of the cabin, in front of several rows of seats, were banks of gleaming computer consoles and high-resolution screens. And bolted to the floor was a massive instrument that resembled a cargo container you might attach to the roof of your car to haul gear on a vacation. This was a laser tool—an altimeter—to measure the height of the ice below.

On the outside, the plane looked ready for the scrapyard. On the inside, it was a fortress of technology. I took another look around. The science team was already logging on to the computers, readying themselves for the schedule of flight missions that the IceBridge team would follow this year. The pilots, with fresh crewcuts, were introduced to us as aces recruited from the navy and air force. Then they, too, excused themselves so they could check the instruments in the cockpit and get ready.

The first flight, I was told, would leave the following morning at eight-thirty sharp. "Be here or be left behind," John Sonntag, the mission leader and a native Texan, told me after we walked back out to the tarmac. Sonntag had been deployed to Greenland more times than he could properly recall. He had an easy manner, a friendly grin. But he meant what he said. The work was too important to accept delays. His team had come here, pretty much to the end of the world, to understand how and why trillions of tons of ice were melting into the ocean. They were not acting on the assumption that they would soon find out, but with each flight and each yearly mission, the data piled up: ice lost; water gained. The goal was to gather more and more evidence in the hope that it would ultimately lead us toward an answer—before it was too late.

Kangerlussuaq is located on the west coast of Greenland, just north of the Arctic Circle. For several years, I came in and out of this village with some regularity to connect with flights that were either heading to towns along the western or southern coast of Greenland, or to link up with scientists who were using Kangerlussuaq as a base for field-work on the central ice sheet, which begins about twenty miles east of the settlement. In truth, Kangerlussuaq is more of an airstrip than an actual town, a place to pass through on your way to somewhere else. Though its population hovers around five hundred, the only real hub of activity is the airport café. Caught in the half-light of transit, visitors eat and drink here as they await arrivals or departures. There are no highways out of town—indeed, there are no highways or arterial road systems anywhere in Greenland. In winter, to get to another village, you either fly by propeller plane or—if the ice along the shoreline is thick enough—take a dogsled. In warmer months, you might opt for a ferry that runs up and down the coast.

Sometimes, I would find myself stuck in Kangerlussuaq for a weekend, with no flights coming in or going out, and to pass the time I would walk for hours on empty trails in the neighboring countryside or explore the deserted roads around the airport. The town dates back to World War II, when Greenland—a convenient halfway point between North America and northern Europe—became a key outpost for American military defenses and a pit stop for refueling planes. There are few trees in Greenland, and only a modest amount of greenery, but in the Kangerlussuaq region there are wildflowers and willow bushes and soft sphagnum mosses; there are staggering mountaintop vistas, too, over rocky bluffs and hidden fjords and lakes too numerous to count. There are caribou in abundance, as well as musk oxen and puffy white Arctic hares. The breezes are clean and otherworldly. The first time I landed here and took a deep breath, I stopped to wonder if the air was made from some other substance altogether.

On the morning after our C-130 arrived, we took off on that first IceBridge flight. Our route from the west coast was plotted across the island, due southeast, toward Greenland's rugged eastern coast, where dark, jagged peaks jut up like huge animal teeth from a prehistoric crust of snow-covered ice. It would be a long ride. Greenland is the world's largest island—about five times the size of California, and three times the size of Texas.[2] Just over 80 percent of the land is cov ered by the central ice sheet. Though it's home to a population of about 56,000 people, most of whom are descendants of the native Inuit, this is the least densely populated nation on earth. Only Antarctica, on the opposite end of the globe (and, technically, not a country) is more barren, and only Antarctica has more ice.

After we took off, we scudded through a layer of thick clouds for a half hour. But the sky soon cleared, and the white world below came into crisp resolution. The strategy for these missions is not to fly high but to fly low. Staying steady all day at fifteen hundred feet is ideal. There was agreement on the C-130 that the ice sheet, at least from our height, tended to look like handmade paper, the kind sometimes used for fine stationery, with visible fibers and textured imperfections. But the technicians on the flight spent very little time gazing out at the scenery. With the clearing weather, they began scrutinizing their computer screens, watching sine waves and radar images and the data streaming in about the ice below.

At that point, I made my way through the main cabin toward the front of the plane. From there, I could hop up a short ladder to the flight deck and watch, through large cockpit windows, as the pilots skimmed over Greenland's frozen interior. For three hours we passed above this pale world, until we at last approached the east coast and began trailing the snaking course of big glaciers—wide rivers of ice that flow from the edges of the sheet, down through mountain valleys to the ocean's dark edge, where they collapse and explode into iceberg-strewn chaos. Without exception, what lay below was a sight of uncommon beauty and uncommon strangeness. Taking in the immense expanse of Greenland from low altitude was like surveying the landscape of some kind of frozen exoplanet. The hard blackness of the coastal mountains, the soft whiteness of the ice sheet—the only color intruding on the scenery was the light blue of the sky and a deeper blue from crevasses in the ice that radiated a luminous, aquamarine glow. Down below there were no people, no houses, no movement. For hours on end, there was only ice and rock, ice and rock, ice and rock. In my notebook I wrote: *Someone would think we'd left no traces here at all.*

Many of the places below had names, though. And during the course of the day and those that followed, I could piece together from my aerial view the history of an island where men and women had spent centuries charting an apparently vast emptiness that had turned out to be anything but empty.[3] Along the coasts, Greenland's peninsulas, capes, and glaciers bore the names of explorers who had passed this way on geographical expeditions in the nineteenth and early twentieth centuries—French, British, Danish, Inuit, Norwegian, German, and American. Many of these people were fairly obscure, all of them were now dead. But down below were also reminders of a more recent age of science. As our plane passed the center of the island, we roared over coordinates that marked historical sites from the 1930s and 1950s—scientific outposts in the middle of the ice sheet where large leaps were made in our understanding of the earth. These camps were now invisible, lost beneath decades of accumulating ice and snow, but near to where they once stood I could discern a place that was still functional: Summit Camp, a research station located in the dead center of the ice sheet, sited at an altitude of about ten thousand feet. A cluster of buildings comprised the camp. Down below I saw a few tractors frosted white. Then all signs of civilization fell away, and our plane was again zooming low over the nothingness of the ice sheet.

I had to remind myself that it wasn't actually *nothingness*. I recalled a story from the early 1930s about a German glaciologist named Ernst Sorge who took one of the first flights over Greenland's central ice— "the white desert," as it was sometimes called then—as a passenger in a small airplane. Sorge had already spent a brutal winter in the center of the ice sheet; he had also traversed it many times by dogsled. But the view from above that day was different than what he had so far encountered. It transfixed him. He would later write: "I said to myself, 'I'm looking at a landscape whose vast simplicity is nowhere to be surpassed on earth, and which yet conceals a thousand secrets.'"[4]

Thirty years ago, in his book *Arctic Dreams*, the writer Barry Lopez put forward the notion that "as temperate-zone people, we have long been ill-disposed toward deserts and expanses of tundra and ice. They have been wastelands for us; historically we have not cared at all what happened in them or to them." Lopez predicted, however, that the value of these places would one day "prove to be inestimable."[5] To a fair degree, this book picks up on his observation and asks whether that moment has now arrived.

Lopez's gift to readers was his ability to render the poetic complexity of the north and explain how its fragile lands and wildlife were coming into conflict with fishing and mining industries. Other writers, meanwhile, have focused less on the ecosystems of the Arctic and more on the resilient culture of the Inuit, who came to Greenland about eight hundred years ago by way of the sturdy sea ice that connected the island with northern Canada. These books can be especially compelling in how they challenge Western conventions. To read, say, the French anthropologist Jean Malaurie's description of how in 1951 he watched a northwestern tribe of native men hunt, butcher, and eat a walrus—the blood was drained into a gasoline can and shared between them as *qajoq,* or soup; the eyes were distributed for snacking; the digested food in the intestine (mussels, mostly) was freed and devoured; and the head, ivory tusks, and the heart, weighing seventeen pounds, were awarded to the hunter who led the kill—is to see how different our twenty-first-century lives, and our twenty-first-century sensibilities, really are.[6]

This book is mainly about Greenland's ice sheet—the vast frontier that "conceals a thousand secrets" and is among the most remote and inhospitable places on earth. More specifically, the pages that follow trace the lives of men and women who sought to understand the mysteries above, around, and within that ice. By describing their work, my intention is to tell the story of how we have come to know what we know. Without that knowledge of the ice—without that story—it seems all too easy to believe that our understanding of the Arctic, and of an endangered natural world, is somehow the product of ideology or opinion, rather than the result of hard-won facts and observations.

There are many ways to tell such a tale. The white expanse that covers the center of Greenland began to form about one million years ago when snow that had fallen during the colder months of the year stayed on the ground and endured through summer.[7] Greenland is a semi-arid environment, which means that snow accumulations are modest. Still, decade after decade, enough snow began to pile up that its weight and pressure formed it into ice. Eventually, that ice became hundreds of feet thick—and later still, several miles thick. We might for a moment imagine these Greenlandic beginnings: A windy, rocky, barren stretch beset by deep freezes lasting tens of thousands of years. All the while, snow piles grow slowly but inexorably, year by year, and fuse together into an ice sheet, which in turn starts to thicken, flow, and spread in many directions, thanks to additional snowfalls and the forces of gravity. And then, finally—only very recently in geologic time—humans begin to wander onto the ice to try and uncover the secrets it might contain. This book starts at this recent moment in history.

The late 1800s and early 1900s are what we might call the waning days of the age of exploration. The men of that era discussed in the ensuing chapters—Fridtjof Nansen, Robert Peary, Knud Rasmussen, and Peter Freuchen—pursued geographical knowledge while testing themselves against the mortal dangers of the ice. In the years after these expeditions, the pursuit of adventure was eclipsed by the pursuit of ideas; or, to put it another way, a point arrived when the age of exploration was transformed into the age of discovery. In 1930, a German scientist named Alfred Wegener led a team of men on a hellish journey to the center of the ice sheet to set up a winter research station. Wegener's colleagues, as detailed here, began to use new technologies to cross the ice and investigate it. And in the decades afterward, hundreds of scientists followed in Wegener's footsteps. In the late 1940s, Paul-Émile Victor, a trailblazing Frenchman, was among the first to perceive Greenland not as the goal of an occasional expedition but as the subject of constant scientific inquiry. Not long after, Henri Bader, a Swiss-American scientist, helped the U.S. military understand the Arctic terrain and began to see the potential scientific value hidden within the ice sheet. This work continues—not only in rarified endeavors such as NASA IceBridge flights, but in ongoing field research that brings teams of scientists to remote locations for weeks or months at a time, often drilling into the ice to collect samples known as ice cores, which contain records of ancient climates, or working to gather other precious information that comes at great financial and human costs.

For many of us living far from the Arctic, it seems reasonable to ask how much relevance we might find in these stories, or in the history of glaciology this book brings to light. Another way to ask that question is to consider whether the past, present, and future of Greenland's ice is important to how we live now. The answer is no longer in doubt. It is difficult to grasp the full implications of climate change without understanding the work done on Greenland, seeing as its ice has told us (and is telling us still) secrets that cannot be found anywhere else. Foremost is the matter of sea level rise: Within just a few decades, the melting in Greenland and Antarctica will afflict hundreds of millions of people who live or work near a coastal region, leading us into a terrible epoch of human dislocation and economic hardship. At the same time, the collapsing Greenland glaciers, coupled with a drastic decline in floating sea ice near the North Pole, may lead to other potential disasters, such as a change in Atlantic Ocean circulations and a global disruption in weather patterns. "The melting of the Arctic will come to haunt us all," climate scientist Stefan Rahmstorf declared in the summer of 2017.[8]

But there's something more here, too—the possibility of encountering in this remote area, and in these stories of explorers and scientists, an essential aspect of our world too long neglected. We have no shortage of historical works about twentieth-century wars, politics, and cultural upheavals. This book offers an alternative history, running counter to the mainstream, that is sympathetic to the idea that understanding this Arctic island is not only useful but crucial. In the history of Greenland—"the world's largest laboratory," as the scientist David Holland once described it to me—we can find not only outsized characters who risked their lives in an unforgiving landscape, but a deeper sense of how scientific discovery happens on an epic scale. In light of this, we might ask ourselves what society does with new knowledge once it is found. How do we build upon it? How should we act upon it? And especially in the case of Greenland, how might it help us let go of the antiquated notion—the notion that concerned Barry Lopez—that the fragile, frozen far reaches of the planet are somehow unconnected to our lives?

I've come to see Greenland's ice as an analog for time. It contains the past. It reflects the present. And it seems capable, too, of telling us how much time we might have left.

On those IceBridge flights, the days seemed to last forever. On our first mission, the engines of the old C-130 throbbed and whined as data about the surface below poured in from radar instruments under the wings and zipped at light speed into the computer servers a few seats ahead of me. The information from our journey would be downloaded and eventually posted on the Web for scientists around the world to see. The afternoon progressed. And in my stiff, military-issue chair I began to sink away: In the dream I was having, we were flying above an endless roll of blank paper sprinkled with Domino Sugar crystals. When I awoke, we were arriving at the ice sheet's western border and Greenland's more populous west coast. Soon we passed over the oil tanks and small houses, red and green and blue and yellow, of the coastal villages that held fast to the rocky edge of Greenland, looking westward over Baffin Bay toward Canada, poised on the rim of a silver-white sea of ice. Our flight had followed a rectangular pattern across the center of Greenland, and eight hours after takeoff we were nearly back in Kangerlussuaq, the place where we started.

It is a truism that we take for granted journeys that today are effortless and safe but that not long ago were extraordinary and death defying. Nowhere is this more apt than in Greenland, where travel and science have been utterly transformed by technology. Many months after my tour with IceBridge, after I returned to the United States and began sorting through my notes, it struck me that to truly understand Greenland's ice—to trace it through past, present, and future—I would need to understand this transformation.

It seemed likely to me that I could locate the start of Greenland's scientific evolution by returning to a premodern age, just before aircraft like our C-130 could soar above the ice sheet with ease. To be sure, that meant going back in time—back to the polar explorers' era of headlong risk, back to a period when the reckoning of direction came by way of sun and stars, back to when communications between parties sometimes involved notes stored for months (or years) inside stone markers built on windswept shorelines. It likewise meant going back to Greenland, in person and in archives, to see how the early interrogators of this frozen world, often enduring suffering on an almost unimaginable level, began to uncover its myths and secrets. With that knowledge, I imagined I could start to observe how our understanding of the ice began. I might likewise grasp how the early work in the Arctic, unbeknownst to the first wave of explorers and researchers, had set later events and ideas into motion.

To start, then, it seemed worth considering a risky notion that took hold of two men on opposite sides of the Atlantic at about the same time in the late nineteenth century: As for this great, impassable field of ice in the center of Greenland—why not try to cross it?

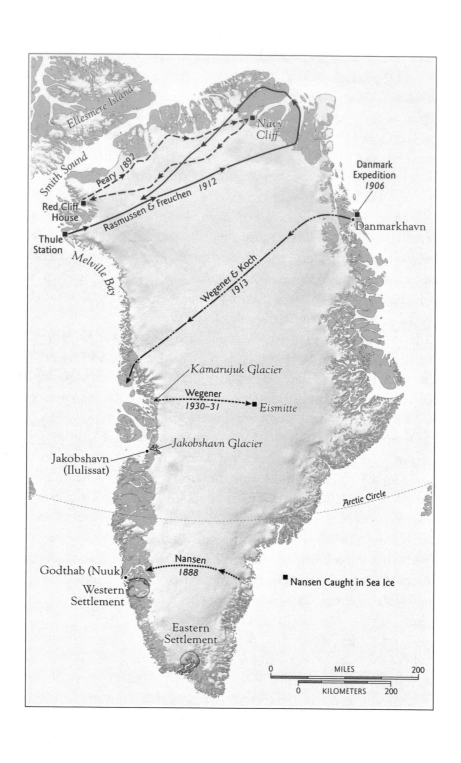

Ellesmere Island

Smith Sound

Navy
Cliff

Danmark
Expedition
1906

Peary 1892

Red Cliff
House

Rasmussen & Freuchen 1912

Danmarkhavn

Thule
Station

Melville Bay

Wegener & Koch
1913

Kamarujuk Glacier

Wegener
1930–31

Eismitte

Jakobshavn Glacier

Jakobshavn
(Ilulissat)

Arctic Circle

Godthab (Nuuk)

Nansen
1888

Nansen Caught in Sea Ice

Western
Settlement

Eastern
Settlement

0   MILES   200

0   KILOMETERS   200

Part I
EXPLORATIONS

(1888–1931)

Fridtjof Nansen (Library of Congress)

1
The Scheme of a Lunatic

On the old maps of the Arctic—the ones drawn by hand, where geographical features were left blank in places where the world was still unexplored—an area along Greenland's southeastern coast was sometimes marked as inaccessible "by reason of floating and fixed mountains of ice."[1] In a region so bitterly cold that winter sometimes lasted ten or eleven months, the dark waters were covered by vast and dangerous frozen fields. During the summer of 1882, a Norwegian sailing vessel called the *Viking* came to this location, only to become entrapped in the icy white sea.

Every morning, a passenger named Fridtjof Nansen would leave his cabin, walk up to the deck, and climb the mast to look out and consider his predicament. Nansen was twenty years old, tall and athletic, with clear gray eyes and a shock of blond hair that often stood straight up. A student in zoology at the university in Oslo,[2] he had signed on for the summer with the *Viking*, a sturdy seal-hunting ship outfitted for icy Arctic waters, to collect marine specimens for his research. From his perch on the mast, though, Nansen could see that the hunting, as well as his studies, would need to be put on hold. All around was the frozen expanse of the Denmark Strait, a body of water that runs between Iceland and Greenland. Jagged white plates, massive and locked tightly together, stretched away for miles in every direction.

Usually, the strategy for a crew locked in pack ice was to wait and hope for good fortune. Wind or sea currents might present a break in the ice and a dark route toward open water—what sailors would call a "lead," which would mobilize them with great urgency to stoke the engines, so the captain could direct the ship as quickly as possible toward safety.[3] If no lead appeared, however, the alternative could be dire. Months could go by. Sometimes even years could pass, in which case food would need to be drastically rationed. And at any time, the icy grip on a ship could tighten.

The literature of the Arctic is rich with descriptions of how it feels to be on a ship fixed helplessly in ice. Sailors often disembark temporarily, to engage in games of strength or skill on the ice floes, or to stage elaborate comedic plays. A ship's officer might wander onto the ice to play his violin under the midnight sun. Frustration, boredom, and insomnia are typical aspects of the entrapment. So is fear. The sound of ice tightening around a ship is said to be so unbearable that it drives some sailors to the edge of psychosis; it pushes against the wooden shell harder and harder, night and day, until pressure bends the planks beyond mere creaking and they begin to scream. In the worst cases, the ship's hull is punctured, seawater rushes in, and eventually the ribs and keel explode into splinters. Nansen was aware of a legendary but true story: In 1777, a century before the *Viking*'s voyage, a dozen whaling ships had all been caught simultaneously in the same east Greenland ice belt that had now snagged his vessel. During that summer, when one ship caught in the ice sank, the sailors would board a neighboring ship or they would make camps on the pack ice—a cruel Arctic variation on musical chairs. The pattern continued until the ice had destroyed all twelve ships, and the men who remained had nowhere to go. They dragged rowboats to open water and headed south, away from the deadly eastern coast. Most tried to row around the bottom tip of Greenland, a distance of several hundred miles, and then up toward the Danish settlements on the island's more temperate western coast. About 150 men survived the journey. But for every sailor who made it through alive, two did not.

On the *Viking*, there was no lead in sight. And as the days passed, Nansen, looking with field glasses from his perch on the mast, began to focus less on the dangers of the surrounding ice—he was not easily, if ever, given to fear—than on the contours of the nearby coast of Greenland. He would later reflect that his curiosity "was drawn irresistibly to the charms and mysteries of this unknown world." By his reckoning, the ship was about twenty-five miles from shore.[4] Nansen asked the captain of the *Viking* if he might be allowed to walk alone over the ice from the ship to Greenland's shore. It was one of Nansen's many peculiarities—the embrace of an idea that seemed so risky as to be almost absurd. The captain denied Nansen's request immediately. But Nansen continued to mull the prospect of the east Greenland coast. From the lookout, he reasoned that the best approach would be to bring a ship close enough to shore, as the *Viking* was now, so a team of men could jump onto the ice, drag a small boat and supplies to shore, and then do some exploring before returning to the ship.

He filed the plan away in his thoughts. And a few days later, when the pack ice opened and the *Viking* sailed away, Nansen became too involved with the ship's regular business to think much more of coastal exploration. Over the course of the *Viking*'s four-month voyage, in addition to collecting birds and fish and small sea mammals to bring home, Nansen estimated that he shot five hundred seals on the ice floes. The fearless zoology student impressed his shipmates by his skill with a rifle, even as he was privately repulsed by the brutality of the seal harvest. The hunt was a literal bloodbath—work that involved shooting and then *flensing* the seal, whereby a crewman would slice the pelt off the animal, capture the rich fat underneath (which was used for the manufacture of soaps and oils), and then leave the gore and flesh on the ice floe as carrion for the birds.

It wasn't until a year later, in 1883, long after he had returned to Norway, that Nansen thought of east Greenland again. A friend was reading the evening newspaper aloud to him. Nansen learned that a Swedish explorer named Erik Nordenskiöld had ventured about a hundred miles into Greenland's interior, the immense expanse covered by massive glaciers and a permanent sheet of ice. Nordenskiöld had long held the belief, not grounded in any set of facts or observations, that at the center of Greenland's ice sheet was an oasis—green and temperate, full of reindeer and vegetation. For Nansen a plan came together at that moment: He would organize an expedition that would cross the ice sheet from coast to coast, a feat that had been contemplated but never completed, and test Nordenskiöld's hypothesis. To be sure, it would be dangerous. No explorer had ever reached the center of the ice sheet, let alone made it across.

His idea had two advantages, though. First, he posited that expedition members could use skis to traverse the ice cap, which would greatly accelerate the crossing. This would be a natural advantage for Nansen, who was among the most accomplished young skiers in Norway. Second, the expedition would not begin near one of the towns on Greenland's western coast, as all other failed expeditions had done. "It struck me that the only sure road to success," he later wrote, "was to force a passage through the floe-belt, land on the desolate and ice-bound east coast, and thence cross over to the inhabited west coast."

Later, he would call his idea "the scheme of a lunatic."[5] But this was not precisely correct. Growing up, Nansen's friends tried to refrain from saying that a particular challenge—a ski jump, a cliff walk—was "not possible," because it would incite Nansen to demonstrate that it was. By nature, he tended to combine logic and forethought with an exceedingly high tolerance for risk. He would eventually become an accomplished neuroscientist, but even as a young man he had an appreciation for incentives that sharpen and direct human behavior. The point of his strategy was that once a team was deposited on the east coast of Greenland they would be unable to survive and unable to turn back; the east was too barren to be sustaining, and the barrier of sea ice along the coast would preclude the possibility of rescue by another ship. So they would have no choice but to cross the inland ice. The plan would prove either lethal or perfect. The motto of the expedition, Nansen said, could be, "Death or the west coast of Greenland."

Probably the earliest recorded mention of Greenland's ice sheet dates back to an Old Norse text written in the early 1200s—a section of dialogue between a king and his son about the island's geography, in which the king declares: "When you asked whether the land was free of ice, or whether it was covered with ice like the sea, then you must know that there is a tiny part of the country which is without ice, but all the rest is covered with it."[6]

The Scandinavian kings knew about Greenland's icy landscape because they had traded with settlers there for hundreds of years. The Greenland colonies were divided into two residential clusters, known as Eastern Settlement and Western Settlement; they had been established beginning in the year 985, after several hundred men and women followed a charismatic warrior named Erik the Red there from Iceland. Banished from Iceland for the crime of murder, Erik the Red apparently chose the name Greenland to make the treacherous journey from Iceland to Greenland seem more appealing to potential followers.[7] His guile paid off. In Greenland, he succeeded in founding what in time became "the westernmost outpost of medieval Christendom."[8] In a region where mild summers were balanced by long, ferocious winters, the Greenland Norse, eventually expanding to a population of perhaps twenty-five hundred inhabitants, set up a network of farms on the lush meadows around the breathtaking inland bays of the island's southwestern coast. There, they raised sheep and cows and supplemented their diet with reindeer and seal meat.

The Norse soon established a shipping route with Iceland and Norway. European traders would regularly visit the Greenlanders, who exchanged the valuable walrus ivory and furs they had reaped from their hunting expeditions (and, on occasion, the twisted tusk of a narwhal, the basis for the mythic unicorn) for fabrics, wood, and iron. By modern measures, and by the standards of medieval Europe as well, the Greenlanders' lives were difficult. Though they enjoyed seasonable summers before the year 1200—after which the climate began to cool, in a period now known as the "Little Ice Age"—the families who lived there were attempting to stake out an existence at a latitude where the practice of agriculture would have been challenging and sometimes impossible. The ruins of these colonies suggest the settlers' deprivations in later years, and their mounting sorrows—crops rotted by early frosts, skeletal sheep, thin soups flavored by seal bones. A visitor today can only try to imagine their isolation, as well as the winter nights of wracking, subzero cold. The Greenlanders' physical distance from Europe constituted thousands of miles of ice-jammed waters during an era when a letter carried by ship might arrive a year or two after a sender had sealed it closed with wax. Still, they remained.[9]

In the late 1300s, the trade routes to Greenland's Norse began to break down.[10] As Scandinavian merchants became more focused on ivory from Russia and Africa, and as increasingly icy and stormy conditions in the North Atlantic made navigation more hazardous, fewer ships made the journey to Greenland.[11] There is little in the way of diaries from this period. We can only surmise from small shards of evidence—bones of animals and humans, mostly, that tell us about what the settlers ate and help scientists discern periods of famine or disruption—about a catastrophe that likely befell this small, stranded civilization. The disappearance of the Norse colonies remains one of European history's supreme mysteries, and theories of its causes abound. The Greenlanders' end could have come from a change in climate that led to mass starvation or a storm that wiped out scores of the colony's best fishermen. Perhaps a deadly skirmish with the Inuit settlers ended things, or a plague that spread rapidly ravaged their ranks. (From archaeological evidence, it seems likely that groups of Inuit settlers arrived on the Greenland coast, via the sea ice connecting northern Canada, around the year 1200.)[12] What we know for certain is that the last ship to sail from Europe to Greenland probably arrived in Greenland in 1406 and returned to Europe in 1410. Also, we have proof that the Greenlandic Norse were alive from a letter, dated 1424, that made reference to a wedding in Greenland sixteen years prior.[13] There was no suggestion that anything at the settlement was amiss. After that, we have nothing.[14]

Even a century after all commerce and communication with the Greenlanders had ceased, Danish kings discussed the colonies with an air of speculation clouded by years of ghostly silence. Could the Norse have survived all this time on the far, forsaken island? Or had they died off entirely? The prevailing belief was that the Greenlanders had indeed survived, and various attempts were made in the late 1500s to visit the island. Some ships that tried to land on Greenland— especially on the frozen east coast—were repelled by weather and the thick pack ice. But those that succeeded in landing encountered the native Inuit and not a trace of the Norse colonies.[15] In 1607 a Danish expedition was sent to find the Greenlanders. Fridtjof Nansen, who would eventually write long, scholarly studies of this period in history, noted that the promoters of the 1607 voyage were so sure of its eventual success "that they even had Icelanders and Norwegians on board, who were there to serve as interpreters when the descendants of the old Norwegian settlers were found."[16] Due to the ice, however, the voyage never made it to shore. Other expeditions were launched later in the 1600s, but those, too, came to nothing. And so in Denmark and in Norway, as the factual details about the Greenland Norse receded and the silence continued, the narrative of the colonies came to resemble myth. In time, the exact locations of the Western and Eastern Settlements were lost.

One day in October 1708, on a remote island in northern Norway, a young Lutheran pastor named Hans Egede went for a walk alone at dusk. Egede, who had just started working in the island's parish, found his mind drifting. On that evening he recalled a book he had read many years earlier, one that included a description of the Greenland settlements—and that in those settlements "there were Christians, churches, and monasteries." He began to ask fishermen in his parish if they knew what had happened to the Greenlanders, but he failed to find more details. He then became obsessed with the question of how they had fared.[17] A biographer of Egede wrote that as the pastor's curiosity grew, he resolved "either to discover the old Norwegian settlements, or to form a new one, and to devote his life to the instruction of the barbarous and uncivilized Greenlanders." His determination must have been remarkable—it took nearly fifteen years of Egede's petitioning before the Danish monarch appointed him as the official missionary to Greenland. (At the time, Norway was part of the Danish kingdom.)[18] With a ship and crew funded by local merchants and the king, Egede departed Europe on May 12, 1721. He arrived safely in southwestern Greenland on July 3 and quickly set up a mission on a coastal island he called the Island of Hope.[19] A few years later the mission moved to the mainland, to a town soon known as Godthab, which in time became the Greenlandic capital of Nuuk.

It was the search for the lost colonies that led to the initial exploration of the ice sheet. Not long after he arrived, Egede, with the help of the local Inuit, encountered signs of the vanished Norse—on the fjords of the coast, he explored the foundations of ruined houses and came upon a roofless stone church.[20] But a common and mistaken belief at the time was that the Eastern Settlement, as its name implied, had been located on Greenland's eastern coast—when in truth the two settlements of the Greenland Norse, Eastern and Western, were located on the southwestern coast, about two hundred miles apart. A few years after his arrival, Egede was asked by his Danish benefactors to cross the center of Greenland, from west to east, and search for the lost Eastern Settlement. Having already glimpsed the glaciers of what was then called "the inland ice," he rejected the idea as too risky.[21] "Nothing was more impossible than this project," he later wrote, "on account of the impracticable, high, and craggy mountains perpetually covered with ice and snow, which never thaws."[22]

Not long after Egede had rejected the idea, however, others thought it worth an attempt. The center of the country was by then understood to be filled by "an appalling tract of ice," yet a trip over the island's middle was considered a reasonable way—perhaps more efficient than sailing *around* the island—to get to the east coast and make contact with the possible descendants of the old Norwegians.[23] This belief led to a failed ice sheet expedition in 1728, and another in 1751. The latter, led by a merchant named Lars Dalager, penetrated about ten miles into the interior. Dalager's group spent a few days on the ice, turned back, and camped at the edge of the sheet, whereupon Dalager, exhausted, drank an entire bottle of Portuguese wine and slept for a day.[24]

In the era before airborne observations, it was only possible to look at the ice sheet by climbing a tall peak on its periphery and, on a clear day, peering toward the horizon, which might give an idea of what was thirty or forty miles ahead. Dalager had barely penetrated the ice sheet; he saw only mountains and more ice ahead. What might be beyond that? More ice? Snowless mountains? A fabled oasis? Dalager conceded that it would be impossible to find out because an explorer would not be able to bring enough provisions for the march across the ice. What's more, the path, from his brief experience, was subject to such severe cold that he doubted "any living creature could exist" within the confines of the inland ice. His words must have had some cautionary effect: Greenland's center was being dismissed by one of the few men who had ever walked within it. In subsequent decades, a few scientists made forays around the edges of the ice sheet.[25] But for a hundred years after that, Fridtjof Nansen would later write, "the interior of Greenland seems to have lost all interest for the world."

One Danish official, stationed in Greenland in the 1850s and 1860s, was an exception. Hinrich Johannes Rink was the first outsider to collect native folktales from the Inuit and create an Inuit-Danish dictionary; he was also the first to try to understand how glaciers flow from the edges of the ice sheet toward the fjords and seas. He considered the ice important—not only as a subject for science, but because it was the source of icebergs that cluttered the North Atlantic and endangered ships. He also perceived that the ice was an intrinsic part of the folklore and tradition of the northern native culture. His book on Greenland, published in English in 1877, noted that "the Greenlanders entertain a sort of superstitious awe regarding the icy interior of their country." The looming, solitary mountains that rise above the ice sheet in some places—*nunataks*, as they're known—"are looked upon as the dwelling place of people who have fled from human society and acquired supernatural senses, quickness, and longevity. Besides these, several monstrous and terrible beings have their haunts upon these lonely hills and roam over the great glaciers."

Rink understood the practical hazards of getting on the inland ice to explore: Venturing in from the shoreline, one would have to ascend the "wall" of the ice sheet and then walk many miles until the ice became "tolerably smooth and level." He had walked on the periphery and observed that in warmer months the ice sheet was pocked with lakes and rivers that emptied into mysterious holes. These rivers, he remarked, move in "beautiful torrents, which rush along their icy beds until they meet a fissure and turn into waterfalls disappearing in the bottomless abyss."[26] He was describing what we now call *moulins*.

Still, in Rink's time, there were perhaps only a few dozen people in the world who had thought deeply about the ice sheet's geological characteristics and mysteries. Nansen would later note that almost as a rule, explorers saw little reason to look inside Greenland as it became increasingly apparent "that the unknown interior in all probability contained no wealth or material treasures."[27] Curiosity wasn't enough. The center was a cold place, a useless place, a dangerous place. The center, as long as it remained encased in ice, was a place that Egede, the eighteenth-century missionary, had proclaimed with confidence would have "no use to mankind."[28]

On June 4, 1888, a brilliant sunny day, Nansen and a team of five other men boarded a sealing ship called the *Jason* in Ísafjörður, a seaside village in the remote reaches of northeast Iceland. Nansen and his men had spent the past several weeks making their way here—a journey that began in Norway and took them to ports in Denmark, Scotland, and the Faroe Islands before approaching the Icelandic coast. Six years had elapsed since Nansen had sat locked in ice aboard the *Viking*, staring out at the dark peaks and ice on Greenland's shore; he was now going back to attempt the ice sheet crossing he had long envisioned.[29]

Nansen was much the same person he'd been as a younger man:

intense, charming, and strung with complex inner tensions. He had developed an unusual talent for striking an easy intimacy with strangers and rapidly pushing a conversation with a new acquaintance beyond mere pleasantries and into debates on heady subjects—recent scientific discoveries, say, or the critical reaction to the newest drama by Henrik Ibsen. One professor who first encountered Nansen in Oslo at about this time recalled: "This man whose name I had never so much as heard until a couple of hours before [we met] had in these few minutes made me feel as though I had known him all my days."[30] His zeal for conversation and camaraderie could suddenly vanish, however. He carried a Scandinavian burden of melancholy. As a teenager, Nansen would sometimes go alone into the wilderness for the weekend and live off the fish he caught and cooked. As an adult, he would ski alone for days, sometimes late into the night on dangerous terrain, testing himself on trails that connected remote Norwegian towns to one another, seeking only a glass of milk from a farmer along the way to regain his energy for the next leg of the trip. It had become a pattern, in fact, for Nansen to follow ebullient bursts of behavior, singing and dancing and conversing late into the night, with brooding periods of solitary retreat into nature. Sometimes it wasn't a physical retreat into the wild but a simple descent into silence—he would sit "without moving" for many hours, a friend would recall—and afterward remain "absolutely mute," again for a span of hours, even when addressed by his companions.[31]

Nansen seemed to understand his contradictions well: hunter and intellectual; athlete and aesthete; extrovert and self-described "lone wolf." A man sometimes inclined to flamboyant displays of masculinity (at dinner Nansen had once convinced a ship's captain to share the heart of a young polar bear he had killed), he was nonetheless an elegant writer who almost always carried a sketchpad with him, so that he could complement his eloquent journal entries with drawings.[32] His writings, done with exacting care during his journeys north, expressed a wondrous appreciation of life that he stitched together with dark, mortal thread. Eventually, Nansen would write of the Arctic: "I found [there] the great adventure of the ice, deep and pure as infinity, the silent, starry night, the depths of Nature herself, the fullness of the mystery of life, the eternal round of the universe and its eternal death."[33]

His plan for the Greenland crossing of 1888 remained similar to the one he had formulated some years before: He and five other men would be dropped off on the ice floes, as close as possible to the east Greenland coast. They would then use two small wooden boats to transport their equipment and provisions over the sea ice and water, and once they made landfall, they would begin their ascent to the ice sheet and head west.[34] The equipment they brought along had involved an immense amount of consideration and engineering. Nansen had experimented for months over what kind of sledges to build (he settled on "a clever and conscientious Norwegian carpenter," who fashioned them out of strong and lightweight ash with steel runners). He experimented with different cookers and fuels, and he tested various wooden skis before settling on ones of oak and birch, measuring seven feet six and a half inches, that were made with a "free-heel" design similar to those used by modern Nordic skiers. He wrestled with the question of whether reindeer hide was a better material for sleeping bags than wool (he decided it was). It wasn't merely performance that Nansen cared about; it was also weight. He had originally seemed partial to the idea that dogs or reindeer should be brought along to haul the sledges, but during the planning stages he saw that the difficulties of bringing—and feeding—animals would be prohibitive. The team, he decided, would pull their sledges over the ice sheet by hand. Thus every pound, even every ounce, mattered.

The men were handpicked by Nansen. During the planning stages, he had received about forty applications from volunteers eager to join him on the journey, but in the end he had settled on three Norwegian men (Otto Sverdrup, Oluf Dietrichson, and Kristian Kristiansen Trana) and two native Laplanders (Samuel Balto and Ole Ravna). All were chosen because of their athleticism and skiing abilities; Sverdrup was also considered valuable because of his experience working as a ship's captain. During the months before the trip, Nansen, Sverdrup, and Dietrichson together made visits to the home of Hinrich Rink, the old Danish-Greenlandic official who had spent decades in Greenland and knew more about the island than anyone in

Copenhagen. Rink gave the men tutorials on the landscape and cul-ture and some basic instructions in the Greenlandic tongue. He also showed them maps and discussed possible routes. The aging Dane quickly grew fond of Nansen, but he worried about his blithe attitude and apparent lack of fear. Rink seemed to think that the trip across the ice would prove far more dangerous than Nansen anticipated.[35]

Just one day after the *Jason* left Iceland with Nansen's team, the ship smashed into chunks of sea ice floating in the water, the first hint of the outer edge of the Greenland ice pack. Nansen would recall that the ship moved south to skirt the ice, then west again, and that over the next few days they encountered herds of seals that put their jour-ney on pause so that the crew could engage in a hunt. On June 11, Nansen caught sight of the mountains of the Greenland coast in the distance, probably some sixty or seventy miles away, but separated from their ship by a thick cordon of sea ice. This was to be expected. And so, in the middle of June, the *Jason* moved near to a cluster of fifteen Scandinavian sealing ships that were stalled in the same waters—often, captains would share dinner or grog together and trade stories as they waited for the ice to open and the weather to improve, so they could proceed with the seal hunt. For the next ten days, in fact, until about June 21, Nansen recalled that his ship "lay knocking about in fog and dirty weather at the edge of the ice, rolling in the swell, and never a seal did we come across."[36]

The days after that weren't much better. The ship moved into fields of pack ice and back out again; seal hunts were launched, but rarely with much success. The *Jason* trembled and lurched as its thick hull knocked constantly against huge hunks of ice bobbing in the water. All the while, however, the ship moved closer to Nansen's goal, and by July 14, Nansen estimated they were about thirty-five miles from an area of east Greenland known as Cape Dan, a coastal island just below the Arctic Circle. And then, on July 17, Nansen awoke and saw from the deck that Cape Dan was only about ten miles away, and that the floes between the ship and the shore looked to be loose and passable. "I saw plainly enough that the landing must be attempted that day," Nansen later recalled. If he didn't act quickly, the ice floes might lock together into a pack and ruin the opportunity.

He dashed off a letter to a Norwegian newspaper that would be brought back to Norway by the ship. "I think the country north of Cape Dan is the very wildest and roughest I have ever seen," he wrote, adding that the area had never been trodden by a European, but that it would be soon. "Behind the mountains we here see for the first time the edge of the 'Inland Ice,' the mysterious desert which in the near future is to be, as far as we can see, our playground for more than a month." Nansen sounded optimistic, and by turns romantic, about the coming journey; he said in the letter that he hoped to be home by autumn.

By now it was seven P.M. With the Arctic sun still high, he and his team rushed to lower their gear into their two small boats. "More confident than ever," Nansen would recall, they made their way down the ladder, pushed off from the *Jason,* and began rowing toward the Greenland shore. "All of us had the most implicit faith in our luck."[37]

The Arctic tends to lure and trap men. Challenges that at first glance seem difficult but surmountable—a short journey by foot over sea ice, for instance, or a brief ascent up a steep hill—somehow, almost invariably, turn ominous. A blue sky blackens, a storm barrels in, and years of careful planning are rendered irrelevant. Diaries of men who nearly died in Greenland, and diaries of men who did in fact perish, are filled with pages of temperature readings so low or wind measurements so high that they seem to defy even the belief of the diarist.

At the start of their journey from the *Jason* to the shore, on the first day, Nansen and his men moved quickly through the leads between the floes, fighting their way to shore through dark waters strewn with chunks of ice. Occasionally they would hit a broad layer of sea ice and be forced to chop an open water path with axes. Other times, faced with a barrier they couldn't overcome, they would simply drag their boats over expansive rafts of ice until they came to open water again.

It began to drizzle and then rain steadily. In the meantime, other obstacles arose. Sea ice is not necessarily a flat, smooth expanse like a skating rink. Often, it contains stretches of hummocks, which are formed when enormous blocks of floating ice smash together to form a ragged landscape of white peaks and valleys. One might imagine the torn-down wreckage of a neighborhood of condemned buildings, pushed into a tight pile and frosted with snow. Nansen and his colleagues had to drag their gear over ice peaks that were sometimes as high as houses, and only when these barriers were overcome could the men reach water again and relaunch their boats.

As difficult as the process was, they nevertheless were making progress. By the hour, Greenland's shore grew closer. "We are so self-confident," Nansen wrote in his journal, "that we already begin to discuss where and when we shall take our boats ashore."[38]

That same night the ice packed together again, and Nansen's boat was pierced by the sharp edge of an ice floe and began taking on water. Immediately, the men pulled the boat onto a floe for repairs—a wooden patch was nailed over the gash. The other men joined them there. As the rain intensified, they all donned their waterproof gear; then the rain began coming down in sheets. At Nansen's suggestion, the six men took a break. They set up their tent on the floe, crawled inside, and went to sleep.

One dilemma of tying your fortune to pack ice is that it moves—sometimes tens of miles per day—and carries whatever is caught within it toward an unknown destination. Nansen awoke after a few hours to find his floe was caught in a larger drift pulling them in the wrong direction. Not only was he now much farther south than he anticipated; he was also twenty miles from the shoreline, which meant he was eleven miles farther out than when he had set out the day before. He wasn't yet daunted. All his team had to do, Nansen reasoned, "is work resolutely across the current, and we must get to shore sooner or later."[39]

At first this resolution seemed to bring results. The men spotted open water, moved their boats into the sea, and made progress rowing to the coast. But the ice packed together again during the day, and to avoid getting crushed, the men once more moved their boats onto the ice. A keg of beer—a parting gift from the *Jason*—seemed to ease their stress; they ate pea soup and coffee, and also snacked on raw horseflesh, which appalled several members of the party, though not Nansen, who enjoyed the meat. (The horse had been purchased in Iceland and brought on the *Jason* for possible use in Greenland—to pull boats and baggage over the ice floes. But when the horse had stopped eating a week before, it had been slaughtered.[40]) Another positive development was the weather: The rain had stopped, the air had warmed, and the sun emerged. That night, after his colleagues had gone into the tent to sleep, Nansen, in the "dreamy, melancholy light," took out his sketchbook to capture the view. He drew "the wild range of jagged peaks" he could see ringing a distant Greenland fjord, along with "the huge expanse of the Inland Ice" that began just inland from the coast and stretched to a broad vanishing point on the western horizon. Before he turned in for the night, Nansen felt a rumbling on their floe that he attributed to a rougher ocean.

In the morning he awoke to find that the floe they had camped upon, originally about seventy yards across, had split into two, the crack having just missed where the men set their tent. Worse, it was rocking violently in the rough surf. "We are being carried seawards with ominous rapidity," Nansen concluded after looking around. By his estimation, they were now thirty miles from the shore. And the farther they drifted, the more intense were the swells and pitch of the ocean.

Almost three days had now gone by. Survival on a sea of ice can often come down to moving adroitly from floe to floe, much like a seal or a polar bear might. As Nansen's ice cake rocked in the waves, it was also smashing and grinding against other floes, so that chunks continued to crack off into the ice-strewn water. In other words, the floe was getting smaller by the hour and likely developing fissures within that would soon shatter it into smaller pieces. The havoc of the situation assaulted the men's senses. It wasn't only the panorama of ice and "boiling water," as Nansen described it; it was the clamor and chaos, too. As they drifted, waves crashed onto their floe, salt spray hissed above and around them, and a wild percussion—pops, thwacks, and booms—issued from huge ice masses that in collision sent pulverized crystals into the air. Nansen spotted a larger platform of ice nearby, a chunk that looked to be large, thick, and flat, and shouted to his team to move their camp and boats over. But after they did, he saw that the current was still moving them out to sea. The coast was receding.

"The breakers seem to be drawing nearer," he wrote in his journal a few hours later, "their roar grows louder, the swell comes rolling in and washes over the ice all around us, and the situation promises before long to be critical." Nansen reasoned they would not survive on an ice floe in the open ocean; it would rock too violently and break apart within hours. That would leave the men to fend for themselves. At the very least, the dream of crossing the Greenland ice sheet would then be over.

Crossing the ice, late summer 1888 (National Library of Norway)

2

Hauling

Eleven days passed before Nansen and his crew made it from the *Jason* to the Greenland shore. On the fourth night, just when their situation had seemed almost hopeless, an ocean current had pulled their ice floe back from the open sea and began moving them toward the coast again. When they landed on the beach several days later, they scrambled out to enjoy the sensation of solid ground under their feet. "We were just like children," Nansen admitted, "and a bit of moss, a stalk of grass, to say nothing of a flower, drew out a whole rush of feelings." They settled in a grassy spot ringed with heather and juniper bushes and sat down to a meal of biscuits, cheese, and jam.

Nansen had promised his men hot chocolate, so he began heating water on his portable stove.

During the week and a half at sea, they had drifted far south of where they wanted to be—far south of the location on the east coast, that is, from which Nansen intended to climb onto the inland ice and begin the long journey to Greenland's west coast. So the day that started with a celebratory feast soon turned to work. At five P.M. the men got back in their boats and, hugging the coastline, began rowing north. By Nansen's estimation, they would need to cover at least 120 miles. This part of Greenland's east shore—the southern region—had been explored just a few years before by a Danish expedition; the land here, as Nansen described it, was mostly low-lying rocks covered in snow, with the tongues of large glaciers sometimes extending out into the water. These glacial tongues could be treacherous. Frequently—and unpredictably—their edges would break and icebergs would fall into the water with a thunderous cracking sound, followed by high waves that could easily tip a small boat.

Nansen had brought with him detailed maps made by the recent Danish expedition and used them to navigate northward. Still, the route was slow and frustrating. "The ice," Nansen would recall, "was closer packed than ever, and inch by inch and foot by foot we had to break our way." On the second day of the trek, after bringing their boats onto shore to rest, Nansen and his men were visited by two Inuit men who approached in kayaks—"the first representatives of these heathen Eskimo of the east coast," Nansen wrote, "of whom we have heard so much." Nansen was charmed by their looks and affability, and impressed by how they handled their boats and harpoons. A few days later, farther north up the coast, he and his men visited an Inuit encampment—a temporary community where the natives lived in large animal-skin tents, each housing four or five different families. Welcomed inside, the men noted that the dim rooms were lit by small lamps that burned seal blubber, making the atmosphere thick with stifling smoke; the natives' faces, many encrusted with dirt, glistened with blubber, too. Nansen's impressions of the men and women there are striking for his reluctance to judge their looks (or their preference for near nudity) by European standards. With an anthropological air, he conceded that his own ways were equally foreign to the Inuit, and "they plainly thought us the most amusing lot of people they had ever seen." He and his men decided to set up a tent for the night near the Inuit camp. In the late morning, they set off once again through ice-choked waters.

They moved north by tiny increments. The days went by—painful, seventeen-hour days, the men often cold and soaked. By August 9, Nansen reckoned "that an ascent of the 'Inland ice' would be fairly easy from any point of the coast along which we were now passing." A day later the men reached a fjord near a small coastal peak known as Kiatak. They camped on a field of smooth gray rocks that slanted down to the water, between the tongues of two glaciers that ran to the sea. When they turned around, they could see in the distance the vast expanse of the ice sheet, a looming ocean of white. They would begin their journey to the west coast from here, Nansen decided. The next morning, his men began to get their equipment into shape—rubbing rust off their skis and off the runners of the sledges that would carry their gear—and taking observations to mark their exact location.

Not all of the difficulties in crossing Greenland's ice sheet are attributable to its brutal temperatures or to its great expanse. A large part of the challenge relates to the dangers of getting an expedition, along with its equipment, onto the surface of the ice sheet itself. The ice sheet is not flat but rather shaped as a kind of gently sloping dome, with its highest altitude in the center of Greenland; in practice, to walk across the ice sheet is tantamount to climbing half of it, and then descending the other half. But adding to that difficulty is the fact that the edges of the ice sheet can be vertical cliffs dozens or hundreds of feet high. Even worse, beyond the steep outer edge the traveler next encounters a dangerous boundary area, sometimes persisting for twenty or thirty miles or more, that is sometimes called the "crevasse zone." The zone varies in width or difficulty from place to place, but on both the west and east coasts of Greenland it is often formidable and marked by a succession of steep ridges or hills. The reasons for this relate to how the ice sheet, as it is pulled by gravity and flows over uneven ground, literally stretches apart at its ends, causing large fissures to open. In nineteenth-century paintings of the crevasse zone, the icescape appears both surreal and terrifying: Men somehow try to make their way over it, dragging sledges and gear through a white barricade of high icy pinnacles and blue, depthless cracks.

Nansen and his colleague Otto Sverdrup conducted a brief scouting mission of the ice sheet's edges on August 11. On that day they came to their first encounter with the crevasses. "Though at first they were narrow and harmless and easily covered in the stride," they soon saw that the cracks in the ice "grew broader and opened a view to depths unfathomable." They could walk around some of these cracks. Others were made passable by bridges of fallen snow that the men could scoot across. "When these bridges were too weak to tread upon," Nansen noted, "we had recourse to a more cautious method, and crawled over flat on our stomachs."[1] What proved to be especially treacherous about the crevasse zone, Nansen would learn, wasn't the largest cracks but the smallest. Many were snow-shrouded and undetectable, so much so that he might be moving along a seemingly smooth surface and plunge down suddenly into a fissure in the ice, saved only by a reflexive urge to outstretch his arms. The fall would leave him up to his armpits in snow, his legs dangling over nothingness.

The team embarked on their crossing a few days later, on August 15, almost one month after they had left the *Jason*. They dragged their boats up from the beach and nestled them, bottom side up, in a protected cleft of rocks; then they surrounded the boats with large stones so they wouldn't be moved about by heavy winds. Before he left, Nansen wrote a quick note about the progress of his expedition so far, sealed it in an empty tin, put the tin inside a bread box he had brought along from the *Jason*, and left it beneath a boat.

Then the men departed, hoping they would never see the east coast of Greenland again.

Six men dragging five heavy sledges. Each of the sledges laden with at least two hundred pounds of gear—tents, sleeping bags, raincoats, skis, snowshoes, cameras, thermometers, barometers, canned food, dried food, tea, sugar, coffee, chocolate, matches, stoves, and fuel. All of the sledges pulled during the night, when the August sun isn't melting the snow atop the ice sheet, transforming it into a soft white bog that makes hauling even more difficult. The skis remain packed up for later; they would be useless at this steep gradient, anyway. Using boots fitted with sharp metal crampons, the men need to hike up—up through the crevasse zone to make their way to what they anticipate be the smooth rising dome of the central ice sheet. The ropes for the sledges burn into their shoulders. The white glare, even at nighttime, when the sun barely slips below the horizon, leaves them nearly snowblind. The frozen hills and valleys seem to ripple on without end. One of Nansen's team will later note that the terrain, in these early days, makes him feel hopeless, as if he is walking upon the waves of an interminable ocean of ice.

On August 17, a storm of wind and driving rain confined the men to their tent for three days. They passed some of the time reading the Bible and books about glaciers, but mostly they lay on their backs, "gazing at the tent roof and listening to the rain splashing overhead and wind tearing and shrieking round the walls."[2] When the rain stopped, Nansen and his men pushed on. According to Nansen's journals, they could usually cover three or four miles per day. By August 23, a week after they began, they had made it beyond the worst of the crevasses; by August 26, they had reached an altitude of around six thousand feet. On good days, it seemed that they moved at a pace of about ten miles; on bad days—days when pulling the sleds through the soft snow cover felt like pulling them through sand—they covered half that distance. When they finally reached the smooth rising dome of the ice sheet they found it easier but still physically punishing. Apart from the treachery of the soft snow that was clinging to the runners of the sledges, the men experienced a constant and unquenchable thirst. In planning the expedition, Nansen had reasoned that each man should fill a flask with snow in the morning and let their body heat melt it to water. But one flask of water wasn't nearly enough to slake their dehydration. At the same time, fuel for their stove was too valuable, and their time too precious, to stop and cook down snow for drinking water. So they would have to wait for the end of each day. And even then it wasn't enough. Surrounded by trillions of tons of frozen water, they awoke thirsty, worked thirsty, slept thirsty.

By the end of August, with hundreds of miles of ice still in front of them, the men on Nansen's team began to question whether their goal was worth such misery. They could not turn back; their motto still bound them—*Death or the west coast of Greenland*. In his accounts of the journey, Nansen seemed to share some of the feelings of hopelessness that began to define their brutal daily routines—"we toiled across an interminable flat desert of snow," he wrote, "one day began and ended like another, and all were characterized by nothing but a wearisome, wearing uniformity which no one who has not experienced the like will easily realize."[3] The men's faces were blistered by the sun and wind; their pulling and slogging continually brought them to a wall of exhaustion after only a few hours; their constant hunger was so intense it even incited one of the men to gobble up his weekly allotment of a half pound of butter immediately after it was rationed.

Meanwhile, as they moved farther up the ice sheet, every night grew colder, and the ice grew firmer. They now began traveling by day.

During these arduous afternoons of thirst and exhaustion, there was time to consider the quest that had brought them here. Five men, seeking adventure and possibly fame, had acceded to Nansen's plan to cross the ice sheet. To some extent these were the goals driving Nansen forward, too. He was crossing Greenland because it had never been done and because he believed he could show the world it *could* be done. The crossing, he would later joke, was "just a little ski trip," but it was a ski trip that could demonstrate humanity's ability to overcome nature's most formidable obstacles.

Yet the trip would have other merits, too. One of Nansen's biographers would later characterize Nansen's belief in the scientific value of crossing Greenland's ice as "humbug," but such a dismissal doesn't accord with the case Nansen put forward in his own writings. Greenland, Nansen pointed out, was still in the midst of an ice age, which allowed visitors a rare glimpse into Europe's geological past—that is, a view of "remnants from the time when the north of the continent and the high regions of the Alps were buried under mantles of ice like that which now forms the great 'Inland ice.'"[4] His argument was based upon the research of a Swiss scientist named Louis Agassiz, who by the mid-1850s had persuaded much of the scientific community in Europe that large tracts of the Northern Hemisphere had in ancient eras been covered by vast sheets of flowing ice, which had since receded. Agassiz's ice-age theory, published in 1841 (and in part built upon the insights of other geologists), replaced the widespread notion that a biblical flood—a flood akin to what prompted Noah to build an ark—had moved massive boulders around the European landscape.[5] In essence, Agassiz forced Europeans to imagine something ancient and secular and true: Long ago, their world had been subject to a deep freeze that sculpted their farm fields and froze out all life. Long ago, in fact, their region of central and northern Europe was much like Greenland.[6]

At the time, the idea was revolutionary. And in the words of one science historian, it mattered not because the recognition of an ancient ice age had suddenly altered Europeans' day-to-day lives, but because it marked "the uneasy passage from medieval to modern ways of thinking about the history of the earth."[7] As a forward-thinking scientist of the late nineteenth century—one who, as it happened, had finished his PhD thesis just days before he left for Greenland— Nansen had embraced Agassiz's insights. But it was likely that Nansen's mentor, Hinrich Rink, had schooled him on Greenland's specific geological importance. Rink was the first to believe that "what the Rosetta Stone was to hieroglyphics, offering the clue that led to the deciphering of the ancient Egyptian writings, the inland Ice-Cap of Greenland" promised to offer for the ice-age era.[8] The ice of present-day Greenland, in other words, could be a key to understanding what had befallen so much of the earth, so long ago. And if Nansen could chart its contours and characteristics, then his trek could be the start of something more important than an adventure. It would be an expedition to lend new and basic insights into geological science—or,

more precisely, into a new subset of geological science known as glaciology.

During his lifetime, Agassiz was lauded for his willingness to do
dangerous fieldwork in the Alps—and for once being lowered, by
rope, deep into a glacial crevasse to observe it from within. Nansen
was learning that extracting the mysteries of a much larger ice sheet
presented a far higher degree of difficulty. His plan had been to cross
the inland ice and end up near the small Danish settlement of Chris-
tianhab on the west coast. But by the end of August, with the trekking still arduous, he began to consider a quicker route that would
take his team farther south, so that they would end up in Godthab,
the settlement founded by Hans Egede in the early 1700s, when he
came to Greenland looking for the lost Norse colonies. As Nansen
trudged up the ice sheet, he turned over the calculations in his mind.
Because he had spent several weeks fighting the sea ice and locating
a good access point to the ice sheet, getting this far had taken him
longer than he had anticipated. Summer would soon be on the wane,
if it wasn't already. "It was now late in the year," he later wrote, "and
the autumn of the 'Inland ice' was not likely to prove a gentle season."[9] Landing farther south might also allow him to catch a ship
bound for Europe before winter locked the harbor up in ice—meaning
he wouldn't need to spend the winter on the west coast of Greenland.
His original plan across the ice sheet would have been about 450
miles; a shorter route bringing them south of their original destination would total about 350 miles. When Nansen brought the idea to
the men, they resoundingly agreed in favor of the plan. Of course they
did. They were exhausted, hungry, miserable. "They seemed to have
already had more than enough of the 'Inland Ice,'" he realized.

The conditions for moving forward had improved slightly by then.
Some of the men tried using snowshoes for traction, with positive
results. And by September 2, as the gradient of the ice sheet eased—it
was now a smooth, very gradual, upward slope—they had all switched
over to skis. Still, Nansen complained in his journal that the cold was
"unusually bad" and that "the going was so unconscionably heavy
that it was only with the exertion of all our strength that we're able to
make any progress at all." One good day of weather and progress—

a cause for hope—was invariably countered by a bad day of driving snow, physical discomfort, slow sledges, and despair. The days wore on. The short Arctic nights—typically a gloomy few hours of twilight— began to lengthen and darken, presenting the men on some evenings with dazzling shows of the northern lights. The ice sheet soon seemed to be leveling out, which suggested that at an altitude of just over eight thousand feet they had reached the central high plateau. Meanwhile, it was growing colder. As they skied, ice crusted their beards so thickly that it became difficult to speak or even open their mouths. Their hair froze to their hats, and frostbite numbed their fingers. Nansen and his three Norwegian colleagues wore coats with hoods that surrounded their faces and came to a sharp peak above their heads, giving them the look of cloistered monks. In photos from these weeks, the men appear as if on some kind of death march; they trudge forward, wearing skis, hauling by rope their heavy sledges, one behind the other, clad in the pointed hoods drawn close around miserable, unshaven faces rimed with ice.

At night, in the tent, the temperature had been ranging between −10 and −20 degrees Fahrenheit. On the night of September 11, Nansen placed a thermometer under his pillow and in the morning took a reading. The instrument's markings, he noted, only went down to −35 degrees Fahrenheit, but the mercury was registering well below that. He surmised the temperature was below −40, perhaps around −50. He later concluded that would make it the lowest temperature ever recorded on earth for that time of year.

The snowstorms that blow in regularly over the Greenland ice sheet— and the powerful winds that accompany them, known by the Inuit as *piteraq*—engulfed Nansen's party in early September. Though on some days the men could make good progress, frequently they would be tent bound, in near whiteout conditions. Sometimes the snow was whipped up by extraordinary gusts and would fly through the crevices of the canvas tent, turning the interior into what Nansen called "a mist of flying snow."[10] For the Greenland crossing, Nansen had personally designed large sleeping bags of reindeer fur—three men to

each bag, so that their body heat would warm one another more efficiently. In the mornings, Nansen related, "when I woke I generally
found my head completely surrounded with ice and rime. This was
inside the sleeping-bag, where the breath had frozen and settled upon
the hair of the reindeer skin." On days of bad weather the men would
stay put and pass a day of cold, crushing boredom. On mornings of
good weather, though, Nansen would rouse himself to start the small
stove for hot water so he could make the men tea or hot chocolate,
which he would serve with biscuits. Inside the tent, the temperature
"was something like –40 Fahr., and the walls, except that on the wind
side, [were] covered with inch-long fringes of hoar-frost."[11]

Usually it was Nansen's duty to manage the food and cooking. Portions were typically weighed out with a scale—just over two pounds a
day for each man—in an effort to make sure they were equal and adequate but not so excessive as to deplete the larder. In addition to
biscuits, the men lived on soup made from peas or lentils, and (as
their main staple) slices of pemmican, a calorie-rich concoction of
dried pulverized meat and berries that they carried with them in
tins.[12] Nansen, apparently suffering from serious tooth decay, would
try to reserve for himself the softer, moldy parts of the pemmican
rather than the tougher pieces. They were easier and less painful to
chew. "The interior of Greenland," he noted, "is certainly not the
place for the fastidious or the epicurean."[13]

On September 12, just after the coldest night of the journey, Nansen wrote in his journal that the ground was "falling well and con-
tinuously." In other words, they were now on the downslope of the ice
sheet, and it looked as though the days of slow slogging might be behind them. Indeed, they soon began making swift progress—not only
on skis but also by outfitting their sledges, much like a boat, with a
square sail made from their canvas tents that harnessed the powerful
wind to push along their gear. Their moods brightened. On September 17, precisely two months to the day after leaving the *Jason*, the
men stopped for a quick meal at one P.M. and were startled to hear
birdsong. Soon enough, a small bird—a snow bunting—flew by. Nansen wrote: "It gave us a friendly greeting from the land we were sure
must now be near." They continued to descend. On September 19,

peering through the mists of a driving snowstorm, Balto, one of the Lapps on the expedition, shouted, "Land ahead!" The men squinted, agreed it was true, and cheered; they paused to celebrate with biscuits, jam, and butter, and then pushed on ahead. As dusk was coming on, while still moving west on skis in the storm, Nansen saw something dark lying in the path ahead. He stopped a few yards in front of it to see what it was, close enough to see that it wasn't an object; rather, it was "a chasm broad enough to swallow comfortably sledges, steersman, and passengers." They had now arrived at the crevasse zone on the opposite side of the ice sheet. They stopped for the night.

In the morning they could see that the crevasses ahead appeared even more formidable than what they had faced a month before—a seemingly endless series of deep cracks running parallel to one another. Farther still, in the distance beyond where the ice ended, they could see a wild and empty country, mountains and lakes and deep gorges. Nansen's problem was twofold: First, he had to get his men and sledges over the crevasses and ridges and make it safely to land; then, when they reached land, there would be another challenge— their route would lead them into a fjord many miles south of Godthab, their destination. They would thus have to figure out how to build a boat, sail it out of the fjord, hug the coast, and make the journey north to the settlement.

Nansen's notes on the final leg of the journey suggest a misery quite different from what he endured on the ice sheet. He was now in an existential predicament. To be sure, the cold was ebbing. With every passing night the men were more comfortable in their tent, which was free of icicles in the morning. But there was still hail and snow and rain. And with land within constant sight—indeed, with their achievement now clear to them as they walked—they saw that it was possible they might glimpse but never reach their final destination.

Over the next few days, the crevasses forced them to advance as if following the strictures of a chessboard. They moved forward, sideways, diagonally, backward—and then forward again to find a strategic passage. It was not a hill they were walking down; once again, it was solid waves of ice, meaning they would climb up and then down and then up again, hauling sledges laden with gear, all the while trying desperately to avoid cracks and chasms. Sometimes they simply could not proceed—when "things became utterly hopeless," as Nansen conceded at one point, and they came upon "a simple network of interlacing fissures, the ice protruding in small square islands from the midst of the blue abysses." Often, such an incidence meant a lost day of backtracking.

Thus the week wore on: up the ridges of ice; down the ridges of ice; over or around the crevasses. For miles and for days, the last leg of the journey stretched on. But on September 24, the men crossed a final steep slope of ice and walked off the ice sheet and onto a bed of gravel. In front of them looked to be many more miles of rocky land that led down to the fjord, all of it difficult terrain they would have to cross. This fact, along with the prospect of having to find a way to get to Godthab—probably ninety miles in distance—did not subtract from their ecstasy. "One thing is at least certain," Nansen wrote, "that we are once more at the sea level, if not exactly at its edge, and are in all probability at the end of our toils and sufferings. A difficulty has been overcome, a difficulty which many, perhaps most, of those qualified to judge have deemed insuperable."[14]

Nansen was wrong about being at the end of his toils. His men needed to build a boat to get to Godthab, and under the instruction of Sverdrup, the former ship's captain, they got to work. The project took several days. They used their tent's canvas floor, stretched over wooden ribs fashioned from willow branches cut from nearby bushes. When they were finished, it looked like half a boat, which is why Nansen called it the "tortoise shell." The boat proved to be seaworthy when tested—as long as every ten minutes the occupants used a cooking pot to bail out the water seeping in. On September 28, Nansen and Sverdrup set out in the tortoise shell for civilization.

Nansen had decided he and Sverdrup would make contact with the Danish officials in Godthab and send a boat back for the other four men, who would camp near the fjord and wait. But Nansen and Sverdrup immediately found that the water in the fjord was too shallow for them to actually use the boat, so they walked along the shore for many miles, towing it instead. It was easier than hauling the sledges through the crevasses of ice—but not by much. Black flies buzzed around the men's heads and bit their faces. The ground was ruinously soft. "Our feet sank deep," Nansen recalled, "and fixed themselves, and wheezed like the piston of an air-pump as we pulled them out again at each step."[15] Several times the men sank up to their hips in mud. Their level of exhaustion was extreme.

The situation improved within forty-eight hours. The water deepened, and the two men got into the tortoise shell. They began rowing furiously, and bailing furiously, heading across the fjord and up the coast. In the evenings, they camped on the shore, where they foraged for berries and used a rifle they had brought to hunt local birds that they would then cook over a fire. Nansen and Sverdrup rested their bodies and exulted in their food and good fortune and the receding memory of what now seemed a strange, frozen sleepwalk. This world, unlike the pale one they had left, was infused with green hues and streaks of dazzling color. One night after dinner, as the moon and stars came out, the two men "sat by [the] sinking fire and talked of the 'Inland ice' as a distant dream." For all the previous discussions about "death or the west coast of Greenland," here they were, bellies full, on a path back to civilization, and very much alive.

Greenland Expedition of 1891–92, with Robert Peary (rear, right) *alongside Jo* (U.S. National Archives)

3

Simple and Easy

On the day Fridtjof Nansen and Otto Sverdrup paddled the tortoise shell into harbor at Godthab, a Danish official named Gustav Baumann walked over to the shoreline to greet the Norwegians, ragged and filthy from their journey. After Nansen identified himself—"My name is Nansen, and we have just come from the interior"—Baumann congratulated him on crossing the ice sheet.[1] All along the western coast of Greenland, they had been waiting for Nansen's arrival.

The first question Nansen asked Baumann was whether there was a ship he could take back to Norway. Baumann informed him that the last boat had left two months before—in fact, it had departed from

Godthab village even before Nansen began his trek across the ice sheet from the east coast. Nansen understood that meant he and his colleagues would be spending the winter in Greenland and returning to Norway in the spring. Yet he wondered if he could at least convey the news of his success to his friends back home. Since Greenland lacked any phone or telegraph connections to the continent, he saw only one possible option: By enlisting the help of several of the best Inuit kayakers, he might be able to get a letter to the *Fox*, a boat that supposedly was soon leaving for Europe from a port in Greenland about three hundred miles to the south.[2] By the narrowest of margins, the kayakers succeeded. This was how a brief sentence about Nansen's successful crossing ("We reached Godthab on October 3, and are all in good health.") was relayed, via the kayakers, to the *Fox*, and eventually to the Danish press, on November 9, 1888.[3] The news arrived in England and America several days later. By that point, the other members of Nansen's team had all been rescued from the southern coast, where they had been left to camp and wait, and brought back to Godthab.

In May 1889, Nansen and his teammates would be greeted with a heroes' welcome in Copenhagen and Oslo, as tens of thousands of admirers packed the harbor docks to hail their return. But until then, as they enjoyed the hospitality of Danish officials in tiny Godthab—hospitality that included cigars, fresh linens, and formal meals with china and silverware—the explorers were cut off from the world's response to their triumph.

During this period, the news of Nansen's achievement reached a young civil engineer living in New York City named Robert Peary. When he heard about the crossing, Peary did not react with surprise or delight; his wife would later recall that he responded "as if he had just seen someone die."[4] To those who knew Peary, the reason was obvious: Nansen had beaten him to become the first man to cross Greenland's ice sheet. Two years before, in the summer of 1886, Peary had traveled to the west coast of Greenland, funded mainly by a five-hundred-dollar loan from his mother, and with a young Danish companion named Christian Maigaard had climbed onto the ice sheet with his sledges. For novice explorers, the two men had done remarkably well: Using snowshoes and skis, and pulling their own sledges, they had penetrated about a hundred miles into the ice sheet. (Nansen had read about their endeavor, but it had not swayed him from his original plan to cross the ice sheet from east to west.) Peary had overcome several harrowing near-death accidents in the crevasse zone and had turned back after two weeks only when his rations ran low. He considered the expedition a prelude to a series of increasingly ambitious polar achievements, and when he returned home to the United States, he laid out in the *Bulletin of the American Geographic Society* a deliberate case for how a complete crossing of the ice sheet (from west to east, and then back again) should be carried out. One of the possible routes that Peary proposed involved crossing over the ice sheet at a latitude that ran across the center of Greenland; another involved a much longer and more northerly route, so that an expedition would traverse the ice by starting on Greenland's northwest coast and end somewhere on the northeast coast, before returning home.[5]

But now Nansen had taken—in Peary's view, had stolen—the central route across the ice sheet. That meant an explorer wishing to distinguish himself from Nansen would need to seek a far more ambitious goal. "I could only fall back upon the other northern route," Peary decided. The uppermost reaches of the ice sheet, still blank spots on the maps of that era, had never been explored; indeed, due to the pack ice that repelled even the most intrepid whaling captains, the northern shores of Greenland had, at least in modern times, never even been *observed*. At the time it was thought possible that if Greenland was not an actual island, then an isthmus of ice-covered land, or perhaps an archipelago of ice-capped islands, might connect its northern reaches with the North Pole. Thus Peary considered the journey "the key to the solution of the Greenland problem." If he could solve the question of whether Greenland was an island, he could likewise help geographers delineate its northern coastline.[6]

In the process, he might also eclipse Nansen's achievement. And as it happened, while the Norwegian was being hailed as a hero in Scandinavia, the importance of his crossing, at least initially, was downplayed in the United States. In December 1888, the *New-York*

Tribune complained that Nansen and his men "endured great hardships, braved great perils, and put forth great exertions: for what?" He and his comrades had only covered a narrow swath of the ice sheet, the *Tribune* pointed out, which meant that the great expanse of ice, this so-called Sahara of the North, was still mostly unknown. "It is evident that their toil has added scarcely anything to the world's knowledge of Greenland, while the difficulties they encountered will probably deter others from seeking further to penetrate the mysteries of that frozen continent."[7]

Peary wrote back to the *Tribune* a couple weeks later to defend Nansen, even though he despised the fact that he had been beaten by him. "Whatever Dr. Nansen may or may not have discovered in minutiae," he wrote, "it appears clear to me that his work possesses much value." Most important in Peary's view was that Nansen had shown that a small group of men, using snowshoes and skis and sledges, could cross a vast frozen wasteland—and that the traditional concept of a successful Arctic exploration requiring large parties and tons of supplies should be discarded. The goal now, Peary asserted, was for someone to tackle the northern route, a journey that would undoubtedly add much more to the sum of human knowledge of this frozen island. In his letter, Peary reiterated the path he had described publicly two years before—he intended to begin on Greenland's west coast, in an area near Whale Sound, and would cross the ice sheet so as to draw in Greenland's mysterious northern border. In essence, he was making a marketing pitch. The key to success, Peary insisted, would be a matter of hiking quickly onto the smooth parts of the inland ice in this northern region, which was almost certainly preferable to the glacial edge in the south of Greenland, which was marked by dangerous ice fields and crevasses. Peary noted his route would likely involve climbing up the ice-covered slopes of coastal mountains and, from the high elevations, moving laterally onto the ice sheet. But in taking this approach, he insisted, the exploration of the ice sheet would be a relatively safe endeavor; in fact, he wrote, "it will be seen how simple and easy it is."[8]

Privately he understood the route he was preparing would likely be far more difficult than what Nansen had attempted.[9] In a memo Peary wrote at the time, he noted that his trek across Greenland would cover nearly six hundred miles of frozen and unexplored territory, as compared to Nansen's trek of about half that length; but in truth his trek would be more like twelve hundred miles, since it would necessitate a return trip back to his starting point. That meant the expedition would need to carry enough supplies, or deposit enough supplies in caches on the ice sheet, to ensure the men could stay alive during the entire ordeal. It was possible they could hunt game on the way, and fish when they got to the northern coast—but that was a gamble, since this was uncharted territory. For now they could only speculate on what they would find where they were going. While interviewing a candidate for his prospective expedition, Peary confided that the trip would prove brutal. "Where the expedition is going we will be as much out of touch with the world as we would on another planet," Peary told him. "Death will be hovering near us always. Some of us more than likely will never return to civilization. I advise you not to go if there is any fear in your heart."[10]

The man signed on anyway. And by the time Peary had finished recruiting in 1891, his North Greenland Expedition consisted of six men—Peary, along with Frederick Cook, Langdon Gibson, Eivind Astrup, John Verhoeff, and Matthew Henson. Informally, it also included Peary's young wife, Josephine—"the pretty, sweet-faced, delicate young matron," as one journalist described her—who planned to accompany her husband and stay at a base camp on the Greenland coast while he traveled across the ice cap.[11]

By February 1891, as Peary's plans and funding were being finalized, it was clear that he wanted the public to believe that something momentous would result from the trip. "Mr. Peary believes that by taking a northeasterly route from Whale Sound he will be able to reach the northernmost point of Greenland," The New York Times reported. "How near the pole this will bring him is not known, and it is not certain that he will be able to go further; but if he finds the sea there covered with fixed ice, he will continue to the north, and may reach the pole."[12] Privately, in letters and memos, Peary said nothing of the pole, which he likely saw as a vanishingly remote possibility. "My object is to locate the northern terminus of Greenland," he wrote to Langdon Gibson, a member of his party.[13]

A few months later, aboard the *Kite*, a three-masted, steam-powered ship outfitted for the icy Arctic seas, Peary began his journal for the Greenland expedition. "At five in the afternoon of June 6, 1891," he wrote, "the *Kite* cast off from the foot of Baltic Street, Brooklyn."[14] The ship sailed under the Brooklyn Bridge, up the East River, and headed northeast along the coast of New England. If things went according to plan, the party would return in a year and a half.

From the vantage point of the twenty-first century, it is easy to conclude that "the Greenland problem" that Peary described in the late 1800s—the challenge of drawing in missing parts of the frozen island's northern border—was not an urgent problem at all. Rather, it was an opportunity for Peary to gain fame, and possibly riches, by following the example of Nansen and others before him. The accomplishments of explorers dominated the media and culture of Peary's era. In the United States and Europe, the public seemed eager to regard almost any Arctic expedition as a compelling and patriotic endeavor, especially journeys that settled hazy questions of remote geographic borders. Indeed, explorers of the north like Peary became great luminaries; their stories, told in detail on the front pages of the country's newspapers, satisfied a mass curiosity for glimpses of the intense internal dramas that seemed to animate so many cold-weather expeditions. "The Victorian reading public," the polar historian Chauncey Loomis later noted, "followed the preparations of an expedition to the North with the same avid attention modern television viewers give to a space flight."[15]

A few newspaper columnists, politicians, and academics asked whether the knowledge gained from these expeditions would actually prove useful to science or to the public good. Was it really worth the risks? But mostly the voices of skepticism were drowned out by the narratives that men like Peary fashioned in publicizing their travels, stories of tough men triumphing over nature and conquering frozen and exotic lands. On occasion, the explorers' justification for their work followed a flimsy, circular logic: The Arctic was worth exploring because the Arctic was worth exploring. (Peary once wrote, "I am after the Pole because it *is* the Pole.") In Peary's own writings and speeches, however, he often expanded on that idea by staunchly defending the "value" of northern exploration, not because it gave "a dollar-for-dollar return for every effort," but because the Arctic was an unknown region that could serve as a focus for our species' curiosity, a place that might satisfy the "irrepressible restlessness of the human animal."[16]

As the *Kite* steamed along the coast of northern Canada in June 1891, Peary—"Bert," rather than Robert, to those who knew him well—would have struck anyone who met him on deck as the pure product of America. In the national tradition of self-creation, he had invented himself out of nothing. Peary's many biographers (and Peary, too, in his letters and diaries) would come to attribute his relentless explorations to an unwavering desire for admiration. He once wrote in a letter: "Remember, mother, I *must* have fame."[17] But Peary's desire for acclaim doesn't adequately account for an inner hunger that during the course of his life pushed him repeatedly to the cusp of death, most notably when he lost eight of his toes in the north (in some accounts, several snapped off his frostbitten feet when a colleague removed his boots after a two-week journey, conducted in −50 degrees Fahrenheit temperatures and near-total darkness, along the coast of northeastern Canada).[18] It seems more plausible that Peary was driven to geographical and endurance extremes not only by what he sought to gain, but by the lonely anonymity he sought to escape.

Peary had spent most of his early years in Portland, Maine. An only child, his father died of pneumonia when he was two years old. His widowed mother did not remarry. As a young boy, he spoke with a lisp, and his mother's insistence that he wear bonnets to protect his fair skin made him a target for bullying. But Peary's athleticism and his willingness to fight when necessary earned him a measure of respect amongst his peers; within a few years, his teenage life came to mirror that of his future rival in northern Europe, Fridtjof Nansen: His love of fishing and hunting, as well as his proximity to Maine's vast wilderness, led him toward the natural world at every turn. By the time he was seventeen, he was taking summer hikes into the mountains that lasted for weeks. Peary and his friends slept on the ground or in barns as they wandered; when they needed money or supplies, they would stop at farms along the way and work the fields before moving farther into the mountains.[19] He thought nothing of walking thirty miles in a day. He would jump without hesitation into a freezing river in mid-winter. Being outdoors elicited within him religious feelings of transcendence. During those years, his diary captures the exultations of a young man's joy at witnessing the distant mountains, the sky brightening at sunrise, and the start of birdsong in morning.

A sinewy, broad-shouldered young man standing a shade below six feet, he moved after high school to Brunswick, Maine, to attend Bowdoin College, and would often hike the twenty-five miles back to Portland to visit his mother.[20] At Bowdoin he focused with ferocious intensity on his civil engineering studies, showing a habit of extreme preparation that would later define his Arctic trips. The young Peary—bright but not brilliant, yet always thorough and persistent—would grind away on a difficult engineering problem for days until he found an answer.[21]

On August 16, 1880, Peary wrote to his mother: "I don't want to live and die without accomplishing anything or without being known beyond a narrow circle of friends. . . . Here I am, twenty-four years old, and what have I done, what am I doing, or what am I in the way of doing? Nothing."[22] He had graduated from Bowdoin a few years earlier and was working in Washington, D.C., as a draftsman for the U.S. Coast and Geodetic Survey. The job, which focused on mapmaking, bored him, so he decided to study on his own for an exam for admission into the U.S. Navy's engineering corps. He passed in October 1881 and was soon commissioned as a naval lieutenant.

For his first few years in the navy he worked on a project to rebuild a pier in Key West, Florida. But then opportunity presented itself to Peary when his supervisor asked him to help oversee a survey in Nicaragua. At the time, the United States was debating how it should finance and build a canal that cut through the isthmus of Central

America. The great question was which route—one through Panama or one through Nicaragua—was better suited. There were cases to be made for each country, and the debate in Washington was fought over the course of several decades in arguments that were by turns political, economic, and technical. While the Panama route ultimately won the support of Theodore Roosevelt and the U.S. Congress for engineering reasons, "the choice was never so clear cut," the historian David McCullough later explained.[23] In fact, in the 1880s, when the potential project involved Peary in several Central American expeditions, it seemed more likely that the canal would go through Nicaragua.

For Peary, the Nicaragua work was often excruciating. But at that point in his life, he had begun to think deeply about the merits of physical endurance, noting in his personal papers that the mind can effect "a direct, conscious painful exertion of the will, saying to the body 'you shall not give up,' 'you must keep on,' 'I will make you.'"[24] He and a team of Nicaraguan men spent months canoeing on remote rivers and hacking with machetes, moving forward a few miles a day, at best, through dense vegetation and uncharted, mosquito-infested swamps. They endured torrential rains, snakes, black flies, and stinging ants. On hot, humid nights, Peary slept under palm trees with "a rubber blanket thrown over a pile of leaves for a bed."[25] As the team charted various routes and made measurements to gauge distance and elevation, their daily toil involved moving ahead incrementally over fallen logs and through soft mudbanks and pools of water. Peary would write that often the water was "to our knees and waists and even necks, [and we were] cutting, lifting, pulling, pushing, swimming."[26]

He did not complain. In fact, he volunteered at one point to do an additional week of arduous fieldwork.[27]

For more than a century, historians have labored to explain Peary's complex and shrouded life, many of them unable to grasp how his choices time and again defied reason and an innate desire for self-preservation. But most seem to agree that his zeal during the Nicaragua assignment pointed to something significant about his character: the first public demonstration of Peary's capacity to endure almost unimaginable misery. It was just after his trip across Nicaragua that he made his first, short visit to Greenland—the 1886 hike onto the ice sheet that brought him a hundred miles in. After returning to the United States, he spent seven more unpleasant months in Nicaragua, again charting a possible canal route through the underbrush. And by the summer of 1888, when the second trip to the tropics was over, Peary was clear about two things: The first was that he should marry, and so he returned north to Washington to wed Josephine Diebitsch, whom he had met at a dance several years before. The second was that his future goals would involve Arctic exploration, and he immediately began planning the northeast crossing of the ice sheet.

As his ship, the *Kite*, sailed toward Greenland in the summer of 1891, he was prepared for almost anything. In the 1880s, while he worked in Washington and Nicaragua, Peary had become an assiduous student of the Arctic, nurturing a dream begun in childhood when he had read thrilling accounts of survival and tragedy by earlier travelers who had sailed ships into the ice-locked waters of the north seeking (and invariably failing) to reach the mysterious top of the world. Ever since, Peary had pored over Arctic books, pamphlets, and heady tracts of scientific observation and conjecture. There was likely not a single account of exploration or fieldwork in the far northern latitudes that Peary hadn't scrutinized; there was probably not a paragraph about weather and hunting grounds and indigenous tribes in west Greenland he hadn't memorized.

But then there came a moment when the *Kite*, after crossing Baffin Bay and sailing up the coast of north Greenland in early July, encountered heavy sea ice. Peary was on deck, standing at the ship's rail, when a jarring impact with an ice block wrenched the ship's wheel from the hands of two men steering it. The ship's heavy iron tiller suddenly swung over and, catching Peary's right leg against a railing, "snapped both bones just above the ankle."[28] This was something he never could have prepared for. In her journal, Peary's wife, Jo, recalled that she rushed over to her husband and saw him looking "pale as death."

"Don't be frightened, dearest," Peary told her. "I have hurt my leg."[29]

*   *   *

The *Kite* landed in northern Greenland two weeks later, in late July 1891. Peary had been cabin-bound for the remainder of the trip, and in such severe pain that the doctor who set his leg had to administer morphine for several days.[30] His injury swept the entire crew into a depression and led some of the members to believe that the mission had failed before even launching. Two scientists accompanying Peary on the *Kite* would later recall: "Several of the party thought it better to abandon the attempt for this year, but to this Lieutenant Peary would not listen."[31] As he lay on his back, relaying orders and opinions, Peary insisted that his injury would heal during the autumn months at a Greenland base camp, and that the leg "will be as good as it ever was" when it came time to cross the ice sheet in early spring. He refused to consider turning back.

The ship's captain agreed to stay the course. And when the *Kite* finally cast anchor near a rocky, flower-strewn hillside alongside McCormick Bay, a small coastal inlet off Smith Sound, Peary was strapped to a plank, lowered into a rowboat, and transferred to shore. The ship's cargo had already been unloaded by that point—food, lumber, rifles, navigational tools, cameras, and enough coffee, milk, tea, and sugar to last several years.[32] What's more, Peary's expedition members had already begun construction on a twelve-by-twenty-one-foot cabin, with eight-foot ceilings, that soon became known as Red Cliff House, based on its proximity to some red sandstone cliffs nearby. They would be living here for many months, through fall and winter, as they prepared for the journey in late spring. From a tent that he shared with Jo near the construction site, Peary did his best to supervise his five-member team—"the boys," he called them. Meanwhile, the *Kite* sailed off, leaving the party alone on the vast, empty, rocky northwest coast of Greenland.

Peary was able to get around on crutches by early August. By that point, the house was complete and the entire expedition team of seven people—six men and Jo—had moved in. Red Cliff House was divided into two main spaces. Peary and Jo shared a tiny bedroom, about seven by twelve feet, while the rest of the men occupied bunk beds in a larger common room furnished with a dining table and chairs. A stove was placed in the partition wall between the two rooms. From the outside, the cottage looked grim: Photographs taken by Peary show a small wood shack, covered with tar-paper, surrounded by barrels and crates full of provisions and gear. The interior of the house, meanwhile, was covered in red wool blankets, for insulation purposes mainly, but Jo noted that the blankets imparted "a warm feeling to the interior," and relieved "what would otherwise be a cheerless expanse of boards and tar paper."[33] Without question, Red Cliff House was better than spending the fall and winter in a tent. In time, Jo Peary—charmed by the house's breathtaking location, nestled between fifteen-hundred-foot cliffs on the one side and the iceberg-strewn bay on the other—came to call it "our 'cottage by the sea.'"

In the months of preparation leading up to the trip, Peary had decided that he would improve upon Nansen's feat not only by taking a longer, more ambitious route across Greenland's ice sheet. His methods of travel would be distinctive, too. Later in life Peary would refer to these techniques as the "Peary System"—a multipart approach to Arctic travel that relied on appropriating the methods of the Inuit (fur boots, igloos, and sledge dogs, for instance) and applying them to his own excursions.[34] "It is only reasonable to assume that these people, having lived for generations under the severe conditions of the Arctic lands, have evolved the best methods of meeting the needs of their daily existence," he once explained to a newspaper reporter. "Everything they eat or wear, and everything in the way of sledges or other equipments which they possess, they have proved by long experience to be the most suitable articles of their respective kinds."[35] As they settled in at Red Cliff House, Peary and his men made contact with the local Inuit tribe of northern Greenland, often referred to at the time as the Smith Sound Eskimo or the Polar Eskimo.[36] For centuries, the Smith Sound tribe had been isolated from Greenland's other Inuit, who lived hundreds of miles away, on the island's eastern and more populous southwestern coasts.

By Peary's count, the Smith Sound Inuit now numbered 253 men, women, and children. For perhaps seven hundred years, they had en-dured on the brutal, frozen shores of north Greenland, a small itiner-ant society that moved from hunting ground to hunting ground depending on the season; they had in the process established elabo-rate customs and a rich oral tradition of folklore and knowledge, all while maintaining a society that did not believe in land ownership, currency, written language, or organized leadership. At the urging of the American visitors, nearly a dozen of the Smith Sound Inuit came to live in close proximity to Red Cliff House, where they were em-ployed, beginning in early autumn, sewing reindeer skins into cloth-ing and sleeping bags, and fashioning animal skins into boots called *kamiks*. The Americans would use these on their ice sheet expedition. The native men—led by a hunter named Ikwa—helped Peary's men hunt for deer and walrus to build up their food reserves. Over the course of a few months, they killed dozens of each.

Josephine Peary would eventually warm to the Eskimo men and women, but her first reaction was ugly and ignorant. "These Eskimos were the queerest, dirtiest-looking individuals I had ever seen," she wrote. "Clad entirely in furs, they reminded me more of monkeys than of human beings."[37] It seemed lost on her that they could be the key to her husband's success.

Peary was the kind of man who would reach into a barrel of biscuits, throw a handful into the air, and then laugh as his Inuit friends scur-ried to pick them up from the floor and eat them.

On the one hand, he was an admirer of the Polar Eskimo, and many of the natives, in turn, would come to revere him—*Piulerssuaq*, the Inuit called him: the Great Peary—for his iron grip of command. Publicly, he would defend his relationship to the tribe, describing it as patriarchal and noting his gifts of guns and food. ("They are a com-munity of children in their simplicity," he wrote.)[38] Yet Peary was ruthless in his exploitation of these families—ruthless not only in what he sought in service to his goals of exploration, but in what he sought for his personal appetites and his imperial desire for control.[39] In addition to his two children with his wife, Jo, he would eventually father two sons during his years in Greenland. Born to a young Inuit woman named Allakasingwah—"Ally"—who was perhaps as young as thirteen when she and Peary began their liaisons, these two boys were kept secret from the American press (though not from Jo) and were eventually ignored by him entirely. Sometimes, Peary would demand that Ally's Inuit husband go on errands requiring weeks away from home so he could woo her.

"With habits and conditions of life hardly above the animal, these people seem at first to be very near the bottom of the scale of civilization," he would write a few years after meeting the Polar Inuit. "Yet closer acquaintance shows them to be quick, intelligent, ingenious, and thoroughly human."[40] Most of all Peary wanted—needed, really—the Inuit's cold-weather skills and commodities, and especially their dogs, which would pull his sledges across the ice sheet. The Greenland dog, bred by natives over the course of centuries, was the most fundamental and powerful technology of polar travel. Wolflike, ferocious, and with a strong resemblance to other northern canines such as the Siberian husky or Alaskan malamute, the breed was seemingly immune to cold and adversity. In the hands of a good driver, the dogs demonstrated remarkable pulling strength, and their extraordinary power belied their compact stature and modest weight.[41] The stamina of the dogs was equally astounding. Eivind Astrup, one of Peary's companions on the Northwest Greenland Expedition, later wrote that "the fatigue and privation that they can endure really borders upon the incredible."[42] Typically, the Greenland dogs were harnessed in straps made of sealskin or bearskin, with each leash, or "trace," leading back to the driver. On his crossing, Peary intended to use a small sledge (holding about two hundred pounds of gear) and a larger, longer sledge (holding about eight hundred pounds).[43] This arrangement would result in the dogs—in teams as small as two or as large as twelve, depending on the load they were pulling—fanning out broadly in front of the sledge, sometimes with one alpha dog, tethered with a longer leash, leading the pack. The driver would use a free hand with the whip, which was about twenty-five feet long. On good ice, and in cold and clear conditions, the sledge would slide along at a brisk clip.[44]

With his leg almost healed—Peary could limp around without crutches by October—he began seeking out the best dogs amongst the locals. He later wrote, "As a result of a systematic series of interviews with the natives who came to Red Cliff, I had, when the spring of 1892 dawned upon us, in my possession information as to the location and ownership of probably every dog in the tribe." He also had calculated how much he would need to offer—usually a knife, some wood, a "trifling present," or in rare instances a gun—to secure a deal with the owner. In all, Peary traded for twenty Greenland dogs that would pull his sledges and supplies in the spring. He was already finalizing the details for his departure a few months hence. His round-trip route still looked to be about twelve hundred miles, and his strategy was not a secret. Dogs need to ingest a tremendous number of calories to pull sleds—far more than men require to be carried along on the ice.[45] He would need to bring a considerable amount of walrus meat for the dogs. But when the stocks ran out, he planned to slaughter the dogs one by one and feed them to each other.[46]

In more severe circumstances, he and his colleagues would eat the dogs themselves.

Peary on Navy Cliff, July 4, 1892 (U.S. National Archives)

4

North by Northeast

The members of the North Greenland Expedition watched the sun dip below the horizon on October 26—the start of the long Arctic winter, and the long Arctic night. The sun would not rise again until February 13. Some years later, Peary would acknowledge the "great night" as the most difficult part of his Arctic travels, a string of woeful months when he "had the blues repeatedly":

> Nine out of every ten people the first time they meet me ask, "How did you stand the cold?" As a matter of fact, the cold of the Arctic regions to a well man, properly fed and

properly clothed, is no more serious than is the cold of our own winters to us here. But the darkness, the months-long winter night! That is different.[1]

Outside Red Cliff house, the men spent their afternoons building and testing sledges in the gloom. The Inuit women, living in igloos nearby and working by lamps burning seal fat, prepared sleeping bags and clothing for the expedition party, a process that involved chewing deerskins for hours to make them pliable before sewing. The process was so difficult that the women could only chew two deerskins per day and needed to follow each work session with a day off, so as to rest their jaws.[2] Jo Peary, meanwhile, had spent late summer setting up the household and venturing out with various hunting parties, rifle in hand, to stalk and kill walrus and reindeer. She now passed her hours with less excitement and less enthusiasm. "My daily routine is always the same," she wrote in her journal in mid-November. "I take my coffee in bed, then get lunch for my family, take a walk afterward, usually with Mr. Peary, then sew or read, and at four o'clock begin to get dinner."[3] With help from Matthew Henson, Peary's personal assistant, she cooked meals for the group on three small oil-burning stoves.[4]

The boys spent their evenings reading books about the Arctic that Peary had brought with him from Brooklyn. His small library—several dozen volumes in all—was organized on a wooden bookshelf in the main room. In Peary's view, his men were looking to these books for good stories and "useful hints" that could help them on their upcoming journey across the ice sheet. Peary often would tell journalists, as well as potential funders of his campaigns, that the dangers of the Arctic were exaggerated, and that its hazards could be avoided with expansive reading and careful preparation. Almost certainly he knew this was not the case. In the Arctic, things often went wrong; one might even say that in the Arctic, things always went wrong. The books he brought to Greenland reinforced the notion that his expedition was spending the winter of 1891 in one of the world's most perilous locations. As they stayed awake reading at Red Cliff House, the men studied tales of Americans who just a few decades prior to their arrival had been brutalized by loneliness, starvation, frostbite, and failure.

Many of these Arctic trials had occurred just a short journey from where Peary's team read by lamplight. Sixty miles from Red Cliff House, for instance, in the fall of 1853, a ship called the *Advance* had become locked in ice off the Greenland coast. Led by a frail young doctor from Philadelphia named Elisha Kent Kane, the crew had journeyed there for two reasons: to find a path to the North Pole, and to seek out survivors of an ill-fated Arctic voyage, known as the Franklin Expedition, which had not been heard from since July 1845. In the winter of 1853 and spring of 1854, Kane and his men did not find a path to the North Pole, nor did they find any trace of the Franklin Expedition. The following summer the men waited for the sea ice around the *Advance* to break up so they could move on. But that summer, the ice did not break up. It stayed fast and frozen, and the ship remained trapped. In his journal Kane wrote: "I inspected the ice again to-day . . . It is *horrible*—yes that is the word—to look forward to another year of disease and darkness to be met without fresh food and without fuel."[5] Two weeks later he called his men together to tell them the news: The brig could not escape the ice. "There is no possibility of our release."[6]

His crew was beset by hunger and weak from scurvy; moreover, they were panicked by their entrapment. Kane's leadership made the situation even worse. He was domineering and often resentful when challenged, leading to fights and festering divisions. And soon enough, some of the men began to discuss mutiny. Running low on coal, the group started to burn the planks and ropes of their ship to keep warm; at the same time, their food supplies grew thin. Kane supplemented his meager diet by eating rats that had infested the stranded ship. Other members of his party, rather than engage in all-out mutiny, split from Kane to set up camp on the Greenland coast. These men ate moss to survive. Only the assistance of the local Smith Sound Inuit, who traded the men seal meat and walrus, kept the party from starving to death.[7]

By May 1855, two years after becoming stranded, Kane's crew

could bear no more. Leaving the ice-locked *Advance*, they began a desperate eighty-four-day trek, which took them hundreds of miles south along the Greenland coast. They were finally rescued by a Danish ship near the Greenlandic village of Upernavik. When he returned to the United States, Kane wrote a bestselling book about the experience, *Arctic Explorations*, which made a permanent impression upon Robert Peary, then a young boy growing up in Maine. The Arctic historian Pierre Berton would later note that in Kane's published account, "the bitterness and paranoia that marked those nightmare months vanished into the background."[8] As a result, the actual, darker lessons of Kane's journey—poor preparation, poor leadership, group discord, and the deathly grip of Arctic weather—were lost on almost everyone who read the book. Kane's seemingly heroic exploits, meanwhile, bestowed upon him fame comparable to that of the greatest politicians, stage actors, and business leaders of the age. When he died in 1857, he was mourned so deeply by the American public that his funeral train attracted audiences surpassed in size only by those who came to see President Lincoln's cortege a few years later.

In the annals of Arctic exploration, however, Kane and his colleagues were actually some of the luckier ones. Charles Francis Hall's ordeal occurred fifty miles from Peary's camp at Red Cliff House. Hall was a portly, bearded newspaper publisher from Cincinnati who initially became interested in the Arctic following the disappearance of the Franklin Expedition in the mid-1840s. In 1860 and 1864 he booked passage on whaling ships so that he could learn the customs and language of the native tribes of northern Canada.[9] By 1870 he had moved on to the idea of reaching the North Pole, and with the help of some funding from the U.S. Congress, he chartered a ship, the *Polaris*, that left New York City in July 1871. In September, near a location on the coast of north Greenland that Hall christened Thank God Harbor, the *Polaris* anchored for the winter. The crew was on edge, with many of the officers no longer hiding their loathing for Hall, who seemed to lack the skills and qualifications to lead a large Arctic expedition. Though well aware of the resentment from the ship's officers, Hall was steadfast about his desire to the reach the North Pole from the ship's anchorage. In October he dogsledded north from the *Polaris* on a two-week scouting mission to plot his spring route to the pole, and when he returned to the ship from the mission he drank a cup of coffee. He then grew increasingly ill. His stomach burned, he complained; he saw blue smoke pouring from the mouth of a colleague he mistrusted. In a hallucinatory state, the historian Berton notes, "He accused almost everyone of trying to murder him." A week later, Hall died, and nearly a hundred years after his death, an analysis of his corpse, still partly preserved in the Greenland permafrost, proved he was a victim of arsenic poisoning.[10]

Hall's death—or murder—made things worse for the factionalized *Polaris* crew, not better. All sense of order collapsed. The ship freed itself from the ice in August, nine months after Hall's death, but an encounter with rough seas in October left the *Polaris* taking on water. In desperation, the captain anchored to an ice floe, but in the chaos of another powerful storm, the crew, fearful of the ship being sunk, dumped food and supplies on the ice. Part of the crew—nineteen in all, including four Inuit men and women and five Inuit children—joined the stock of supplies on the ice near the ship. And then suddenly, unexpectedly, the ship broke free from the ice floe and drifted out of range. Over the course of the next five months, these nineteen men, women, and children floated for approximately two thousand miles on a block of ice that measured about a mile across. Usually at the brink of starvation (often, too, on the brink of murdering one another) they survived mainly because two Inuit men, Hans Hendrick and Joe Ebierbing, succeeded in hunting seals from the floe. The *Polaris* exiles, emaciated and freezing, were rescued in April 1873 by a whaling ship near the coast of Labrador.[11]

The stories of Kane and Hall were the stories that shaped the contours of Peary's imagination; the sites where these men had anchored their ships on the north Greenland coast were landmarks of near-mystical significance that Peary would visit during trips to the region, as if on a pilgrimage, so as to gather rusted souvenirs (ships' fittings, bolts, hooks, and hinges) to bring home.[12] Above all, the stories of Kane and Hall contained details that Peary and his colleagues at Red Cliff House would discuss, deep into the night, in an effort to discern between bad luck and error. Was it poor weather or a twist of fate that tended to destroy the plans—and sometimes the life—of a Greenland explorer? Or was it bad judgment?

The temperature dropped to –16½ degrees Fahrenheit on Thanksgiving Day at Red Cliff House. A celebration was held in the cabin's warm main room. The expedition members enjoyed broiled local wild birds, green peas, venison pie, plum pudding, coffee, a whisky cocktail, and some wine. Peary did not eat vegetables, and almost never drank alcohol, but he feasted on the meat. A few days later, a heavy snow began falling. "Red Cliff," Peary noted, "was sinking into a huge drift that almost buried it from view."[13] Over the course of the next few weeks, he and Jo hosted dinners for Christmas Eve, New Year's Eve, and New Year's Day; the latter included several dozen Inuit guests, some of whom dogsledded two hundred miles to join the Americans to eat reindeer legs and engage in athletic games outside— hundred-yard dashes, run forward and then backward, in the cold polar darkness. On January 2, 1892, Peary wrote in his journal, "The holidays have come and gone at Red Cliff House and we have entered on the new year. Will it bring a fruition of my hopes? The year itself will tell. It seems to me as if everything is favorable."[14]

The "white march" began at the end of April.

With two sledges and twelve dogs, three members of Peary's party (Cook, Gibson, and Astrup) set out from Red Cliff House to cart supplies to a camp nearer to the ice sheet, a journey that took about seven hours by dogsled. They were assisted by three Inuit men and two boys. Three days later, on May 3, Peary and his assistant, Matthew Henson, took a large sledge, pulled by eight dogs, to meet the group. The team spent the next forty-eight hours moving to another camp—"Cache Camp"—located 2.5 miles away, on the edge of the ice sheet, at an elevation of 2,525 feet above sea level. Looking into the distance, Peary could see "that my old friend the Inland Ice was evidently preparing its usual reception for me; the leaden-grey clouds massing above it giving every indication of an approaching storm." Soon enough a furious squall blew in, and over the next week, Peary's attempts to make any progress proved futile. At one point, after being awake for sixty-four hours straight, Peary slept for twelve hours, only to wake and find that the snowstorm outside was so fierce that he couldn't leave the igloo for another entire day.

Other unpleasant surprises arrived. Peary's recently mended leg began to ache terribly. Also, his trusted aide Matt Henson told Peary that his heel seemed frostbitten, so Peary sent him back to Red Cliff House, thus reducing the party of Americans to four.

Above all, there was the problem of the dogs. The relationship between a Greenland dog team and an Inuit driver is not casual; it is nurtured through months of training, the firm cry of verbal commands, and the frequent sting of a sealskin lash. As Peary began his trek, the unruliness of his dog team, and the difficulty of driving in stormy conditions, exceeded his worst expectations. "Restless under their new masters, and fighting constantly among themselves, these brutes gave us not a moment's peace," he later wrote. While at camp, the dogs would chew through their harnesses and break free—leaving the men to tackle the snapping, growling animals in the snow, "lasso and choke" them into submission, and then start the harnessing process all over again. In the course of the chaos, Peary and his colleagues suffered multiple bites; some were so severe that Cook, the doctor, needed to suture them closed. What unnerved Peary more, though, was that some of his dogs seemed not merely aggressive but increasingly feral. Indeed, he watched helplessly as several began to suffer from an illness that Peary, adopting an Inuit word, called *piblockto*. The sickness was characterized by frenzied behavior that either led to a dog's quick natural death or such violent paroxysms that Peary felt forced to silence the dog with a gunshot to the head. By the time he began the ascent of the ice sheet, Peary was down to sixteen dogs.

As the weather cleared, the men moved forward, following a northeast direction. In the planning stages of the expedition, Peary had predicted to friends that the journey would be "simple and easy," but he seemed to have willfully overlooked what history had taught him. Here, on the periphery of what he liked to call "the Great Ice," he reckoned with its difficulties.

First-time visitors to the Greenland ice sheet soon discover it is not what they imagine—that is, a smooth glistening sheet. Even where it is not riddled with crevasses, the ice on the margins is often rough, sharp, and bumpy. In colder months, much of it is covered by drifts of snow; in warmer months, large areas are covered in dirt and dust and soaked in places by meltwater lakes, anywhere from three to thirty feet deep, that pool atop the ice and form hazards for the traveler.[15] Without trees, without greenery, without hunting grounds or any life or sustenance, it was logical that the Inuit avoided the ice sheet whenever possible. They considered it a forsaken land of death and spiritual danger. At the start, in a blowing snowstorm, Peary's men and dogs climbed and skidded in an exhausting fifteen-mile journey over hills and rises in an attempt to get to the smoother parts of the inland ice. It was the same up-and-down exhaustion that had led one of Nansen's men, four years before, to say he felt as though he were walking on the waves of an interminable ocean of ice.

After a week the four men had made their way to smoother terrain, and once there the pace—twelve to twenty miles a day—increased. A snowstorm confined the men to an igloo they had built for two days, but afterward they made another twenty miles. "We were now one hundred and thirty miles from the shore of McCormick Bay," Peary said, noting the distance from their starting point. It was May 24, and he declared that it was time to split up: "I told the boys that this was our last camp together, that after we had slept two would return and two go on."[16] His reasoning was that a smaller party would fare better in terms of carting enough supplies, and that he sought the maximum number of dogs and the minimum number of men. He then asked for volunteers.

All three—Cook, Astrup, and Gibson—asked to continue with him across the ice sheet. Peary chose Astrup, the youngest of the group and almost certainly the best skier and athlete. He noted that Cook, the doctor, would be useful at Red Cliff House and that Gibson could hunt game back at the cabin to bolster the food stocks. He directed the two men to return together, and in the morning he gave them precise directions, a compass, a chronometer, two dogs, and provisions for twelve days. Peary and Astrup, meanwhile, kept twelve hundred pounds of rations on their own sleds for the crossing. Before the party split up, the four men shook hands and exchanged good-byes.

"There was a certain solemnity about this hour of separation," Astrup wrote in his journal. Peary then watched Cook and Gibson begin their journey back home, and he continued to watch as they grew smaller in the distance and then disappeared behind a far hummock of the ice sheet.

Like Fridtjof Nansen before him, Peary eventually wrote an account of his adventures on the Greenland ice. It was an effort to explain the challenges he'd experienced during his first few trips to the Arctic, while glossing over its discomforts, interpersonal tensions, and his private fear of failure.[17] Like Nansen's account, Peary's drew heavily on his own journals and was published in two volumes; it was well over one thousand pages in length and included long digressions on Inuit culture, sledges, weather, food, and equipment. It is perhaps unsurprising that each man's personality comes across in his prose. Nansen is witty, engaging, and scientifically insightful. Peary is aloof, authoritative, and scientifically incurious. It is also the case that Peary seems to come alive whenever the subject at hand is the brutality of the natural order. In describing how he had to shoot one of his uncontrollable dogs in the head or the bloody hunt of a pack of walrus, he seems electrified by being so close to life and death, of witnessing the precise moment when one becomes the other.

Peary's continuing journey with Eivind Astrup over the ice sheet was similarly fraught—a protracted struggle between life and death, fought over the course of several freezing months. Yet Peary refrained from describing it as such, even when matters on the ice grew desperate. In the account of his first journey to and from Greenland's north coast (he would ultimately make the same trek two years later, under far more difficult circumstances), Peary's descriptions are simply those of a man immersed in work, detail oriented and cautiously optimistic, always behaving as the persistent engineer as he pushes himself and his dogs across a barren wasteland, knowing all the while that if something were to go wrong—a serious injury, a shortage of food, a loss of his sledges—he would be well beyond the reach of any rescue party. As he traveled northeast, in fact, Peary's journals focused mostly on snow conditions, broken equipment, and how much distance he could cover in a day. Ten miles was a disappointment. Twenty miles was cause for contentment.

The men passed landmarks—Petermann Fjord, Humboldt Glacier—which one hundred years later scientists would fly to with ease, and with increasing urgency, as they conducted their research on glaciers crumbling in a warmer climate. Peary was not immune to the beauty of the white panorama or the strangeness of what he called "the eternal ice." But mostly he and Astrup discussed the dogs—several more were shot and fed to the others—repairs to the sledges, and the weather. When they camped, the men shared an igloo, and their meals were spartan. "Breakfast," Astrup would later write, "consisted of tea, six biscuits, from four to eight ounces of pemmican, a small frozen lump of unsalted butter, and as a dessert, a cup, or half a cup of water."[18] As the two set out for a day of travel, their sledges were often so far apart as to preclude conversation. On good days of smooth ice and decent weather, Astrup would sing to himself and Peary would whistle. But beyond these strains and the gusts of wind and the bark of the dogs, there was the loud and constant rasp of the sledge runners against the ice, "crisp and resonant" and audible from three quarters of a mile away.[19] Astrup later remarked that the monotony of these hours was sometimes debilitating. Snow and ice, hills and hummocks, snow and ice, hills and hummocks. His "imagination ran riot" with thoughts of old acquaintances, living and dead, until he ceased to think at all "and the nightmare of emptiness possessed us." He had prepared for the physical trials of the ice sheet. The psychological burdens of the dogsledding—seeing only white beyond white, nothing beyond nothing—were just as difficult.

By early June, the men were enduring a host of discouraging days. A powerful storm—"the wind howling past us down the slope . . . and the blinding drifts of snow hissing and whirling"—kept Peary and Astrup imprisoned for two days in their igloo; a few days later, warm weather and wet snow slowed the progress of the sleds to the point that they sometimes had to stop altogether.[20] On a route through a glacial basin, the men found themselves one day in a field of deep, deadly crevasses with a thick fog rolling in, unable to see more than three feet ahead. Stranded within a crystal labyrinth, with a few steps in either direction likely to end in a deadly plunge, the men and dogs stayed immobile for eighteen hours, until the fog finally lifted and they could make their way over ice bridges that spanned the chasms. Peary would later record this and other deadly trials—dogs falling into a crevasse, the near loss of most of his food supplies, a lost spyglass—with a tone that approached nonchalance. Here, mortal challenges became modest inconveniences. And every day was merely another problem to be solved in the service of the only thing that actually mattered to Peary: the heroic completion of the journey.

By June 26 they sensed a change in the ice—a perceptible descent. A few days later Peary could see Greenland's snowless northern shore beyond the edge of the ice sheet to his left side. The landscape appeared as brown and red cliffs: "deep valleys, mountains capped with cloud-shadowed domes of ice, stretched away in a wild panorama, upon which no human eyes had ever looked before."[21] On July 1, he turned toward the coast and brought his dogs and sledges down a steep gradient of ice. Ahead of him, at the bottom of the slope, he could see rushing rivers of meltwater flowing off the ice cap and pouring into lakes. In the distance, perhaps five miles away, he saw a mountain that he thought he could climb to survey the landscape of the north coast. He and Astrup had now reached a moraine—a hill of ice and snow and stones—at the edge of the ice sheet. Peary ordered Astrup to camp there with the dogs and then told him he would go farther ahead on his own. "A mile or more of slush, a two-hundred foot slide down the nearly forty-five degree slope of the extreme edge of the ice, and my feet were on the sharp, chaos-strewn stones which cover the iceward borders of this land of rock."

Two months after leaving Red Cliff House, in other words, he had made it over nearly six hundred miles of ice. He had landed, finally, on an unexplored part of the northern coast, where no American or European had ever been before.

*   *   *

In Greenland, visitors are often deceived by the tricks of the Arctic light. Landmarks that appear proximate are somehow refracted by the crystal air and seem to be closer than they actually are. Peary's mountain turned out to be twelve miles away rather than five, and though he would have willingly made a longer walk, his shoes—Inuit-style *kamik* boots, sewn from soft sealskin and suited mostly for ice—were being torn apart by the sharp stones of the landscape.[22] He eventually turned back. By his own estimation, he had been walking for twenty-three hours straight when he returned to where Astrup had camped with the dogs; exhausted, he ate a meal of tea, pemmican, and biscuits and collapsed into sleep. In the morning, when the two men awakened, Peary declared they should pack supplies for four days, harness the dogs, reinforce their boots, and try again as a team.

As they stepped off the ice sheet that morning, Peary was mostly concerned with finding a lookout so he could survey the northern coast. But he was also intent on hunting. The day before, he had found tracks of musk ox, a large, wooly denizen of Greenland, more closely related to goats than to oxen.[23] If Peary could kill a musk ox or two, he knew it might ensure success for his return trip. He and Astrup needed the meat—provisions were running low—and so did the dogs. And so, mile after mile, the men and dogs walked through endless fields of sharp stones—"a region of such utter barrenness I never saw before," Peary remarked.[24] All around, the men saw the skeletons of musk ox, along with a few blooming Arctic poppies. The dogs panted in the heat.

In time, Peary and Astrup came upon a herd of musk ox and ultimately killed five. They skinned the animals and cut up the meat, and Peary threw the dogs a carcass so they could lick the bones clean. The two men cooked musk-ox steaks for themselves over a small stove, took a short nap, and set out to climb a distant cliff to survey the land. This time they succeeded, reaching a breathtaking lookout point about twenty-six miles from where they had left the sledges on the edge of the ice sheet. In Peary's sight was a field of red and brown rocks several thousand feet below him. The rocks stretched for miles and ultimately led to another series of bronze cliffs and an immense, fan-shaped river of ice that collapsed into the frozen bay. Here, for

Peary, was the culmination of all his efforts of the past year and the end of an extraordinary journey. The date was July 4, and from the height of Peary's observation point, with the prerogative that goes to the first explorer, he would name the natural features in the panorama before him.

First, he and Astrup took sips of brandy from a small silver flask. Then Peary christened the frozen bay in the distance as Independence Bay, in honor of his country's birthday. The escarpment on which he stood he called Navy Cliff, in honor of his employer, the U.S. Navy. And the glacier to his right he named Academy Glacier, in honor of the Academy of Natural Sciences in Philadelphia, a supporter of the expedition. He and Astrup then proceeded to build a cairn, a pyramidal, shoulder-high pile of stones, and flew the American flag from a pole above it.

Long before the advent of electronic communications, tradition dictated that an Arctic explorer would write a note, seal it in a canister or bottle, and place it within the cairn in a location that would be visible from the shoreline.[25] The cairns could signify a marker for direction or an SOS message; for parties who were stranded or were about to do something involving great risks, cairns would often mark their last conveyance of information, an explanation, sometimes found years or decades later, as to the causes and specific details of their deaths. Peary considered his cairn to be "the silent record of our visit there"—his way of saying, *I was here first*. On his note he wrote:

July 4, 1892, latitude 81°37'5".
Have this day, with one companion, Eivind Astrup, and eight dogs, reached this point, via the Inland Ice, from McCormick Bay, Whale Sound. We have travelled over five hundred miles, and Astrup, myself, and the dogs are in the best condition. I have named this fjord 'Independence,' in honour of that day, July 4th, dear to all Americans, on which we looked down into it. Have killed five musk-oxen in the valley above, and have seen several others. I start back for Whale Sound to-morrow.
R. E. Peary, U.S.N.

On the back of the letter he requested that whoever discovered it should return it to the Secretary of the Navy in Washington, D.C. Peary then placed it inside the bottle, sealed it tightly with a cork, and put it inside the cairn. Next, he placed a capstone on the pile of rocks. He and Astrup took a photograph of their handiwork, turned away, and then began the long journey back, over more than twenty miles of sharp stones, to the ice sheet.

All through the spring and summer of 1892, Jo Peary waited for her husband at Red Cliff House. "Never in my life," she wrote in her journal, "have I felt so utterly alone and forsaken." Cook and Gibson, who had split off from Peary and Astrup in the middle of the ice sheet, returned to the house on June 3, their faces burned and blistered by the sun. They informed Jo that her husband and Astrup were doing well and had continued together, as planned, to the northern coast of Greenland. Jo was relieved by the news. But soon more doubts crept in. "I feel as if the chances were almost even as to whether I shall ever see my husband again," she wrote. Her Inuit friends assured her in a friendly manner that Peary would never return—a reflection of their tendency to equate the ice sheet with death. "My instinct revolts against this judgment, but it makes an impression upon me, nevertheless."

She suffered in silence. By mid-June, her daily regimen of long walks, sometimes involving journeys of fifteen miles at a stretch, grew more difficult: The snow had softened and the terrain around the house had turned to slush and mud, and food stocks were limited to canned and dried provisions brought over from Brooklyn the year before. She longed for some of the men in the party to succeed in their reindeer hunts—a break from the regimen of seal meat, which left her nauseated. Then, on July 24, there was a welcome surprise: The *Kite*, the same ship that had brought her to Greenland, arrived with the intention of bringing the expedition party home. The problem was that Peary was still not back. Bert had told Jo he would return by August 1, but when he did not show up on that date, an expedition party left the *Kite* and set out for the ice sheet to look for him, a journey that could take up to a week. The ship could not wait forever; if it did not leave within a few weeks, it would risk being locked in the ice and stuck for the winter. Jo's Inuit companions consoled her— Peary had died, they told her gently. One local man informed her that he'd had a dream in which only one white man returned from the ice.

Jo wondered: *Did that mean Astrup, or did it mean Bert?*

In fact both men were struggling, but both were alive. On the crossing to reach the north coast, they had stayed close to Greenland's coastline, but on the return home, Peary decided to take a more direct route, straight across the ice sheet, which entailed fewer miles but a more arduous trek at higher elevations. The weather was difficult: hot one day, cold the next, with conditions often leaving the snow soft and soggy. On the way, they endured a whiteout storm that confined them to a hole in the snow for sixty hours. Also, three more dogs had died, and Peary, with disappointment but not hesitation, fed their butchered bodies to the other dogs. Most worrisome was that just over two weeks into the journey, Peary realized he was running out of food. The musk-ox steaks had not been enough, and Peary had left himself no margin for error. Like all explorers who challenged the ice sheet, he did a quick calculus, plugging in all the relevant variables. *Food and men and dogs and time and distance: What could be accomplished?* "I found I had only ninety pounds of pemmican on which to feed two men and six dogs till our journey was ended," he wrote on July 21, "and at the rate we were travelling it would take over twenty days."[26]

Still, the pace was picking up—twenty-mile days soon became typical, and in one instance Peary and Astrup covered thirty-five miles in a day. On August 3, as he neared the edge of the ice sheet, Peary and Astrup came over a cliff and saw the search party of a half-dozen men that had come up from the *Kite*. Recognizing them, he gleefully shouted to Astrup, "The boys are out looking for us." Peary could also hear that "a faint cheer came across the white waste to our ears."[27] They quickly met up, and after excited greetings, the party eventually made its way, over the next few days, to where the *Kite* was at anchor.

It was now August 6. Peary and Astrup had been gone for ninety-three days. By that point, Jo had left Red Cliff House and was sleeping on the *Kite*. She was agonizing over the prospect of whether she would have to leave with the ship if Peary did not return soon.[28] Jo recalled that on that evening she woke up "at the sound of oars and loud talking." Slowly, she came to realize that the search party had returned from the ice sheet. She heard someone jump over the rail on the deck just above her head. And amidst the commotion, she recognized a familiar footstep coming down the hallway toward her room.

Knud Rasmussen (Library of Congress)

5

A Pure Primitive Realm

Robert Peary returned to New York City in late summer of 1892 with the air of a victorious field general. He embarked on a series of paid talks to raise money for his next expedition—lectures delivered on a stage shared with several sled dogs, which howled disruptively as Peary and his colleague Matthew Henson, sweating inside their Arctic furs, explained their Greenland expedition.[1] The showmanship was in keeping with Peary's new, self-styled mystique: He was America's fearless Arctic pioneer, who had returned from the end of the earth to tell scarcely believable tales about Eskimo tribes that hunted whales and ate polar-bear steaks.

Peary had a legitimate claim to greatness now. He had traveled farther over Greenland's inland ice than anyone in recorded history and reached a part of the world that had previously been unmapped; he also seemed confident that he had proven, for the first time, that Greenland was an island. In subsequent years, visitors to the northeast coast of Greenland would discover that Peary had made errors in his geographical observations of the shoreline, especially those made from the high lookout on Navy Cliff.[2] Yet for the moment, at least, the apparent triumph of his ninety-three-day journey gave him a reputation as the most courageous of a new generation of explorers. His Greenland excursion, as The New York Times noted, was "a story of gratifying success."[3] In late September 1892, not long after Peary came home to Brooklyn, he received an admiring letter from Fridtjof Nansen. "As I am one of those who have seen a little of the inland ice and who have followed your expedition with the keenest interest and sympathy," Nansen wrote, "I hope you will not think it impudent when I now send you my most heartfelt congratulations with your wonderful achievements and grand results."[4]

The men would apparently never meet in person. Over the next decade, however, they went on to pursue an identical goal—the conquest of the North Pole—with differing approaches, like two wary predators circling the same prey. The Pole is located on a thick and expansive crust of floating sea ice that in the winters of the late 1800s covered most of the Arctic Ocean. When he wrote to Peary, Nansen was finalizing the details of a journey that involved a profound idea. A student of ocean and wind currents, he was in the midst of overseeing construction of a ship, christened the Fram (Norwegian for "forward"), which had been built with the intention of getting trapped in the polar ice for as long as five years. Nansen's belief was that the natural, circular drift of the sea ice could eventually bring the Fram's passengers near enough to the pole to allow him access to it by foot or by dogsled.[5] By designing the ship with a curved wooden hull and sloping sides, and by reinforcing its planks with an extraordinarily dense wood called greenheart, Nansen predicted the ship—"round and slippery like an eel," as one newspaper described it—would be lifted up smoothly by the jaws of encroaching ice, rather than be crushed by it.[6]

Nansen left Oslo for his journey on June 24, 1893. At the helm of the ship was Otto Sverdrup, his old comrade from the Greenland expedition. It turned out that the smooth curves of the *Fram* worked splendidly in accepting the crush of pack ice without damage, but its ice drift over the course of the next two years did not bring it near the pole as rapidly as Nansen hoped, prompting him and a colleague to leave the ship in March 1895 and attempt to reach the top of the world by dogsled. The men made halting progress, and within three weeks they were running low on food and facing enormous ice blocks piled in their way. "These ridges are enough to make one despair," Nansen wrote in his journal just before turning back, "and there seems to be no prospect of things bettering."[7] In making a retreat to the south, the two men eventually ended up on a remote archipelago of uninhabited Arctic islands known as Franz Josef Land. There, stranded for eight months, they lived on polar bear meat and walrus blubber and resided in a subterranean hut, built from stones and moss, which they roofed with walrus hides. "On the whole, we had quite a comfortable time," Nansen would recall, not very persuasively, in one of his memoirs.[8] In fact he slept on a couch of rough, cold stones. By luck the men were rescued in June 1896 by a British explorer who was passing nearby on an expedition, and when Nansen returned to Oslo he was celebrated as a hero for reaching a point farther north than any man had gone before. What's more, as a grace note to his success, the *Fram* made it safely back to Oslo at almost the same time as Nansen did, carrying a healthy crew and multiple volumes of scientific observations on polar ice and weather.

Peary, meanwhile, clung to a plodding and exhausting strategy. Dog teams and sledges had brought him over the ice sheet to the uncharted northern coast of Greenland, and dog teams and sledges would bring him over the sea ice to the top of the world as well. There is little doubt that after returning from his Greenland expedition Peary thought "the prize," as he liked to describe the North Pole, was something he could claim within a few years' time; it was this notion of *nearness* to the pole—proved false over the ensuing decade, as Peary faced a demoralizing succession of failures—that drove him onward.

During those dark and difficult years, he made frequent visits to the Smith Sound area of Greenland, the same area where he and Jo had built Red Cliff House in 1892.[9] Some of these visits were made with larger ambitions in mind: Two years after his crossing of the ice sheet, for instance, he repeated the same journey with the notion that when he got to the far side he might be able to continue on to the North Pole. The mission proved calamitous, however, and he barely made it home alive.[10] In other years, he journeyed to Greenland simply to exploit its resources and its people, behaving much like a feudal lord overseeing his barony.[11] He brought home valuable fox furs that his Inuit acquaintances traded to him; in return, Peary gave them guns, ammunition, wood, and useless bric-a-brac. He brought home human cargo too six Inuit, four of whom later died from infectious disease, and several Inuit skeletons, essentially robbed from Inuit graves, for scientific study.[12] Also, Peary pried several massive meteorites from the ice and dirt of an island on Greenland's northwest coast. With a team of Americans and some paid Inuits, over the course of two summers he loaded the massive stones onto ships, brought them back to the United States, and delivered them to the American Museum of Natural History in New York City, where they remain on display one hundred and twenty-five years later.[13] For this contraband, which the Inuit considered sacred and for many centuries had been their only source of tool iron, Peary earned fifty thousand dollars, a sum that helped him finance other expeditions, and led eventually to his final assault on the North Pole in 1909, which he proclaimed to be successful.

These later Arctic journeys—excursions Nansen and Peary conducted after their Greenland work, when each was more famous and more ambitious—are considered the apexes of their careers. But their earlier efforts on the Greenland ice, racking up hundreds of miles as they crossed over what Peary described as "that awful frozen desert," created a more important legacy.[14] Indeed, the conviction of Peary and Nansen had originally been that the challenge of "the great ice" was not to study it but to get across it; in the process, they would thereby prove to the world that man could be victorious over the most formidable obstructions of nature. But the course of history, as well as the course of science, was tipped by these pioneers in a new direction. Subsequent events would show that the crossing of the ice, heroic as it might have seemed, was in fact less important than the ice itself.

With their journeys, the two men defined the ice sheet's contours and characteristics in ways that had never been done before. They demonstrated that it was navigable, that it appeared to have no central "oasis," and that Greenland could be the subject of deeper investigation by scientists who were inspired by their example and who soon began tagging along on all types of Arctic expeditions. With the help of Nansen and Peary, the young field of glaciology was given a significant boost.[15] Just as crucially, the ice sheet expeditions of both men, as well as their close contact with the Inuit, led to events and endeavors that neither could have imagined.

As Nansen was journeying over the ice sheet in 1888—trudging from east to west with his exhausted colleagues—a nine-year-old boy in the small town of Jakobshavn on Greenland's west coast awaited his arrival with an excitement that was nearly uncontainable. Knud Rasmussen, the boy in wait, had never met Nansen, and local officials in his village could not say with any certainty where (or whether) Nansen would arrive. Many decades later, when he had become a lauded explorer in his own right, as well as Nansen's good friend, fate would effect a strange reversal when Rasmussen was given the honor of delivering a eulogy at Nansen's funeral. For now, though, such a future was unimaginable. Somewhere out on the ice sheet, many miles east of his tiny village, this man named Nansen was walking in the boy's direction.

A local newspaper, the *Greenland Commercial*, announced a small cash prize to the first person to spot Nansen. Rasmussen and his friend Jørgen Brønlund decided they would try to win. On a summer evening, when the sun lingered in the Arctic sky, the two boys walked east of town for many miles, until they could no longer see the settlement behind them.[16] They reached a high point with a lookout over the desolate region of rocks and lakes that led toward the ice sheet, but they saw no sign of the Norwegian explorer. Eventually, the boys turned back—a wise decision. Nansen, having altered his route in the midst of his trek across the ice, would not be coming to the village of Jakobshavn after all—or to any village nearby for that matter. On that evening, he was three hundred miles to the south, on his way to the settlement of Godthab.

The news of Nansen's successful crossing likely reached Rasmussen's village within a few weeks. Jakobshavn, located about halfway up the western coast of Greenland and presently known as Ilulissat, its Inuit name, sits on granite slopes that descend steeply toward the dark, cold waters of Disko Bay. A fjord cuts into the coastline on the southern edge of the village—a narrow fjord, one of the deepest on earth, which was once believed to slice through the entire center of Greenland. If the fjord had been as long as the old mapmakers liked to imagine, the idea of using it as a ship's passage would nevertheless be fanciful, since any sailor entering the fjord would see that it is rendered impassable, just a dozen miles inland from Jakobshavn, by a spectacular wall of ice that rises several hundred feet above the water level and descends nearly four thousand feet below.[17]

The wall is the terminus of a glacier about six miles wide, which flows out of the central ice sheet and into the fjord. Known by the Inuit as Sermeq Kujalleq, this river of ice is more commonly known to visitors as Jakobshavn Glacier. In warmer months, massive icebergs break—or, in the language of glaciologists, *calve*—from the glacier's front wall, which is known as its calving front; these bergs, larger than the size of medieval castles, in turn float westward for dozens of miles—moving, in effect, from the glacier's wall, into the fjord, into Disko Bay, and then finally into the North Atlantic Ocean.[18] The journey from calving front to sea, which can take years, is so slow as to seem nearly indiscernible. At the early stages of their journey, the icebergs from Jakobshavn Glacier pass by a small church and a red, high-gabled, three-story wood frame house that looks over the frozen waters at the southern edge of Jakobshavn. This is the house where Knud Rasmussen grew up.

Rasmussen would later recall his childhood near the ice-strewn fjord as idyllic, but it only vaguely resembled the experiences of European and American children of his era. School was only an incidental part of his education—indeed, he could barely tolerate long days in the classroom, and mathematics in particular was agonizing. On the other hand, he loved language, loved the outdoors, and loved the ice; he had a knack for raising and driving sled dogs, and was ready to feast with the village hunters whenever they came home with a narwhal or walrus. At an early age, Rasmussen learned his way around the Greenland coastline by accompanying his Danish-born father, the local missionary, on long trips by dogsled where the elder Rasmussen would christen parishioners in distant villages. For Knud, a web of friendships thus began. Though newspaper accounts from the era would sometimes describe his mother as "full-blood" Inuit, she was actually one-fourth Inuit, which made Rasmussen one-eighth Inuit; Danish and Greenlandic were spoken in the family home, and the boy was fluent in both. His native heritage gave him a kindred link with the tribes who would eventually serve as the focus of his life's work. Many years later, Rasmussen would write of his childhood: "From the very nature of things, I was endowed with attributes for Polar work which outlanders have to acquire through painful experience. My playmates were native Greenlanders; from the earliest boyhood I played and worked with the hunters, so that even the hardships of the most strenuous sledge-trips became pleasant routine for me."[19]

Though small in stature at five feet five inches, Rasmussen would astound friends with his strength and stamina—once, during an expedition as an adult while he was starving and sick, he marched up the side of a steep glacier for seven hours without resting while pulling a heavy dogsled by a sling tethered around his forehead.[20] Just as impressive was his ability to drive a sledge on the ice, a talent that derived from his skill with his dogs—he communicated not with reprimands or angry lashes, but with firm direction and whispery persuasion.[21] A friend would recall, "He could always get his dogs to persevere, to do what they had to do, no matter how exhausted they were."[22]

By his teenage years, Rasmussen had the bearing and good looks of a movie star. His Inuit heritage lent an exoticism to his appearance that made him seem not quite European, yet not quite Greenlandic, either. Apart from his physical grace, two aspects of Rasmussen's personality eventually led those who met him, even many years later, to lament that his buoyancy and charisma (or more precisely, his *magnetism*, a word used by friends, again and again) couldn't fully be described through anecdotes or photographs. Most famously, he could celebrate—usually in the form of parties that tended to go all night or last several nights. The partying might involve frenzied dancing and liquor and tobacco or—in times of privation, which was not unusual in Greenland—merely the discovery of a few sweet biscuits or the meat of a rotting but still edible seal. "In this marvelous land," he wrote later of Greenland, "you can hold a perfect bacchanalia on a few cups of tea and a little moldy bread."[23] The celebrations, always inclusive, brought countless friends into his orbit. They also led a steady procession of women into his bed.

But he wasn't merely an extrovert. Rasmussen could listen—could listen attentively to someone for hours, listen patiently without interrupting, especially as the elders of the Inuit tribes, sometimes "doubled up with rheumatism," crouched in animal skin tents where the air was choked with burning seal fat, the fat they used for heat and light, and explained local legends and traced their own arcane history and phantasmagoric visions. Over the course of his life, Rasmussen would dutifully search out, listen to, and memorize these stories. Then he would go outside, or into another room, to write them down.

Rasmussen's childhood in Greenland was cut short by his family's move back to Copenhagen when he was twelve years old. Leaving the pristine emptiness of Greenland (a country, then, of about eleven thousand inhabitants) for the bustle of turn-of-the-century Copenhagen (a city, then, of about four hundred thousand) left him with a nagging ache for home. But he was adaptable to almost any social situation and made a rapid adjustment to city life. His talent for making friends stood in stark contrast with his academic frustrations, especially with mathematics. Rasmussen barely passed his classes. He eventually moved on to study at the University of Copenhagen, where he once again excelled at socializing. When his father announced he could only afford to send one of his children to college, Knud suggested that he drop out and that Christian, his younger brother, be the one to enroll. "Christian is much better fitted for a disciplined study career," he told his father. "Let him choose a profession, for I'll be able to take care of myself."[24] By that point, Rasmussen was anyway more interested in his bohemian coterie, which included writers, actors, political activists, and painters; they would partake, nightly, in impassioned discussions on art and politics, and argue deep into the morning hours. But he remained unclear about what he should do for a career. For a while, he considered being a stage actor; later, an opera singer. Then he decided he might try journalism, after he had signed on to a trip to Iceland organized by Ludvig Mylius-Erichsen, an acquaintance from a university club. With the help of his father, Rasmussen successfully pitched the idea of a travelogue about Iceland to a local newspaper, and his writing career officially began.

From this point onward, Rasmussen would alternately be known as a writer, explorer, anthropologist, entrepreneur, filmmaker, and to some extent a Greenlandian diplomat. Also, for much of his future life, he enjoyed the status of being a hero and celebrity in Denmark. The core of his work relates to how he repeatedly returned to the land of his youth, twenty separate times during his life, and chronicled its landscapes, people, and character in popular and elegantly written books, before Greenland was subsumed by the encroaching modernization and militarism of the twentieth century. One of Rasmussen's biographers, Stephen Bown, notes that among the legends of extreme exploration who rose to prominence in the early 1900s—Nansen, Peary, and Roald Amundsen, a Norwegian who was the first to reach the South Pole—Rasmussen was unique, in that "his goals were cultural as well as geographical."[25] To put it another way, the vanity and desire for achievement in exploration that drove Nansen and Peary were not precisely what drove Rasmussen. He sought out people, rather than places.

It seems likely that the idea of chronicling the indigenous people of Greenland occurred to Rasmussen after his trip to Iceland. He accepted another offer from Ludvig Mylius-Erichsen, his companion from the Iceland trip, to join an expedition to Greenland, where Mylius-Erichsen intended to write articles, and eventually a book, about the island's geography and native communities. This excursion eventually numbered five men and included Rasmussen's childhood pal, the Greenland native Jørgen Brønlund. In the spring of 1902, before leaving, Rasmussen visited Oslo, where he met for the first time with Fridtjof Nansen, who Rasmussen thought might be able to help his group overcome various Danish regulations that forbade most travel to Greenland.[26] Nansen was charmed by Rasmussen's enthusiasm and willingly supported his effort. Bown, the Rasmussen biographer, notes that the Norwegian also "suggested that Rasmussen would make an excellent ethnographer."[27] His advice would prove to be prescient.

The men left from Copenhagen by ship in late May 1902 and landed in southern Greenland several weeks later. They made their way north to the village of Jakobshavn, where they began acquiring sledge dogs and rations; their intention was to stay in the village through the winter and then travel up the frozen western coast, skipping from settlement to settlement on their way to Greenland's far northwest—Cape York and Smith Sound, the home of the Polar Inuit. Smith Sound was the same region where, ten years before, Peary had set up his Red Cliff House and launched his excursions across the ice sheet. The Inuit population there still numbered around two hundred and fifty people, but not in any sizable concentrations; the men and women lived in about a dozen small settlements scattered over a territory of hundreds of square miles and frequently moved in response to seasonal weather and hunting opportunities— the migrations of musk ox, caribou, narwhal, seal, walrus, and various bird colonies.

When he arrived in Jakobshavn, the village of his youth, Rasmussen was twenty-three years old. He was welcomed back into his old community almost instantly and found it easy to rekindle relationships with friends and relatives that he hadn't seen in a decade. Yet his plans, as he described them to his fellow Greenlanders, seemed fantastical. He had not ever visited the place he was going—the far northwest of Greenland. Indeed, no one he knew, and apparently no one who lived on the more populous and developed southwestern coast of Greenland, had ever been to the northwest, either. With the exclusion of Peary and a few explorers who had come before him, the land of the Polar Inuit was terra nova: It was separated from the other Greenlanders by a cultural gap and by hundreds of miles that comprised the wide, forbidding—and mostly frozen—expanse of Melville Bay. And no one in Rasmussen's understanding had ever dogsledded *over* Melville Bay. "When I was a child," Rasmussen would later write, "I used to often hear an old Greenlandic woman tell how, far away North, at the end of the world, there lived a people who dressed in bearskins and ate raw flesh." As Rasmussen grew older, he noted, "the thought of them was always with me, and the first decision I came to as a man was that I would go to look for them."[28]

Rasmussen and his colleagues left Jakobshavn in February, when the sea ice along the coast was thick and good for travel. In March, after an arduous journey of about six hundred miles up the coast and across Melville Bay, Rasmussen and his men finally arrived, hungry and exhausted, at the first encampment belonging to the Polar Inuit. It was a small cluster of stone houses near the frozen sea, ghostly in the cold daylight, with fresh sledge tracks in snow leading out from the camp to the north. After inspecting the camp, Rasmussen seemed to think the settlement had been only recently abandoned. He squeezed himself through a tiny entrance in one of the stone houses. Inside, he caught the scent of raw meat and fox. He later recalled:

> The first time one sees a house of this description one is struck by the little with which human beings can be content. It is all so primitive, and has such an odor of paganism and magic incantation. A cave like this, skillfully built in [the] arch of gigantic blocks of stone, one involuntarily peoples mentally with half supernatural beings. You see them, in your fancy, pulling and tearing at raw flesh, you see the blood dripping from their fingers, and you are seized

yourself with a strange excitement at the thought of the extraordinary life that awaits you in their company.

If Rasmussen had any doubts about his calling, these kinds of encounters would end them. With the Polar Inuit, the most northern tribe in the world, he believed he had arrived at humanity's purest and most primitive realm. In this world, a seeker—first Peary, now Rasmussen—could learn the secret methodologies of endurance. Not long after he happened upon the abandoned settlement, he sledded farther north and made his initial contact with the tribe, who offered him by way of hospitality "a frozen walrus liver, raw, which was our first meal."[29] Over the next eight months Rasmussen would sled all around the Smith Sound region to live with the natives and listen to their stories, which were often related during long sessions with the *angakoks*, or shamans, who explained local customs to him. He would come to understand that the people here, isolated and removed from all modern economies, survived not in spite of the ice and cold, but because of it. Sled travel, hunting tools, kayaks, igloos, sleep patterns—everything had been adapted, over the course of nearly a thousand years, to the Polar Inuit's pitiless environment. Food was often scarce; winters of killing famine were not uncommon; and so celebratory feasts and constant nourishment in seasons of abundance were the norm. Rasmussen—who as a native Greenlander avoided vegetables—would soon explain to fellow Europeans: "To eat, for example, eight huge meals of meat with coffee to follow, one after the other, is a thing to which you only grow used after living for some time among the Greenlanders."[30]

Also customary in the far north were fermented foods—delicacies such as rancid walrus meat whose hot pungency could make men cry but that Rasmussen came to love; or *kiviak*, another Rasmussen favorite, which consisted of tiny auk birds putrefied to a gelatinous consistency over many months in the airless, hollowed-out stomach cavity of a seal.[31] Helpful for survival, useful for celebration ("banquet food," one American traveler observed in 1914), the fermented dishes were also ones that lasted and could be tapped during seasons of scarcity.[32]

But sometimes they weren't nearly enough. If winter descended and everything—the *kiviak*, the gathered bird eggs, the rotten seal meat, the pickled walrus liver, the whale meat "green as grass," even the dogs—was gone, there were few options. With the temperature hovering around 40 degrees below zero and the terrible clamp of hunger tightening, it could sometimes lead to suicide, or to murder. The far northern world of paganism and "magical incantation," as Rasmussen came to see, could be so cruel as to lead to unspeakable acts. In terrible times, for instance, the practice of infanticide could be quietly sanctioned among the Polar Inuit, since a father's death, or a nursing mother's, could make the support and feeding of young children almost impossible. In that case, a mother or father would suffocate the child.[33]

To the cultured men and women of Europe, Rasmussen would eventually try to explain that this was a place where time mattered less than seasons, where days of endless night or nights of endless day merged, fugue-like, into dreamy states of disorientation, and where it seemed that "years would glide by" in the midst of the continuing struggle for existence, which he often described as a constant struggle for meat. Life was meat. Once, when he was out hunting with a group, Rasmussen asked an Inuit who seemed to be enjoying a moment of deep reflection, "What are you standing there thinking about?" The man laughed at his question. "It is only you white men who go in so much for thinking." The hunter said he was only mulling about whether he would have enough meat for the long, dark winter. "If we have meat enough, then there is no need to think."[34]

Soon enough, Rasmussen began to ask these natives of their experiences working with Robert Peary, who had stopped visiting Cape York and Smith Sound so as to undertake journeys farther north in the Arctic. He met and befriended some of the same men and women mentioned in the pages of Peary's accounts of his ice field exploits. Rasmussen was especially curious about how Peary's presence—his employment of the Inuit; and his dispersal to them of guns, ammunition, and other goods—affected the culture and economy of the tribe. What he discovered from his conversations led Rasmussen to conccive of a scheme that would make this small province his second home for the rest of his life.[35]

When he returned to Denmark in November 1904, Rasmussen began writing two books about the Polar Inuit; their publication in Denmark a few years later made him a literary celebrity.[36] He married a Danish woman, Dagmar Andersen, who was the daughter of a well-known businessman and helped edit Rasmussen's writing. The couple bought a house outside Copenhagen and had a daughter (and would later have a son and another daughter together). But Rasmussen was averse to having a settled domestic life. By 1909, he had made several more trips to northern Greenland and was planning to return again, this time to begin something more ambitious.

The practice of ethnography, one of the essential research tools for anthropologists, involves the systematic gathering of information about the customs of a particular culture or group of people. Rasmussen's ethnographic research on the Inuit, begun on his first trip to northern Greenland and conducted over the course of the following three decades, involves tens of thousands of pages of meticulous interviews and observational impressions. It stands as the matchless record of a vanishing, pre-modern culture. From his work, we unravel the human mysteries of Greenland—the food habits and dress of the Polar Inuit, their death rites and creation myths, their myriad rules and superstitions (in the event of a death, anyone who touches the corpse "must remain quiet in tent or house for five days and nights"), and the legends whereby vengeful souls assume the form of polar bears, cracks rippling through glaciers transmogrify into living beings, one-handed women live at the bottom of the sea, and men turn into dogs and back again into men.[37] In regard to the breadth and significance of his research, a Danish academic review once concluded, "no one has even approached the scope of Rasmussen's accomplishments."[38]

Rasmussen once described his own role, in the Greenlandic tradition of self-deprecation, as modest in comparison to explorers like

Nansen or Peary.[39] His achievements, however, bridged a number of crucial gaps that were cultural as well as scientific. He was too late to cross the Greenland ice sheet first, and unlike Peary or Nansen seemed to have little passion for conquering either pole. As he put it, he "came to lift the stones which the others had let lie."[40] His assets—his broad social network, his mixed ancestry, his fluency in a number of languages, his gifts as a writer, his deep curiosity about people and places, and not incidentally his fearlessness—made him a figure of admiration and trust, in Greenland as well as Europe. Ultimately, Rasmussen connected the northern part of Greenland with the southern part and linked the deep but unknown history between all Greenlandic natives. Meanwhile, his writing and informal diplomacy connected Greenland more closely to Denmark.[41]

In the process, his work had the ultimate effect of connecting Greenland more closely to the rest of the world. Partly he did this by bringing a number of social and physical scientists into his orbit and helping to introduce them to the world of ice and the Arctic. And partly he did this by the magnetic power of his celebrity, which in time prompted a variety of outsiders interested in Greenland—movie directors, U.S. military officers, even the aviator Charles Lindbergh—to seek him out.

He did not do it alone, though. In 1908, while on vacation in Bergen, Norway, Rasmussen met a huge, handsome, good-natured man who was nearly a full foot taller and several years younger. Peter Freuchen, with unkempt hair and a full and ragged beard, was on his way back from Greenland, where he had been working for two years as a member of the *Danmark* expedition. In many respects, the *Danmark* group, which took its name from their sturdy ship, was the first endeavor that included scientists who aimed to investigate east Greenland's meteorology, geology, and ice. Freuchen had been a junior member of the team—just nineteen years old when they sailed from Copenhagen in 1906. During their first meeting, Rasmussen was impressed by his zest for life and adventure, along with his evident toughness.

At one point during his stay on east Greenland, Freuchen had been asked to run a tiny weather station near the ice sheet—one that was situated about forty miles away from the *Danmark* base camp. For six long months, in a cramped, nine-by-fourteen-foot hut, with an interior encrusted in ice a foot thick due to moisture from his breath, Freuchen lived alone. He had no way to communicate with the outside world. His job was to make barometric and temperature measurements several times a day. Much of his energy, though, went to keeping his spirits up in the dark winter months (he sang to himself constantly) while making heroic efforts to stretch a meager supply of food and coal from one month to the next, when a colleague from the main camp would bring a fresh load of supplies by dog sledge. The cold, the dark, and the loneliness were all unpleasant. But most unpleasant of all were the wolves. "One after another they ate all seven of my dogs," Freuchen would later recall.[42] For much of his stay, in fact, Freuchen worried about how to keep from being attacked and eaten, and strategized how to trap the wolves that surrounded his cabin, or sometimes howled on his roof, on most days and nights.

What connected Rasmussen to Freuchen, the men discovered when they met, was a complicated web of friendships as well as a recent tragedy. "When I finally returned to the ship," Freuchen recalled of the day during the *Danmark* expedition when his six months in the weather station ended, "I learned that three of my companions were dead—Mylius-Erichsen, the leader; Lieutenant Hagen; and an Eskimo, Jørgen Brønlund."[43] Mylius-Erichsen, Rasmussen's friend from the University of Copenhagen, had been the leader of the Literary Expedition, which just a few years before had introduced Rasmussen to the Polar Inuit and had set the course for his adult life. Brønlund was Rasmussen's childhood friend from Jakobshavn, the boy with whom he had walked toward the edge of the ice sheet to wait for Nansen.[44] It had not been the weather or wolves that killed these men. Rather, Freuchen explained that they had made a springtime journey up the eastern coast of Greenland to map the coastline of the region. But the weather turned warm. With the ice melting, dogsledding was impossible. That they were stranded seemed bad enough—but they soon discovered there was little game to hunt in their vicinity. The following winter, a search-and-rescue party from the *Danmark* discovered Brønlund's body in the snow at a prearranged rendezvous point. He carried a note explaining that Mylius-Erichsen had died two weeks earlier. He had starved to death on the catastrophic journey back to base camp.[45]

Freuchen and Rasmussen parted ways in Bergen. Freuchen returned to Copenhagen. Before his journey north a few years earlier, he had been studying medicine at the university there. But things seemed different now. Still only twenty-one years old, he was a bona fide Arctic explorer who had experienced hardships in a part of the world few of his contemporaries could imagine. "I found it difficult to accept the boys in my classes as equals," he would recall. "They understood nothing but school pastimes and pleasures." To pay for his room and board, Freuchen wrote melodramatic fiction for pulp magazines—"each one was the story of a beautiful girl who fainted on a lonely marble balcony," he recalled—and gave lectures about the Arctic. First he gave the lectures alone. And then, with his new acquaintance Knud Rasmussen, he formed a seriocomic duo (one tall, one short; one serious, one silly) that began traveling around Denmark to give talks, with the help of a recently acquired slide projector, about life in Greenland. "Soon," Freuchen said, "instead of my lectures interfering with my studies, my work at the university was interfering with my lectures." He took a final examination in chemistry and then said goodbye to university life.[46]

In the course of their lecture rounds, Rasmussen had asked Freuchen if he would like to join him in returning to Greenland. His idea was to set up a business in the Smith Sound region and trade goods to the local Inuit in exchange for fox pelts. Though Denmark had colonized southwestern Greenland, the land farther north was not yet under the jurisdiction of any country and there remained a lack of clarity over whether the United States might have a claim (because of Peary's expeditions) or whether Norway might have intentions (because another explorer, Nansen's old comrade Otto Sverdrup, was setting up a seal-hunting station nearby). Rasmussen believed he and Freuchen should move quickly to claim the region for Denmark.

Freuchen thought about the proposal for several days. He was about to turn twenty-four years old. He understood that this was the kind of choice that could divide an ordinary life from an extraordinary one. He knew, too, that the job—manager of the trading station—would mean saying goodbye to family, friends, and girlfriends. He would be living in a place so remote he would be lucky if more than one ship a year came bearing mail and newspapers (there was no telegraph, no telephone), and where life would in some ways resemble the months of depthless solitude that characterized his *Danmark* expedition. He could stay in Copenhagen and become a physician, or perhaps work as a sailor, seeing as he had grown up in a seafaring family. Journalism and writing appealed to him as well. "I might do any number of prosaic things," he reasoned.[47]

But then, after ruling out the alternatives one by one, and after perceiving the possibilities of a risky, wayfaring life, Freuchen told Rasmussen, *ja*. Of course, he said. He would be happy to join him in the frozen north.

Peter Freuchen (Danish Arctic *Institute*)

6
Thule

Rasmussen and Freuchen sailed for northern Greenland in the spring of 1910.

As they neared the island and made their way up the western coast, their ship was whipsawed by gale winds. With a broken tiller and propellers, the vessel limped into a harbor at North Star Bay, just south of where Peary had set up camp nearly two decades before, and made landfall on the rocky coast, near where a small settlement existed already—a Christian mission, along with eight Polar Inuit families. The location had a name, Ummanaq, which meant "seal heart," in Inuit. It was recognizable for miles around by a high, flat-topped mountain at the edge of the harbor. In Rasmussen's view, it was not the ideal location but would suffice, though he believed the trading station needed a new name. There is a lack of clarity as to who suggested it—both Freuchen and Rasmussen would later take credit, and because Freuchen outlived Rasmussen, he had the last word—but the settlement was soon christened Thule station. The name, pronounced TOO-lee, derived from the 330 b.c. voyage of Pytheas, who had sailed north from Greece and reached the pack ice and glimpsed land, the land of the farthest north, that he called Ultima Thule.[1]

The establishment of Thule was a concerted effort to expand on work that Peary had begun two decades before. Peary's primary goal, of course, had been to cross the ice sheet and possibly find a route to the North Pole from the island's as-yet-unexplored northern shore. What mattered to Rasmussen was that during the course of Peary's visits he had transformed the tiny economy of the Polar Inuit. Peary sometimes traded with them for pelts to bring home as gifts—his Brooklyn apartment was literally covered in animal skins garnered from his travels. But mainly he employed the Inuit for making boots and fur clothing for his exploration teams, and for helping him transport goods.[2] Also, he requested from the Inuit their best dogs to pull his sledges. In return, Peary disbursed guns, knives, lumber, and other assorted trifles made of metal or wood, which were difficult to procure in a country without industry or trees.[3]

It seems reasonable to conclude that in this arrangement the balance of power tilted toward Peary and that the Inuit were not adequately compensated for their work, even as they delighted in the prospect of owning rifles and sharp knives. Peary considered the exchange as mutually beneficial—that he was acting not as an agent of exploitation but instead giving the Inuit better odds in their efforts to stockpile winter meat. Rasmussen tended to agree with this view. He had come to believe that the Polar Inuit would soon be forced to confront the modern world and that Peary's trade with them was a significant first step. Modern goods had improved their capacity to hunt and avert winter starvation; and because of Peary's presence, Rasmussen wrote, the Polar Inuit "had jumped from the stone age to the present time in their technical situation."[4] Yet therein lay a problem. As Peary had moved on to different pursuits, no other trader had stepped in to furnish the natives with goods and tools.[5] Rasmussen therefore discerned both an opportunity and a necessity. By setting up this trading post with Freuchen, he could acquire from the Inuit fox pelts and walrus tusks that would almost certainly fetch high prices in the European market. Meanwhile, the trading post could continue to supply the Inuit with food staples and tools, including steel fox traps, which they now lacked.[6]

In many respects, the story of the Thule station, at least in its early years, is the story of a deep friendship between two men seeking meaning and adventure in their lives at a place at the end of the world. But at the start neither Freuchen nor Rasmussen saw it that way; they perceived Thule in terms of its missionary aspects—not in a Christian sense but in how their efforts could service the Polar Inuit with staples while potentially fulfilling two other goals. First, Rasmussen intended to use Thule as a reason to secure the land in this remote part of Greenland for Denmark and put it under the Danish flag, thereby joining it to the more developed region of southwest Greenland. Second, he had the idea that Thule would function as a research base—a place for geographers and archaeologists and other expeditions to use as a launching point for work that involved efforts to understand the Inuit and Greenland itself.[7] What Rasmussen and Freuchen could not anticipate was how their seemingly modest and eccentric efforts to follow Peary at Thule would play a far-reaching role, both in Cold War military endeavors and in late-twentieth-century glaciological research.

The two men quickly settled in. Upon their arrival, Freuchen and Rasmussen unloaded goods from the ship—a heap of packaged food, tools, and a prefabricated wood house that Freuchen was tasked with assembling. He would later quip that the house—a simple rectangular structure with a double-door entrance and a peaked roof running lengthwise—was neither large nor comfortable and that he stayed inside as little as possible. But this place would prove sturdy enough to serve as the men's home, and as the Polar Inuit trading station, for decades. Indeed, the Polar Inuit in the region, most of whom knew Rasmussen already, began traveling frequently to Thule station, especially in winter when sturdy sea ice made travel easier, to drink coffee and trade furs. Several older Inuit women soon came to live in the house with Rasmussen and Freuchen, and Rasmussen would listen to their stories and incantations late into the night and transcribe them. Semigak, for instance, a conjurer who heard the voices of spirits and claimed to plumb connections with the dead, would spend the winter at the Thule house, three and a half months when darkness lasts all day and night, amassing any boxes she could find.

"I collect shadows and darkness," she explained to Freuchen one day, "so that the world will get light again, and I keep it all locked up here in these boxes."

There were other practices, too, that the men found not only poetic but worthy of respect. Freuchen learned that when one returned from a journey where a member of the party had perished, the survivors were obliged by tradition to pause by the harbor with their backs turned to the settlement, their eyes facing the icy seas, so as to communicate the tragedy in silence. Also, upon seeing the sun for the first time in months after a dark winter, you were to remove your mittens and hood and face the glow. "You should not laugh at us for this," Freuchen once heard an old Polar Inuit, Ulugatok, telling a skeptical European scientist during a dogsled journey over Melville Bay. "We think that if we do this we shall not die at least until the sun returns next year. Even if it does no good, we enjoy life so much that we do anything to keep it."[8]

The skeptical European who refused to partake in the ceremony that welcomed the sun died later that year, of starvation and insanity, on the north coast of Greenland.[9]

In his small house by the harbor, Freuchen began to feel his old self, his Danish self, slipping away. "I did not realize that a revolution was taking place in me," he would later write. "Slowly I was developing a great reluctance against ever going home." In adopting some of the conventions of the Polar Inuit—"borrowing" the wife of an Inuit friend, for instance, for a long sledging trip—he would say, "I turned more and more into an Eskimo."

He grew his hair long; his beard thickened and lengthened. At this far-flung outpost, Freuchen began to discern an utterly free way to live, a northern Bohemia. He had moved well beyond the reach of radio communication, and had cast off the conventions of society. He was now able to reject the hidebound notions of a workweek and day job, both of which seemed absurd here, anyway, when the struggle for survival was constant and the need for celebration frequent. "Freuchen has definitely renounced civilization as being unfit for man," another Greenland explorer, Donald B. Macmillan, noted around this time, after spending New Year's Eve with him in a tiny Inuit village. "He has married an Eskimo girl and has settled down for life at the top of the world among ideal socialists."[10]

The name of Freuchen's new wife was Navarana. She was smart, independent, and a far better sledge driver than Freuchen. Her radiance, captured in photographs taken at Thule, belied the misery of her childhood. She had been raised on an island off the northern coast where an epidemic had wiped out everyone except for herself, her mother, and her young brother. The family ate everything they had in storage and then turned to butchering the dogs; after that they ate old clothing and sealskin dog leashes. When her breast milk stopped, Navarana's mother hanged her young son. Navarana then turned to eating grass and rabbit dung. By chance, a man named Uvdluriark came to visit the island from a local village and rescued the two women. In time he became Navarana's stepfather.[11]

Freuchen had seen Navarana around the settlement and courted her in an awkward manner. One day, she came to stay at his house to accompany an older woman who was visiting. That night, Freuchen asked her to move from her sleeping ledge across the room to his ledge. She agreed. Freuchen recalled that in the morning, he asked her not to leave, and she agreed to that, too. "That was all the wedding necessary," he decided. When Rasmussen heard the news after returning from a hunting trip later that day, he ordered a celebration for the entire settlement, which lasted several days and nights.

"I did know that my marriage to an Eskimo girl made a final breach with the world I had known as a young man," Freuchen would write some years later. "But I had already left that world far behind."[12] Later, Freuchen would explain how deeply it pleased him to be at Thule during those early years. "We lived contentedly in our little house, and never shall I be so happy again. Perhaps most men could not have endured the isolation, but I had everything I wanted—friendship, trust, and a busy, active life."

The adversity made his life feel *valuable*. A few years after his marriage, when Freuchen traveled to Copenhagen with Navarana, he noticed that she quickly grew tired of the culture. Life in Greenland was hard; it was impossible to argue otherwise. Yet in Copenhagen, Freuchen remarked, "She had everything she needed . . . but all things came too easily." Navarana grew depressed. "She had no feeling of living when there were no difficulties to surmount."[13]

To visit the site of the Thule trading station now, a century after its inception—a cluster of abandoned and worn wooden shacks, close by the shores of a silvery bay, and near the flat-topped mountain of Ummanaq, which stands close to the vestiges of stone houses from an Inuit hamlet that vanished more than a half-century ago—is to understand just how far from the world these two men had staked their claim. In the face of such isolation, though, the trading station succeeded in creating a modest cultural and economic hub in northwest Greenland, where none had existed before.

To Rasmussen's pleasure—but not his surprise—the station quickly began to turn a profit. But at Thule, as Freuchen soon discovered, buying pelts and selling goods didn't always follow typical European business practices. One day at the trading station, an Inuit man arrived from a long distance to make some exchanges. Among the items he desired was a knife. He took out five fox pelts and gave them to Freuchen.

"You are mistaken," Freuchen told him, explaining that a knife didn't even cost one pelt.

The visitor waved him off. "I have been without a big knife for a whole year and have been missing it terribly," he said. "That is why I give you so many skins."

Freuchen kept reasoning with his visitor, trying to get him to pay less, but he wouldn't budge. "In Eskimo logic," Freuchen later explained, "goods possess a value according to the need of the buyer rather than the scarcity or abundance of the supply." In time, Freuchen made an effort to alter the Inuit approach, which had historically left the tribe at a disadvantage—for instance when they had traded with visiting whalers, or with Peary's team. He and Rasmussen manufactured thirty thousand "kroner" coins out of aluminum so that they could introduce the idea of currency and fixed prices to the Polar Inuit. The coins had no value outside of the small world of Thule, but they created a modest economic order to the business. Freuchen and Rasmussen could offer the coins for pelts, and the coins would in turn be used by the Polar Inuit to buy goods.[14]

Freuchen and Rasmussen had untold hours to fill. Even when they lacked food, they luxuriated in time. About his first visit to the Polar Inuit region in 1903, Rasmussen had written: "It is a strange life that we live up here; no program of arrangements is ever drawn up; the days bring their own diversions and their own work. The only thing laid down beforehand for each day is the allotted portion of walrus flesh that we have to eat to keep our bodies in fit condition for work. But, should a happy suggestion occur to any one, it is acted upon at once."[15] Now, years later, the strange life had become the norm. The difference between work and leisure didn't apply to life at Thule, where days tended to be long and unstructured and the hectic pace of society, thousands of miles away in Europe, was beyond comprehension. On any given day at Thule you might fix your sled, or mend your house, or feed your dogs, or merely eat a piece of meat and go back to sleep. There was a small library of books to read if you pleased.[16]

If someone were to tell Rasmussen at this stage of his life that there would eventually be statues of him all over Denmark and Greenland, the notion would have likely seemed preposterous. And yet, to fill his time, and to assuage his curiosity, he and Freuchen somehow transformed themselves during this period into the last great explorers of Greenland. Sometimes the two men left Thule station to conduct trips on nearby Ellesmere Island, or to mount a dogsled trek southward across Melville Bay, where Rasmussen and Freuchen would visit other trading posts to bring back tobacco or coffee. These excursions could take weeks, or even months, but traveling helped them make contacts among the Polar Inuit and ensure the success of the trading post. Freuchen would explain that another reason for his longer excursions was to break the monotony—"even certain peril was better than sitting about with no prospect of change in routine ahead of us."[17]

But an idea eventually arose for a trip that was far more ambitious than anything Rasmussen and Freuchen had attempted before; it seemed as much about testing their limits as avoiding boredom. And it had to do with the inland ice. What became known as the First Thule Expedition was a response to news the two men had heard during their local travels about a missing explorer named Ejnar Mikkelsen, who in 1909 had traveled from Copenhagen to east Greenland on a ship called the *Alabama*. In the summer of 1911, Freuchen and Rasmussen learned that Mikkelsen had never returned to Copenhagen, and they resolved to go look for him.[18] From their understanding, Mikkelsen had traveled from the east coast of Greenland, where his ship was anchored, to the north—the very top of Greenland, approximately the same place Peary had gone when he crossed the ice sheet. They planned to follow the shoreline north from Thule, and then go east along the coast, until they met Mikkelsen "or came upon traces of him," Freuchen would later explain. "We would return, well, when we returned. We might be forced to spend the winter on the other side [of Greenland], but what of that?"

Just before the men were to leave, Uvdluriark—one of two Inuit men who would accompany Freuchen and Rasmussen—looked at a map of Greenland and seemed skeptical of the route along the shoreline.

"Why can't we go straight across?" he asked. "It looks as if it would be a short cut."

Freuchen explained that such a route would take them over the ice sheet. There would be nothing to hunt.

"The ice cap is only a road without rough ice," Uvdluriark said.

After some discussion, the men agreed it would make sense to go over the ice. "If you can navigate us across," Rasmussen told Freuchen, "we'll look out for the food."[19] Freuchen agreed, and preparations for the trip began.

*   *   *

"I started my expedition on the 6th of April, 1912," Rasmussen would later write. "The expedition consisted of four sleighs, Peter Freuchen, Uvdluriark, Inupitsok, and myself. We had big teams—in all, fifty-three dogs."[20] The early days of the journey were difficult; the ascent onto the ice sheet was steep and wearying, yet to Freuchen's relief they seemed to avoid the worst of the crevasse zone. By the end of the first week, they realized that the ice sheet was different than anything they had tried before, and the arduousness of the sledding exhausted them. As they made their way to the center of the ice, gusting winds and fine drifts of snow poured over them, a relentless dusting of white crystals. "We looked like ghosts driving ghost dogs," Freuchen recalled. When the sky was clear there were other challenges. At an altitude of six thousand feet, in a glittering landscape of ice, Freuchen would remember that daylight hit his eyes like sparks from white-hot metal. Meanwhile, it was so cold that the men's faces both froze and burned. Skin hung off Rasmussen's face "in ragged splotches," and Freuchen's own nose was raw and bloody—though he was more concerned by his advancing snow blindness. Later, he recalled, "Unless a person has experienced it, he cannot appreciate the torture. Your eyelids feel as if they are made of sandpaper." His dilemma was that he served as the expedition's navigator. Several times a day, he was required to make observations with an instrument called a theodolite, which involved looking at the sun. So every day was more torture, more sparks of white-hot metal in his eyes.

Rasmussen tried to boost the group's spirits. In the mornings he would get up first in the tent—or in the igloo, if they had succeeded in cutting snow blocks—and sing silly songs as he made tea. At night, even as their food supplies dwindled and they began to reach the point of starvation, he read luxurious recipes to the men from housekeeping magazines he had brought along. Then he would get them to laugh.[21] Years later, Freuchen would remember these moments—the harsh weather, the dwindling provisions, and Rasmussen's inexplicable good humor. "Nothing draws men closer than to hunger together, to see death in each other's eyes," he wrote. "Lying together in snow huts during snowstorms of many days' duration, waiting for better weather, and seeking to drown our hunger by each telling the other everything he knows—then you pour out your life."

Their journey across the ice took about three weeks. But as they came within sight of the rocky, northern coast, they were beset by other troubles. Near what was now called Peary Land—named in honor of the American explorer—they arrived at a sheer cliff of ice that dropped fifty feet; to descend it, they belayed down a sealskin line. During Freuchen's turn, a harpoon he was carrying punctured his thigh, leaving him with a deep wound and pants soaked with blood. So as he recovered from snow blindness in his tent during the following days, he also limped around. A few weeks later, as Freuchen was recovering, Rasmussen became ill—so sick and feverish he didn't even desire to smoke his pipe.

For the next three months, the four men ambled about on the northern coast as their health mended. In a place with no other humans and no interruptions, their only ambitions were to hunt, eat, sleep, explore, and record all they saw about the region's geography, flora, and fauna. Even with their injuries and illness, they killed dozens of musk ox that sated their hunger and were packed away to be used on the return journey. They didn't find Mikkelsen, the Danish explorer—but Rasmussen discovered evidence that his party had been there and had departed.[22] More curiously, in the gravelly landscape that characterized the north coast, Rasmussen discerned what looked like ancient tent sites, which suggested to him that Inuit settlers had traveled this way many centuries before en route from the western to the eastern coast of Greenland.

By early August, when the expedition party decided to head home, they agreed they had met with various successes, along with one significant achievement. They had explored new areas of the coast and had corrected some of Peary's geographical maps of the area. And during their time in the north, while Rasmussen was ill and recovering in his tent, Freuchen had hiked over to Navy Cliff and climbed it just as Peary had done two decades before. The stone cairn at the summit was still intact. No one had been there since. Awed by the moment, and by his predecessor—"If it had not been for Peary's work and experience we should never have been able to do what we did," Freuchen later wrote—he stood for hours on the precise spot, lost in the strange loop of history, connecting himself to an obscure achievement that he had read about many times before.[23] Eventually, he dug inside the cairn, and among other mementos, he discovered the note that Peary had left in the brandy bottle. In keeping with custom, Freuchen apparently took Peary's record with him (the practice was to return it to the original writer) and left his own note in its place, explaining that he had visited the cairn twenty years after Peary.[24]

He had added some details about his and Rasmussen's own expedition, in case they didn't make it back alive to Thule.

With a few drops of cooking fuel left, and with one flank of musk-ox meat remaining, the four men straggled into Thule, exhausted, in mid-September. To celebrate their conquest of the ice sheet, Rasmussen sponsored a week of parties, where he served whale meat and *kiviak*. In the months following, he and Freuchen slowly returned to their old life. Whatever might come next, nothing at Thule happened too quickly. The men spent leisurely days at the trading station; they embarked on hunting trips for seal and narwhal. In the spring, they made weeklong journeys to collect bird eggs on the coast.

Rasmussen journeyed to Copenhagen to see his family and commence work on a book. From Europe, he sent a letter to Robert Peary, whom he had never met. He explained the discovery of Peary's note in the cairn at Navy Cliff and wrote: "Having recently returned from an expedition in North Greenland, I herewith take the liberty to bring you a greeting from the country and the people which for so many years have been near to your work and your heart. I am glad to testify that old as well as young among the 'arctic highlanders' keep your memory high in honor, and that they remember the time when they lived with you on your famous expeditions with delight and enthusiasm."[25] It was undeniable that Peary was remembered in Greenland, but Rasmussen was also being courteous. It seemed doubtful that Peary was remembered with anything like delight.

The start of World War I made life at Thule even more difficult. In peaceful times, the infrequency of contact with Europe meant it was not unusual for Freuchen to read a year's worth of old newspapers after a supply ship arrived. But the war forced the men to endure long stretches without much news from the continent, and without much in the way of merchandise to sell at their store.[26] So a year passed. And then two. Rasmussen returned from Europe. And a steam ship finally arrived with supplies and mail—"four whole years of newspapers to be read," Freuchen would recall, "with a complete, day by day account of the war." There is a photograph taken inside Thule station during this era: An enormous man, ruggedly handsome, with long hair and an unkempt beard, wearing pants of polar bear fur as he sits in a wicker chair at a rolltop desk, reading a Danish newspaper by lamplight.[27]

The First Thule Expedition had instilled ambitions in Rasmussen to write popular books about his exploits, but it also led him to start thinking about how to legitimize his trading station as a scientific base. In 1917 he embarked on what became known as the Second Thule Expedition—another trip to the north coast that would involve covering long distances over the ice sheet on the return. Freuchen was asked to stay behind and run the trading station. The Second Thule members included Lauge Koch, a young Danish geologist, along with several Inuit travelers and a botanist named Thorild Wulff; it was the first of Rasmussen's expeditions with a serious effort at research. The ambition was to map many of the unexamined fjords and glaciers of northern Greenland and collect information on the area's geography, geology, and natural environment.

The team began traveling in mid-April, choosing a route that ran along the top of Greenland's serpentine coast. A few days after setting out, Rasmussen stumbled upon a food depot from an old British expedition, one led by George Strong Nares. Dating from 1875, it was six boxes that included several nine-pound tins of Australian mutton. The men ate the mutton. Rasmussen would later say, "We were thus able to live grandly on food originally meant for Arctic colleagues who had travelled here before any of us were born."[28] The journey then continued, a steady push toward the northwest, with Rasmussen and his men trying whenever possible to use the "ice-foot," a shelf of ice that clings to the rocky shore of Greenland and can often serve dogsled travelers much like a smooth highway. In various harbors Rasmussen encountered the graves of explorers from decades past. He also came upon a large cairn that contained a note Peary had written on June 8, 1900, seventeen years earlier. It had been stored in a brandy bottle and mentioned that Peary was passing through on his way to the Canadian Arctic.[29]

They reached their destination in mid-June—the northern fjords, 620 miles from Thule. But the days here turned into misery. Though Rasmussen and Koch succeeded in mapping previously unknown parts of the coast, they found getting around exceedingly difficult, and at times the men were forced to wade up to their waists in freezing water for twelve hours a day. Their lack of food became critical. Rasmussen had planned this as what he called a "hunting expedition," meaning that the meat would be procured at their destination on the north coast, with "emergency rations only being carried."[30] The men succeeded in hunting musk ox in various locations, yet it soon became clear that Rasmussen had been counting too much on hunting seals successfully along the coast. In late June, they turned back toward home. Taking a shortcut over a portion of the ice sheet to speed the journey, Rasmussen had a recurring dream, sharpened by his hunger, night after night: His mother back in Denmark, having just finished baking two cakes, was cutting several large slices and saying to him, "There you are, my boy; eat as much as you like!" Then he would awaken to the reality of a tent on the ice, and a day where he would have to slaughter and eat another of his precious dogs for sustenance.[31]

The Second Thule Expedition ended in chaos and panic. Wulff, the botanist, died from starvation in August as the men got close to home, a tragedy that followed the disappearance and death of an Inuit member of the team, Hendrik Olsen. Freuchen was up in his room reading when Rasmussen returned—he had been away for six months. According to Freuchen, a local conjurer had predicted earlier that evening that the expedition party would return soon but would be missing two members; and then Rasmussen poked his head in Freuchen's bedroom doorway to greet his old friend. It had been a terrible summer, he told Freuchen, one "too horrible to think of."

Freuchen looked closely at Rasmussen. He read the summer's events in his friend's thin, bedraggled state. "The look of the icecap was upon him," he wrote, "months of starvation and hardship written on his face."[32] Then he began to listen to his friend's story about what had happened.

In many ways, that night marked a turning point in the blissful experiment of Thule. Rasmussen told Freuchen the details of the expedition, adding that he had missed him "every, every day." [33] Yet in the period immediately after the Second Thule Expedition—the years between 1918 and 1920—the two friends drifted. Tracing the details of these years can be difficult: A superb storyteller not beholden to precise chronologies, Freuchen, in the course of several memoirs, was largely untroubled by rearranging actual sequences of events—or, sometimes, details. Reading over his accounts of those long ago days in the north, in fact, one becomes less sure about the precise truth even as one becomes more enchanted by the poignancy and colorful daring that characterized his adventures. There was, for instance, a story of how on one journey Freuchen killed a polar bear atop an iceberg, and how its blood, flowing down from above, hot red liquid melting into solid white ice, shattered the berg and nearly killed him. He seemed to be taking the myths that his friend Rasmussen was collecting and turning them into his own life story.

What seems certain is that after World War I, Freuchen was relieved of his job at Thule. It was probably due to some kind of lingering depression related to his isolation that affected his behavior and judgment. *Never shall I be so happy again*—he had believed this to be true about Thule. But the sentiment was more a reflection of his early years there, rather than his later years. Over time, life with the Polar Inuit, nine years in all, had worn him down.

He left Greenland for Copenhagen in 1919. He was now the father of two children with Navarana, a boy named Mequsaq and a girl named Pipaluk, and all came with him to visit the city of his birth. It was during this visit that Freuchen became sick—a victim of the influenza epidemic that ravaged Europe just after the war. His illness brought him close to death and confined him to a hospital bed for several months. "I was isolated," he later explained, "and in a ward reserved for dying patients."

In all, his recovery took nearly a year, and when he returned to health he felt a need to return to Greenland. He asked Rasmussen if he could join his next endeavor, a massive research project that would become known as the Fifth Thule Expedition.[34] It was slated to begin in 1921 and would in retrospect be seen as one of the most ambitious social science endeavors of the era. Led by Rasmussen, and beginning with a dozen members (not all of whom would travel together, or go the full distance), the expedition goal was to fan out across the entire span of northern North America by dogsled, covering at least ten thousand miles in the process, moving from Greenland to the Hudson Bay to the Bering Strait to gather geological, historical, and ethnological information. When Rasmussen agreed to include Freuchen as the group's cartographer, Freuchen eagerly set sail for Greenland.

It proved a disastrous decision. Navarana contracted the flu on the west coast of Greenland and died shortly afterward. Freuchen, in a state of despair, decided to continue on the expedition anyway. His problems then multiplied. During the expedition, while venturing out in a snowstorm, he became lost and nearly died. He later explained that he was caught in a blizzard and dug a sleeping hole, covered it with his sledge, and was frozen tight there. He couldn't get out. Sometimes in telling the tale he also liked to add an apocryphal touch: That he considered defecating and then cutting himself from the ice with a frozen knife made of his own feces. In any event, Freuchen got out, and when he stumbled back to camp, the toes on his left foot were so badly frostbitten that gangrene set in, and he had to leave the expedition. Later he would say he cut several of his toes off himself with a tool used to pull nails, and then passed out. What is verifiably true is that a few years later he had the entire foot amputated, marking a definitive end to his exploring days.

Rasmussen suffered no such troubles. "Luck followed him everywhere," Freuchen would say of his friend. After he finished the Fifth Thule Expedition in Alaska, on August 31, 1924, Rasmussen traveled from Seattle to Washington, D.C., where he gave a number of speeches at various geographical societies and accepted an invitation to meet Calvin Coolidge at the White House. In New York City, Rasmussen also met with a number of journalists to discuss his journey. Amidst the bustle and the skyscrapers and the automobile-choked streets, he seemed to radiate curiosity and calm. One writer who met him at the time noted that

> the explorer is a man of medium height, slender and sinewy, whose outdoor life and battles with the elements have developed muscle. His hair is straight and black as an Indian's and his skin is bronzed by wind and sun. From under shaggy brows his eyes gaze forth with a steady clearness. They are a greenish-gray blue—and look as if they had taken to themselves the tones hidden in the depths of frozen water.

Rasmussen explained to the reporter that he had traveled in total some twenty thousand miles. He had visited regions and tribes that few outsiders had ever penetrated. He had collected about a hundred songs and two thousand interviews on his travels. And he had gained a far deeper understanding of the spiritual beliefs, family customs, and methods by which the northern tribes passed down oral histories. "Tradition is the foundation of their lives," Rasmussen said. " 'We do as our fathers did before us,' is their creed. 'Have you anything better to offer?' they ask. 'We are willing to listen.' "

"I found among the Eskimos an admirable quality," he added, "a humility of spirit which causes them to admit that man knows very little."[35]

In time, what came out of the expedition was an acclaimed book by Rasmussen, known in English as *Across Arctic America,* along with ten volumes of scientific observations, comprising thousands of pages—ethnographies; cultural studies; geographic, archaeological, and geological research; and botanical and zoological compendiums.

But by the time he made his visits to America's East Coast, he had been away from home—he had been *traveling*—for three and a half years. He felt a need to get back. In November 1924, he boarded a ship that sailed for Copenhagen. But it stopped first in Oslo. And when it arrived there, among the crowd at the dock, waiting to greet him and celebrate his return, were three aging and familiar faces: Fridtjof Nansen, Otto Sverdrup, and his old friend Peter Freuchen. For a short while, at least, Greenland's wanderers all stood still, together in one place.[36]

Alfred Wegener (Alfred Wegener Institute/Archive of German Polar Research)

7

TNT

Rasmussen's last great expedition resembled a cinematic finale to an age of exploration—the ultimate success before an encroaching era of technology arrived in the Arctic and the era of dog sledges and rugged adventurers faded away. But in truth, these seemingly distinct eras—exploration and science—had overlapped. Peter Freuchen's career was a case in point. Before he lost his foot, before he lost his wife, and before he lost his job at the Thule trading station, Freuchen's life had proceeded from a single momentous decision: In 1906, as a young college student in Copenhagen, he had dared to sign up with the *Danmark* expedition, the journey to the Arctic that required spending two long years on the isolated east coast of Greenland.[1] It was still a few years before he would begin his adventures with Knud Rasmussen; in fact, the two men had not yet even met. Freuchen's job on the trip, we might recall, entailed taking weather and temperature readings, a task that at one point forced him to live many miles from base camp, in a tiny hut with ice-encrusted walls where he was hounded night and day by wolves.

His official title on the *Danmark* journey was scientific assistant. And his supervisor was a young researcher from Germany named Alfred Wegener, who was designated the chief scientist of the expedition. There exists no record of the first conversations between these two men. Wegener's Danish language skills were poor, which isolated him from the rest of the *Danmark* team for many months until he became fluent enough to converse. From Wegener's journal entries and letters, though, it seems clear that he and Freuchen got along well—so well that their friendship remained intact many decades after their initial work together. Both men were affable, quick-witted, and technically adept. More crucially, as part of the *Danmark* group, the two shared an experience that could hardly be fathomed by anyone outside a tiny circle of explorers. The course of their lives was shaped by the fact that as young men they had lived through a Greenland journey so intellectually and spiritually powerful that each felt compelled to go back to the island again and again, for reasons of adventure and science, as long as they were able.

Wegener was twenty-six, seven years older than Freuchen, when he joined the *Danmark* group. Born into a highly cultured, middle-class household in Berlin, he'd spent much of his early life preparing for a career in academia, and by his early twenties he already seemed a stereotypical member of the scientific elite. Scholarly, athletic, charming—his fellow students and instructors would later remember how his penetrating blue eyes made an indelible impression on almost anyone he conversed with. Above all, friends and family would remember Wegener's *serenity*. Even in the face of terrible circumstances and overwhelming workloads, he came off as quiet and unflappable.

At this point in his life there was little indication that he would be regarded as one of the most important thinkers of the early twentieth century. Wegener's American biographer, Mott Greene, wrote of his subject that he had "that special form of perseverance that allows one to stay at exacting tasks for long periods of time, without succumbing to boredom or restlessness in a way that makes work impossible."[2] It was undoubtedly true—by habit the young professor worked constantly on his research, late into the night and through the weekend, oftentimes forgetting meals and family engagements; in his study, as he toiled through endless revisions and calculations, he puffed constantly on cigars or a long pipe stuffed full of tobacco. His dedication to work obscured a crucial aspect of his personality, however. He actually disliked being deskbound and had a recurrent craving for intense physical experiences—hiking, sailing, mountaineering—that led him to leave the city and test himself against nature. Indeed, by his early twenties, Wegener had moved on from the study of astronomy, his first interest, to physics and meteorology, in part because he had decided that "astronomy offers no opportunity for physical activity."[3] Meteorologists of Wegener's era were often engaged in invigorating outdoor fieldwork—constantly sending up and retrieving balloons and kites to collect atmospheric data in difficult weather conditions. Such work allowed Wegener to merge his love of science with his love of the outdoors, while also allowing him to explore his fascination with technological innovations, such as new gauges, instruments, and devices that were becoming crucial in his field. The stratosphere, the layer of thinner air above the more turbulent troposphere (the place where storms raged and clouds sailed by), had only recently been discovered. It was Wegener's good fortune that his career coincided with the start of what looked to be a golden age of atmospheric research, an era when the forces that created weather patterns could be deeply investigated and measured for the first time. It seemed possible, in fact, that better weather measurements would even make possible long-term weather predictions, which up to that point had proven difficult.

Wegener's first job after college was at the Aeronautic Observatory in Lindenberg, near Berlin, where he sent unmanned balloons into the sky to collect data—seven days a week if the weather permitted it. In the parlance of the era, he worked as an *aerologist* with an oc-casional interest in working as an *aeronaut*—that is, someone who liked to pilot a balloon himself. In 1906, he and his brother Kurt participated in an international competition known as the Gordon Bennett Balloon Race. Taking off in a large, hydrogen-filled balloon from a point in central Germany, the brothers mistakenly left their winter coats, along with most of their food and water, on the ground. Cold, hungry, but otherwise exhilarated, they drifted for several days over Germany and Denmark at altitudes ranging from three hundred to eight thousand feet. By the end, the Wegeners had flown for fifty-two and a half hours, an apparent world record, and likely would have stayed aloft longer if they hadn't become so worn down by fatigue and frigid temperatures. As Greene recounts, by the end of the flight they were slowly losing altitude and in such a weakened state "that when they tried to drop ballast, neither of them could push a sack over the rim of the balloon basket."[4] They landed softly and without a scratch, however. And two months later, Alfred left for the *Danmark* expedi-tion to Greenland.

In his late teenage years, he had read of Fridtjof Nansen's exploits on the ice sheet, and ever since Wegener had dreamed of going to the Arctic. When he finally arrived there, he found that his own mental and physical endurance, considerable by any standard, was severely challenged. Here in the field, far from the lecture halls and laborato-ries, he discovered a place of tremendous risk and isolation. On long, lonely walks along the ice at night, he was astonished by the lack of human presence and the stark ringing silence all around him. "Out here, there is work which is worthy of a man," he wrote in his journal on Christmas Day, during the first year of that two-year trip. "Looking nature in the eye and testing one's wit against its puzzles, this gives life an entirely new meaning."[5]

Was he more an explorer than a scientist? The distinction—largely a contemporary one and often used to distinguish men of ego and ob-session from men of research—is perhaps beside the point. Wegener was both. Many of his journal entries from the *Danmark* trip regard

his stoic acceptance of an Arctic expedition's daily discomforts—
miserable weather, frostbite, dreadful food, the long polar night, the
challenges in getting along with a small party of gruff men in trying
isolation. He wrote about how to endure the Arctic chill and how to
shoot encroaching polar bears. On a typical day, though, Wegener
and Freuchen also made painstaking measurements—still consulted
for the historical record today—with weather balloons and compiled
voluminous records of air pressure, wind, and temperature. These observations sometimes included ruminations on the beauty of the
northern lights and meditations on the techniques of dogsledding,
which he worked hard to master and at first approached as a scientist
rather than a sportsman. ("It is very interesting," he noted after his
first driving experience, "this incredible desire which animates them
all to pull the sleds along."[6]) But repeatedly in his journal he would
write: "On future expeditions . . ." And he would then make lists of
what kinds of foods or gear he might bring along if he returned to
Greenland. Also, which colleagues would he enlist to help him? "A
professional photographer should come along on the expedition," he
noted. "A professional cartographer is necessary as well. A meteorologist. A magnetician. An aeronaut for kite and balloon deploy-
ments . . . a geologist, a botanist."[7]

It became something of an obsession of Wegener's during the
winter of 1906 and 1907 to think ahead to what he might do in
Greenland *next* time.[8] The impression that comes through his diaries
is that he had discovered a region that had been explored but not yet
examined by the tools of modern science. And the mysteries here—
high above, in the atmosphere; and far below, within the depths of
the ice sheet—seemed so complex as to necessitate more time, more
men, more thought. By his return to Germany in 1908, he seemed
resolved to make another visit as soon as possible.

Of course there was a question: *Why* go back? Three men perished
during the *Danmark* expedition; during a long journey to the northern coast, they had become stranded and eventually died from starva-

tion. So Wegener knew the dangers were real. Why, then, leave a comfortable life in Germany and a promising career as an academic? Why leave behind his fiancée, Else? Why spend another year or two in what he would later call "the great white wasteland, the most absolute and lifeless desert on earth"?[9]

Two decades before Wegener arrived in Greenland for the first time, Fridtjof Nansen had declared that he would not only be the first man to cross the ice sheet, he would also be the first to make a detailed examination of the "meteorological phenomena" of the island's center. What Nansen meant was that he would take readings for temperature, wind, and precipitation in an icy region that no human being had ever visited.[10] Such information might have seemed inconsequential. Yet polar explorers of Nansen's era were routinely challenged, often by skeptical newspaper editors, about whether they would acquire on their travels knowledge of any real and practical value. In respect to researching the weather on and around the ice sheet, Nansen insisted on its worth: "Every single section of earth's surface stands in intimate and reciprocal relation to its neighbors," he wrote. And the inland ice cap—"so huge a tract of ice and snow"— must therefore have an as-yet-unmeasured influence on the surrounding climate.[11] Nansen's point about the ice sheet's effect on global climate was prescient. He was likewise correct in his larger assertion, which was that to understand the natural systems of the earth one needed to appreciate the characteristics of its individual regions. The interior of Greenland was a significant part of "the planet on which we dwell," Nansen remarked, and unless we came to know it, we could never truly understand our world.[12] The heir to this idea—that earth's ice, seas, and climate share a tangle of important and unexplored links, and that Greenland was a key to understanding a much larger complex system—was Wegener.

His first expedition there, as Wegener saw it, was merely a prelude to a long career of Arctic research. When he returned to Germany in 1908 after the *Danmark* trip, he resumed his teaching and academic work. He published a book that drew heavily on his atmospheric measurements in Greenland, and meanwhile began to develop a theory—a theory unrelated to his Arctic work—which proposed that all the earth's great land masses were once clustered together (later described as a supercontinent known as Pangaea) only to move apart over billions of years. His idea on the origin of continents and oceans, which in time became known as the theory of continental drift, or continental displacement, was surprising not only in its assertions, but in who its author was: An atmospheric scientist with an abiding curiosity in Greenland who had little training in the geological sciences. And in fact, even as Wegener was writing his momentous paper on the continents, he was considering his next trip to the Arctic. The stated goal of the *Danmark* trip had been to chart unexplored parts of Greenland's northern coast; the goal of the next expedition that Wegener was planning—what was eventually called the "Koch-Wegener" expedition, and which included Wegener's colleague J. P. Koch from the *Danmark*, along with Vigfús Sigurðsson (from Iceland) and Lars Larsen (from Denmark)—was to return to the same east coast harbor where the *Danmark* had anchored. From there, the men would spend the winter on the ice sheet—the first group ever to do so. And in the spring they would make a long trek across the ice to the west coast, not unlike Nansen had done years before, where they would take a boat back to Europe.

This time Wegener was intent on making detailed studies of Greenland's glaciers, upper atmosphere, geology, and northern lights. It would be a science expedition, pure and simple.

Wegener arrived on the island's east coast in July 1912. He began to keep a diary, as did his colleague Koch, the official leader of the expedition. Together, their journals offer a meticulous, day-by-day account of their journey. What happened was this:

The men brought with them sixteen sturdy Icelandic horses, rather than dogs, to pull their sleds, believing from experience that the horses would do better at carting the men and their twenty tons of supplies up the steep icy inclines on the edges of the ice sheet. When their ship landed in Greenland, though, thirteen of the sixteen horses ran away. The men succeeded in recapturing ten of them. They now had thirteen horses left.

A few weeks later, while crossing a glacier, Wegener fell and broke a rib. He nearly blacked out from the pain. "I slipped on the smooth, new ice," he scribbled in his journal that night, rigid and in agony in his tent. "I fell on my camera, which I had been carrying on my back, and it bored a hole in my back, just above my pelvic bone."[13] After allowing a few days for recuperation, the expedition kept moving, with Wegener bent over and hobbling in discomfort.

From the coast they tried to reach Queen Louise Land, a region located on the eastern side of the ice sheet, where they intended to set up camp for winter. To get there, they were forced to build a bridge over five deep crevasses and cut a passage through a glacier. In one account, the glacier suddenly calved, "and a block of ice a third of a mile wide fell into the fjord where the expedition had its horses and supplies."[14] Half of their supplies were destroyed, and four of their seven sledges were broken beyond repair. Abandoning the goal of Queen Louise Land, they set up winter camp in a small, prefabricated hut on the eastern edge of the ice sheet. There, they slaughtered several horses and fed the meat to the other horses. They now had five horses left.

J. P. Koch was out crossing a snow bridge not long after; when it collapsed he dropped forty feet into a narrow crevasse. As the team rushed back to camp to get a rope ladder, Wegener feared Koch might die. Koch did not. But in the fall he had broken his leg and lost at the bottom his theodolite, a crucial navigation instrument.[15] "For three months," Koch would recall, "I was confined to my bed and we could not proceed until March."[16] The temperature in camp hovered as low as −58 degrees Fahrenheit. In photographs from these months, a bearded Koch lies on his bunk, eyes closed, intently smoking a pipe and surrounded by books, his lower right leg swathed in bandages.

On April 20, 1913, they set out on a southwest course for the west coast—about seven hundred miles away. Winter was ending, but the weather proved worse than their worst expectations. Staggered by winds and blowing snow, the men often huddled in their tents, waiting for the blizzards to cease. Yet even on travel days they were desperate—walking sideways, walking backward—to avoid the scouring gales. The sun and wind and airborne ice crystals ravaged their skin. Wounds flowered on their cheeks and chins. Wegener noted the tip of his nose had frozen solid "at least 10 times" and that his face, resembling a leprosy victim's, was missing patches of skin and covered with "ulcerous yellow spots."[17] Temperatures bounced between –22 degrees and –30 degrees Fahrenheit. The horses, now almost totally snowblind, plodded forward in numb exhaustion.

Slowly, Wegener was growing aware, too, of how in Greenland the desire to survive so thoroughly overwhelmed the desire to think. Stuck in his tent during a snowstorm on the ice sheet, he noted that

> one would think this forced rest would encourage my mind to ponder, solve scientific questions, and to concentrate on things that I know I think about constantly when I am back home. But only once in a while do I find myself coming up with some unimpressive beginnings of ideas; all these problems, that of the volcanoes, the cyclones, the blue strips in the ice, the daily fluctuation of the barometer, the rotation in the solar system, etc. are always with me; they are always sitting, so to speak, right in front of me, yet my imagination does not make it through, and instead chooses other paths. It persistently returns to two things, back and forth, and both are of a shamefully material nature.

He wondered about the life he and his fiancée, Else, would begin next year in Germany. Also, he wondered about supper: "What kind of food will we cook?"[18]

The situation was becoming more desperate in late May. Realizing that several of the horses could walk no farther, the group agreed the sickest animals should be executed. That left three—Lady, Red, and Grauni. But Lady and Red, Wegener confided in his journal, were "half-dead," and Lady collapsed soon after. "And we still don't know if any of us will reach the west coast alive," he wrote.[19]

They were now nearly halfway across the ice.

On June 11 one of the last two horses, Red, was put down. That left Grauni. The men rallied to the sorrowful cause of the faithful gray horse. This last horse, they decided, had to be saved. And so they fed Grauni portions of their own scarce food. They pulled Grauni on their own sleds to ease its suffering. When they spotted ice-free land several weeks later, on July 2, they realized they had reached the western edge of the ice sheet. A year before, they had arranged to have a depot of food left just a mile away. But that looked to be on the far side of crevasses and streams covered by fragile ice. It seemed treacherous enough for men to get there—but a heavy, sickly animal? "Therefore," Koch would later say, "we were obliged to kill our last horse, our good friend in need."[20] More precisely, Wegener noted in his journal that several of the men made it to the depot and returned to camp with some bread for Grauni, only to find the horse lying on the ground, dying. "Our last bullet was used to end his suffering," he wrote.[21]

In truth the four men were nearly dead as well. They had only four hunks of black bread, some condensed milk, and a small companion dog, Gloë, who Wegener had bought just before the trip. On July 13 the men reached a fjord on the west coast, and by their estimation they were now close to Proven, a small coastal settlement. But hiking along the coast involved going up and down steep terrain, and to compound their difficulties they hit rain, snow, and fog. To stay alert they sniffed drops of camphor, but eventually even this couldn't subdue their exhaustion. Starving and soaked through—and now in a state of continuous nausea and dizziness—the four built a shelter out of rocks and moss, curled up inside, and rested for thirty-seven hours. "I was appalled," Wegener would later recall of these thirty-seven hours. "After surviving such a long and dangerous journey, and less than 2 miles from the colony, we were to die here like animals? In July? Was there even a trace of logic to that? Everything inside me rebelled against this thought, and I concentrated all my mental strength on this: I want to live."[22]

And so, acting on the notion that a final meal might give the men strength to continue, they agreed to slaughter Gloë, their small dog, and use the meat to make stew. As they divided the food and sat down to eat—at the very *instant* they sat down to eat—Wegener thought he saw through the fog a sailboat off the coast. He peered through a telescope to make sure he hadn't mistaken an ice floe bobbing in the water for a boat. It was July 15. The men had been in Greenland for almost exactly one year. After signaling and yelling desperately to the ship, they watched it veer toward the shore. An hour later they were aboard, eating eggs and fresh bread, sipping coffee, and puffing on pipes, now stuffed with fresh tobacco given to them by an obliging crew. Their rescue could never make for a good novel or film; the timing, all true, was too perfect to believe.

Wegener would later consider the expedition a success. Somehow he could put aside the string of calamities—events that would leave most men to imagine they had been cursed by the gods—and conclude that he had not only completed a monumental crossing of the ice sheet but had pushed the cause of Arctic science forward. He wasn't convinced he had yet discovered anything of deep significance. But he and his colleagues returned with a rich written and photographic record of their trip, some of it chronicling their science endeavors (studies of the northern lights, the ice sheet's topography, and Greenland's weather) and some of it chronicling the journey itself. Indeed, the record of the journey would later be featured in Koch's published journal of the crossing, *Through the White Desert,* which captured the stark scenery of the ice sheet without quite conveying its unceasing misery.[23] Also tucked into Koch's journal was a photograph of a pit, comprising descending steps that extended to twenty-three feet in depth, which the men had dug into the ice as an experiment.

The pit was used to study the temperature of the central ice sheet. "This was hard work today, to dig 7 meters deep," Koch wrote on June 12. "From 5 meters downwards the *firn*"—the name for deeper, older, more granular snow—"became so hard that a single powerful spade stroke went only half a decimeter deep."[24] But by measuring the snow temperature deep in the "big hole," as Koch called it, the men were able to determine the temperatures from past years and could estimate that the annual average temperature of central Greenland was just below −30 degrees Celsius. What's more, their pit revealed a layering of the ice sheet—how each year a winter's snowfall had piled on top of one from the year before, which in turn had piled atop one from the year before that, so that each year's snow had pressed down and added incrementally to the enormous bulk of ice they were crossing. Wegener didn't remark on whether (or how) this might serve as a future point for inquiry. Still, he wrote in his journal, "We are pleased with the good results."[25]

Koch's journal of the crossing was published in 1919, six years after the men had returned to Europe. In the intervening time, Wegener had endured several busy and difficult years in Germany. After Greenland he had returned to university life in Marburg, Germany, and married Else, his fiancée; they soon had a daughter and would eventually have two more children together—all girls. In 1915, *The Origin of Continents and Oceans*, his book-length exposition of his theory of continental drift, was published, meeting with a response that ranged from enthusiasm to incredulity. But Wegener's return also coincided with World War I, during which he served as a soldier and was shot twice—once in the arm and once in the neck. And then, in the postwar gloom of Germany's defeat, as his country's economy descended into a perilous free fall, Wegener found he could barely survive as a scientist. He worked as hard as ever, eking out a living on small stipends from teaching and modest paychecks for writing assignments. His family scavenged in a nearby forest for wood to heat their apartment. Wegener's biographer Greene relates that just after the war ended, in 1918, Wegener, at thirty-eight years old, "was cold, exhausted and broke."[26]

He had the good fortune, in 1919, to secure a research and teaching position with the German Marine Observatory in Hamburg. And while he still struggled financially in Hamburg—he and his extended family used large tracts of their yard, along with some nearby lots, to grow vegetables to feed themselves—what ultimately changed Wegener's life was a growing appreciation for his atmospheric research and his theory of continental drift, along with an appointment in 1924 to a professorship at the University of Graz, in Austria. In Graz, Wegener was at last financially secure. Between teaching and research, he could afford weekend ski trips in the Alps with colleagues and summertime hiking tours with the children.

"We had almost settled down to a life of middle-class comfort," his wife, Else, would later write. "It seemed as if this would go on indefinitely."[27]

But he began thinking again about the Arctic. In 1928, one of Wegener's former students, Johannes Georgi, wrote to say that he was planning to set up a scientific base in the center of Greenland's ice sheet to launch aerial balloons and research the jet stream, which was barely understood at the time. Wegener seemed electrified by the idea, and it made him think back to his first expedition to Greenland in 1906. He wrote back to Georgi: "You mention also the question of a station on the inner ice cap. That is an old plan of Freuchen's, Koch's, and mine! If only the War had not happened, it would have been carried out long ago. But meanwhile Freuchen has lost a leg, Koch is in the hospital, and I too have had my own trouble and am no longer a young man."[28] This much was true—his old journeys to Greenland had occurred decades ago and seemed almost to have been part of another man's life.[29]

He was now forty-eight. He smoked heavily, but still appeared healthy—barrel chested and athletic, standing a fit five feet ten inches. If he joined Georgi in Greenland, it would mean leaving his wife, daughters, and professorship for at least a year, maybe more. Should he? This would almost certainly be his last chance to go north again.

What pushed Wegener's interest into a full-fledged commitment was a visit to Graz in the spring of 1928 by a scientist named Wilhelm Meinardus, who told Wegener of a new technique that a colleague of his had developed for measuring the depth of glaciers.[30] After triggering an explosion with dynamite or TNT, Meinardus said, a researcher could rig a seismograph to gauge how long it took for an "earthquake" wave to travel from the ice surface, through to the bedrock, and then back again to the surface. In this manner, they could determine the thickness of the ice. Meinardus suggested Wegener try the technique in Greenland, and Wegener saw immediately that it might solve a difficult question: How deep was the ice sheet in its center? Within a

few months, he formally proposed that Germany fund a massive Arctic expedition, led by him, that involved setting up not one but three
scientific stations in Greenland—one each on the west and east
coasts, and one on the ice sheet's center. Some of Wegener's goals
were now glaciological: to measure the thickness and interior temperatures of the ice sheet, to determine whether the ice sheet was
growing or shrinking, and to understand the flow and movement of
glaciers. The other goals incorporated Georgi's ideas to study weather
and climate. The center of the ice had never before been observed
during winter. Thus the researchers at the three stations, which would
be built in a line along latitude 71° north (the Arctic Circle is at
66°33′), would make continuous meteorological observations for at
least a year, so as to better understand the seasonal climate patterns
surrounding—and influenced by—the ice sheet.[31]

As Wegener began planning the project, he also was corresponding
with his old scientific assistant Peter Freuchen who was at that point
working as a writer near Copenhagen. Twenty years had passed since
the two men worked together in east Greenland. Wegener visited
Freuchen, and he related his intent to create a station in the center of
the ice. He also explained that his team of scientists would be large—
ultimately, it would number twenty-two—and that he would no
doubt bring a substantial amount of gear, food, and fuel.[32] In effect, he
was facing the same problem that every explorer of the ice sheet since
Nansen had had to face. There would likely be sheer cliffs of ice to
overcome on the west coast, which was Wegener's preferred place for
a base station, and almost certainly deadly crevasses. Wegener asked
Freuchen: *Where, Peter, would be the best place to climb onto the ice?*

Time in Europe—time that advanced with the chimes from the central clock tower in Graz; time that progressed by the meeting of dead-
lines for scientific research papers; time that moved ahead in family
dinner conversations and, afterward, in the agreeable curl of pipe
smoke—was not like time in Greenland. On the coasts of Greenland,
where quiet villages softened the stark landscapes, time moved more

slowly, more strangely, marked by hunting seasons and cold spells and iced-in harbors rather than hours and days. Rasmussen and Freuchen understood this especially well. Calendars and clocks were of little use. Time in Greenland was indifferent to man, and often cruel toward him, and seemed deferential only to nature. And for that reason, on the occasion of a final opportunity to study Greenland, time became Wegener's adversary.

In the spring of 1929, he sailed to the island to investigate the coastal geography and picked Kamarujuk Glacier, about halfway up the west coast, as the point where his expedition would begin its ascent onto the ice sheet the following year. He brought along three German scientists he had chosen as the main investigators of the expedition: Johannes Georgi, the former meteorology student; Ernst Sorge, a teacher with an expertise in glaciology; and Fritz Loewe, another meteorologist. Kamarujuk, they agreed, was fairly steep, but seemed superior to other entry points Wegener's team had seen.[33] The glacier was two and a half miles long—a smooth river of ice that flowed down a narrow valley from the central ice cap. At the top, where it joined the ice sheet, the altitude was about 3,280 feet above sea level; at the bottom, the glacier ended near sea level about a quarter of a mile from the coast, which was where a ship could unload supplies. "The main difficulty of the undertaking," Else Wegener would later write, "would be the wearisome and lengthy job of transporting the scientific instruments, winter huts, fuel, and stores for man and beast up on to the ice-cap." Indeed, the complexity of the task was exhausting even to contemplate. Kamarujuk was the ice road Wegener's team would follow, lugging several thousand boxes and containers that weighed, all told, 240,000 pounds. Up the ice it would go—by horse-pulled sled, by dog-pulled sled, by rope and winch, by hand.[34] Some of the supplies would be directed to the west coast research station near the top of the glacier, which was known as Scheideck; the rest would be carted about two hundred and fifty miles east, to the center of the ice sheet, to a station Wegener christened "mid-ice," or Eismitte.[35] An east coast station—the third leg of the expedition—would be supplied by a separate effort.

Wegener returned home to Germany to plan and pack during the fall and winter of 1929.[36] In the spring of 1930, he set off again by ship for Greenland. Before he left, a journalist asked Wegener why he kept returning to the Arctic—now for the fourth time. He said, "It is one goal and one alone that pulls us into the frozen wastes: the joy of battling with the white death. Vast expanses of the earth are still closed to mankind by barriers of ice and snow. It is our mission to open these tracts and conquer the forces of nature, bending them to our will."[37] Johannes Georgi would later say that Wegener's deeper nature was that of "a lion behind the lamb"—and indeed these remarks suggest a glimpse of swaggering bravado. But usually Wegener was a man of careful and thoughtful temperament. What's more, just a few weeks later, when he arrived on the southwest coast of Greenland, it became obvious to him that nature was not bending, that nothing was being conquered, that willpower was almost irrelevant. What mattered was time.

The area where Wegener chose to ascend onto the ice sheet is situated in a landscape of islands and inlets and small peninsulas that, like most of the Greenland coast, was sealed tight during the cooler era of the 1930s with pack ice, which held fast to the coast from midwinter until late spring. Wegener was hoping he could bring his ship to unload supplies at the foot of the Kamarujuk Glacier in mid-May. But the ice in the fjord remained frozen, and so he waited on the ship, many miles south of the unloading point, knowing that each day his chances of success—his chances of getting all the gear onto the ice sheet and building a central station—would diminish. The last weeks of May 1930 slid by. His team tried to break up the ice with explosives, without much success. And then June began. Even for researchers of today, blessed with a passel of modern technologies to ensure their safety, the science season in Greenland is short. Just as for Wegener, it tends to run from May through September, reflecting a keen awareness that weather and darkness make travel in the late fall difficult, and sometimes impossible. As Wegener waited for the ice to break up near Kamarujuk, his journal entries became edged with panic. "Our expedition's program is slowly being seriously jeopardized by the obstinacy of the ice," he wrote on June 9.[38]

At last the harbor ice broke up. Wegener finally landed his supplies on the rocky shore in front of Kamarujuk Glacier on June 16—thirty-eight days later than he'd planned. But his anxiety increased as the magnitude of the job ahead became clearer. Wegener hired teams of Greenlanders to help his men move gear up the glacier. Silent film footage taken during these weeks chronicles the abject misery of the work—lugging heavy boxes up the ice, by dogsled and horse team, slipping and sliding and hopping over small crevasses in the process. As the weeks progressed, transporting supplies became even more difficult: The ice began to melt, and creeks and gullies cut into the glacier and made climbing perilous. The men began to move the gear at night, when temperatures were cooler and the mosquitoes less ferocious. All the while, Wegener watched and worried. "I'm afraid, really afraid, that we're not going to make it," he wrote in his journal on August 5. "It seems that we are slowly moving towards an increasingly unpleasant predicament."[39]

His biggest worry involved getting supplies to Eismitte, the camp in the center of the ice. The scientific validity of the entire expedition depended on building a functional station there before mid-September. Georgi had already gone ahead in mid-July to set up the station and begin his research program; two deliveries of supplies, made by dogsled teams, had followed. But the outfitting of the station had not been done with Wegener's usual forethought, and a prefabricated hut, radio transmitter, and a full load of food and fuel had not yet been delivered.[40] And now summer was almost over. "Providing supplies to Georgi's station is getting harder and harder," Wegener wrote on September 3. "If we are faced with autumn weather and strong winds from now on, everything will go even slower." By September 6, he was inconsolable. "Disaster has struck. It won't be possible to equip [Eismitte] in the way we had planned."[41]

One of Wegener's great hopes at the start of the trip was that technology would make this research expedition different, as well as easier. He had brought to Greenland two large, bright red, propeller-driven sleds—rather than snowmobiles, they resembled small fighter jets—that would carry men and cargo from the edge of the ice sheet to the central station. To Wegener, the machines would eliminate the difficulties of leading—and feeding—dogsled teams across the ice cap. All he would need were some cans of fuel. In his journal, he wrote optimistically, "I have the strong feeling that we are approaching a new era of polar exploration."[42] But it was turning out that the sleds didn't work well in windy conditions or on steep terrain. Also, they struggled to move in deep snow. Peter Freuchen, on a writing assignment in Greenland, came to meet his old friend Wegener in early September. "With some difficulty I managed to climb the glacier with him," Freuchen recalled of his time at Kamarujuk. The steep hill of ice, crisscrossed with crevasses, was difficult for the big Dane, who was hobbling with his wooden foot. Yet Freuchen succeeded in making it to Wegener's western camp on the ice sheet and stayed for two days. Wegener showed him the motorized sleds. He even took Freuchen on a ride, and the latter seemed to think the machines signified a fantastic innovation. "With a speed of fifty miles an hour we zoomed across the ice," he wrote of the sleds, "and in one hour we covered a greater distance than Knud [Rasmussen] and I were able to travel in one day."[43]

Even so, by mid-September, after a demonstrated lack of reliability, the team had abandoned the idea of using the motor-sleds to get to Eismitte. And by that point, Wegener was weighing whether to make a final two-hundred-and-fifty-mile journey by dogsled to the central station. He was certain that Georgi and Ernst Sorge, both now living at Eismitte, needed more food and fuel; without another delivery they might not survive the winter.[44] As leader of the expedition Wegener understood their fate rested on him. He deliberated for a few days, and by September 18 he decided to set out for Eismitte, accompanied by thirteen Greenlanders and the meteorologist Fritz Loewe.[45]

It was late, very late, in the season, he knew, perhaps too late. It was likely that no human had ever ventured so far onto the ice sheet at this time of year. On the morning of September 21, 1930, Wegener nevertheless left camp on the west coast with fifteen sledges and one hundred and thirty howling dogs. He was carrying four thousand pounds of supplies, heading east, into the wind that rushes down the ice sheet's western slope. He was intent on saving his colleagues as well as his grand experiment, his caravan moving toward the center of Greenland, toward Georgi and Sorge, toward the island's coldest place.

Georgi and Loewe under the ice (Alfred Wegener Institute/Archive of German Polar Research)

8
Digging

Eight weeks before Wegener began his journey to Eismitte, on July 30, 1930, Johannes Georgi had stepped off his sledge and ordered his dogs to halt. By Georgi's estimation, he and his companions—a German scientist named Karl Weiken and four Greenlanders—had traveled two hundred and fifty miles over the ice cap and reached an altitude of 9,850 feet. Traveling from the west coast station, it had taken them fifteen days. In accordance with Wegener's wishes, this was the place to set up the mid-ice science station. What the men actually did was dump a tremendous number of crates and packages haphazardly on the ice, along with piles of clothing, a pair of skis, and a tent. They helped Georgi set up a barometer and thermometer station. Then five of the expedition members turned and headed back to the west coast with their dogs. Georgi looked around—it was a splendid day, 18 degrees Fahrenheit, "bright and calm," the blue sky streaked with cirrus clouds. The world stretched away from him in all directions to white vanishing points. He took off his clothes and washed himself in the snow. He wrote his wife a letter. He was alone in the center of the ice sheet.

He could soon see the weather here had wild swings—blowing and blizzardy one moment; sunny and still the next. Inside his tent the temperature seemed to shift from subzero cold at night to uncomfortable warmth, sometimes up to 50 degrees Fahrenheit on a sunny afternoon. Georgi therefore decided on his second day to begin digging out steps and a small room under the surface of the ice cap to house some of his weather instruments; his goal, he later explained, was "to protect them as far as possible from the sharp fluctuations of temperature."[1] He was digging into firn, the crusty snow that comprises the past few decades of accumulation on the ice sheet and has not yet compacted fully into ice. He used a shovel, a saw, and a long knife. And in the course of his digging Georgi decided not only to create a room beneath the ice, but a small round castle above it. "I cut the ice out in great blocks," he would later recall, carried "them up the steps to the surface, and [used] them to build a wall as a protection against drifting snow, and, later, an observation tower." Eismitte was initially planned as a cluster of huts and tents, but it soon began to appear from afar as a scaled-down medieval fortification.

Months before, Georgi had insisted to Wegener that he would be fine staying alone at Eismitte. Almost from the first day he was left behind there, though, he was beset by an ache of loneliness. His weather observations, done religiously three times a day, kept him somewhat busy. He was an obsessive photographer, so he also set about chronicling mundane jobs and documenting passing cloud formations. On many afternoons, he would put on skis and explore the perimeter of mid-ice for several hours. Sometimes he would do so wearing no clothing at all, save for ski boots. And all the while, there were meals to cook and eat: rye crackers with butter and sardines, or soups of pemmican, the dense cakes of dried beef and fat, that he mixed with canned vegetables. And yet his anxiety lingered. Each time Georgi would step out of his tent, he looked to the west, where black flags on tall stakes marked a track toward the coast, two hundred and fifty miles away. He was hoping to discern dots on the horizon moving toward him. The plan had been to have three more sledge teams visit Eismitte with supplies before winter began. Where were they?

On August 9, Georgi noted in his journal: "I often imagine to myself what the column of sledge[s] will look like in the distance." A few days later, on August 11, he admitted: "I looked at least twenty times along the track to westward, down which my comrades must come." On August 17, his third Sunday, he confided to his wife in his journal that he wished she were there, just to see his predicament. "I understand that my words give no real idea of it, nor even photographs," he wrote. "The contrast between the boundless wastes all round and the tiny scrap of individuality cannot be reproduced. Tent, man, instruments—from three miles off not a trace of them can be seen."[2] A sledge party finally arrived the next day, five Greenlanders and Fritz Loewe, hauling eighteen hundred pounds of supplies. Georgi was ecstatic. But after dropping off the cargo, the party headed home twenty-four hours later, and he was alone again. He returned to his work—measuring temperature, air pressure, and wind, and sending up balloons to get a reading on conditions in the upper atmosphere. More meals were cooked and devoured. The sun was setting earlier in the day now, the weather turning colder. His prefabricated hut had not arrived from the west, and living in the tent was becoming unpleasantly cold. Meanwhile, he kept digging into the firn. His tower of ice blocks reached a height of nine feet.

He resumed his habit of looking constantly to the west.

A third sledge party arrived four weeks later, on September 13. Led by the glaciologist Ernst Sorge, the team brought three thousand pounds of provisions and fuel. "My solitary spell is over!"[3] Georgi would later exult in his journal. What he meant was that when this sledge party returned for the west coast, Sorge stayed behind. The plan was for him to remain with Georgi for the winter, doing experiments on the ice and in the late spring or summer doing seismic work involving explosives to gauge the thickness of the ice sheet. As Sorge put it, "The snow-line was the line of demarcation" between the responsibilities of the two men. The meteorological work ("everything above" the snow) was investigated by Georgi; the glaciological work ("everything below") belonged to Sorge.[4]

Sorge, thirty-one years old and recently married, was energetic and optimistic. Georgi, forty-one, was chatty and intense, and his angst seemed not to interfere, at least not yet, with his affability. The two men got along well—and both, as it happened, liked to dig. Within a few days of his arrival, Georgi noted that Sorge had begun "digging [another] large room in our subterranean world" to house the seismic equipment he would use to measure the thickness of the ice.[5] When storms blew through the camp in late September, the men began spending time in this new underground cave as a sanctuary: It was warmer than their tent up above, they realized, and protected from the savage wind and snow. "My digging experience had shown that only six to nine feet down the ice was very firm and strong," Georgi would later write. So he suggested they excavate "a fairly large space and erect our summer tent in it, far below the surface."[6] Eventually, the duo disposed of the idea of bringing the tent down. They simply sculpted bunks out of the firn, on which they laid their fur sleeping bags. With a ventilation hole and a judicious use of their stove, they discovered they could keep the temperature in the room between 23 and 32 degrees Fahrenheit.

On October 5, 1930, Sorge and Georgi officially moved in, six feet under the ice. Almost immediately they considered it an immense improvement in comfort over surface living. Eismitte, now a subglacial bunker, was entered by descending a set of steps next to the castle tower and passing through a door made from discarded sacks and furs. The strangeness of the new digs was not lost on the two scientists. Sorge would later write of the ice cavern that "our first and strongest impression was that we were lying in state in a crypt. Everything was white like marble, and our sleeping-places were clean-cut and rectangular like the marble base of a sarcophagus." When daylight filtered down through the ceiling of ice, its colors were refracted so that it illuminated the room with a surreal blue glow.

The men now had time to ponder their dilemma. They had given up hope of getting a delivery of the prefabricated winter hut; and they had not yet received the wireless receiver they had brought to Greenland, so there was no way to contact the west coast station. Originally they had calculated they couldn't possibly survive the winter without enough fuel—so they had sent a note back to Wegener when the third sledge party departed, on September 14, telling him that if they didn't get a fourth and crucial delivery by October 20 they would set out for the west coast camp and return by foot.[7] But then October 20 arrived. By that point Sorge and Georgi had been living in their under-ice cavern for two weeks, and the men had come to the conclusion they could likely manage their small supply of fuel through the coldest months, even as temperatures at Eismitte were now dropping on some days to –50 degrees Fahrenheit, accompanied by whipping winds. They doubted a fourth delivery would make it to them before winter. To make a 250-mile journey in this weather—*to* or *from* the west coast—seemed unthinkable. He and Sorge would therefore stay here.

For now, the men followed their scientific regimen and, during idle moments, forced themselves into a state of chilled hibernation. Three times a day, Georgi—or on Sundays, Sorge—would go above, clad in furs, to take in some fresh air and conduct weather readings. With the hours of daylight waning, this often meant working under moonlight and bright, starry skies, regardless of the temperature. On some evenings they stood under the glowing aurora borealis—the northern lights—and gaped in wonder. "We were gradually coming to understand Nansen's enthusiasm for the Arctic night," Sorge later recalled.[8] When they weren't working they wrapped themselves in reindeer sleeping bags down below, trying to conserve their energy as well as their meager fuel supply. They were in their ice cave on the afternoon of October 30, on their bunks and ensconced in reindeer fur, when Sorge heard the rasp of dogsleds and some muffled voices on the surface of the ice above. It was –60 degrees Fahrenheit outside. "They're coming!" he shouted, both in shock and excitement—at which point both men leaped from their bunks and ran up the staircase carved from ice.

*   *   *

Wegener's 250-mile journey from the west coast to Eismitte had begun on September 21 and was supposed to take sixteen days. It had taken forty. Soon after he arrived, he told Georgi "that it had been the worst journey in Greenland he had ever heard of."[9] He had begun with his colleague Fritz Loewe and thirteen Greenlanders, but as temperatures plummeted and conditions worsened, many of the Greenlanders, fearing they would never reach the center of the ice alive, turned back. At the ninety-four-mile mark the party was whittled down to three: Wegener, Loewe, and the Greenlander Rasmus Villumsen. The two Europeans were lucky to convince Villumsen to continue with them. Partly it was his experience as a dogsled driver. But also it was the fact that they could rely on the native Inuit people's uncanny ability to *see*. Nearly a century after Wegener made this journey, scientists could still be astonished by the Inuit's skill at identifying objects in their native landscape that are wholly invisible to outsiders—a polar bear clambering on distant sea ice, for instance, or a group of narwhals swimming miles out. Indeed, western researchers sometimes struggled to explain an acuity of perception that was almost superhuman.[10] When Villumsen went ahead to lead the three-man team, Loewe was awestruck that he could pick out in the twilight black flags marking the route—"only a fraction of a square inch showing in a small snowdrift scarcely to be distinguished from any of the others." Without his gift, they might never have made it.

The journey was not only difficult physically. It was difficult emotionally. Not long after he left the west coast, Wegener had received the note that Georgi and Sorge had written—it had been handed off to him, en route, by the third supply team, which crossed his path as it was returning from Eismitte. The note informed him that if the men didn't get more supplies and fuel they would set out by foot for the west coast on October 20. Wegener believed such an act at this time of year was tantamount to suicide. And while the note had not *prompted* Wegener to make the journey to Eismitte, it had hardened his resolve not to turn back. "Whether Sorge and Georgi will be able to remain there or will come back with us remains to be seen," he wrote during this journey in late September. "The whole business is a big catastrophe and there is no use in concealing the fact. It is now a matter of life and death."[11] As Wegener's team had progressed along the route to Eismitte they kept a constant lookout for Georgi and Sorge, who they anticipated might be heading in their direction on foot. They never saw them, of course, because the men had never left Eismitte.

The fact that Wegener had endured such an awful journey so late in the season was something of a miracle. But there were other aspects of his arrival at Eismitte that made an impression on Georgi and Sorge. The first was that Wegener had exhausted his own food on the way and had been forced to dump all the other fuel and supplies he was transporting. Every morning, Wegener had made the same kind of calculations that Peary and Rasmussen and Freuchen had made years before in considering their odds against the ice sheet on any given day: *Men and dogs and distance and food. What can be accomplished?* In the end, the urgent journey to bring cargo to Sorge and Georgi had become a desperate race for his own survival.

What also struck Georgi and Sorge was that Wegener looked to be in robust good health. He seemed amazed by the subglacial cave the two scientists had carved out and paced excitedly around the strange alabaster chamber. "You *are* comfortable here!" he exclaimed over and over again. Descending into a room of calm air at 25 degrees Fahrenheit from a wind-buffeted world of –60 degrees Fahrenheit apparently made the new quarters feel almost tropical to him. And he seemed particularly enthused that the scientific viability of Eismitte could be maintained through the winter. Wegener spent hours talking with the men about their weather measurements and taking notes as they described the building of their underground station. The men meanwhile snacked constantly and sipped coffee. Finally, they slept: Four German scientists and a Greenlander, for two nights, crammed together into the ice cavern.

The most consequential aspect of Wegener's arrival involved Fritz Loewe. Toward the end of the journey, Loewe had lost all feeling in his toes. Every evening and morning in the tent, in fact, Wegener had spent hours massaging the other man's feet in an effort to restore the circulation. And to no avail. After arriving at Eismitte and conferring on how to proceed, the group agreed that it was too risky for Loewe to return to the west coast. He would remain here with Georgi and Sorge, while Villumsen and Wegener would return together by dog-sled.[12] The idea of all five staying at the station wasn't seriously considered. There wasn't enough food or fuel for everyone.

On the morning of November 1, it was time to go. Georgi took some photographs and film footage. Eighty-five years later, the images of that day convey a strange optimism, along with a striking poignancy: Wegener and Villumsen, clad in thick furs, shaking hands and getting ready to leave . . . Wegener putting on his thick gloves and waving . . . Wegener getting on his sled . . . and then Wegener cracking the sealskin whip as his dogs begin pulling.

Loewe was downstairs in the snow cavern; with his frozen toes, he was unable to walk. But Georgi and Sorge watched as their visitors departed and disappeared toward the west. If Wegener's journey to Eismitte had obviously been made too late in the season, they understood the risks of his return journey were beyond reckoning. They could only take comfort in the fact that in traveling west, Wegener and Villumsen would be going downhill on the ice sheet and would likely have the wind at their backs. In addition, the temperatures had crept up a bit. On the day of their departure it was slightly overcast and −38 degrees Fahrenheit. "Splendid sledging weather," Sorge would later recall. Also, it was Alfred Wegener's fiftieth birthday. The men had celebrated the night before by each eating a frozen apple.

Several years later, Fritz Loewe would share a story about a conversation he had with Wegener during these strange and difficult weeks. On the sledge journey toward Eismitte, Loewe and Wegener had waited for a day at mile 204—meaning they were still forty-six miles from Eismitte. It was forty degrees below zero, and dusk was falling. The wind, which had been blowing steadily in their faces, quieted. The men pitched their tent and strolled around the vicinity, as was now their habit, to see if they might intercept Georgi and Sorge, walking westward home. "But nothing appeared," Loewe would later recall. Wegener then took the discussion in an unusually candid direction: He "spoke openly (as he did but rarely) of his belief that there is purpose behind man's evolution, and that mankind will ultimately be liberated by the growth of knowledge." To contribute to the sum total of human understanding, he said to Loewe, was his ideal in life. Loewe accepted it as a kind of explanation. It was why they had taken this nearly impossible journey. It was why Eismitte existed at all.

At the time of Wegener and Loewe's trek, much of Greenland's geology and geography had been charted or was in the process of being studied. This was the dividend of the arduous expeditions conducted by Nansen, Peary, Rasmussen, Freuchen, and Wegener, who between 1888 and 1920 systematically explored remote coasts and glaciers, and who built upon research by the Danish scientist Hinrich Rink as well as a number of other Scandinavian and Swiss adventurers.[13] In the early 1920s, moreover, the Danish geologist Lauge Koch, a former colleague of Rasmussen's, spent several years on Greenland's eastern and northern coasts, doing exhaustive geological and geographical studies. All this cartography was no mean feat. To look at a detailed modern map of Greenland is to confront its dizzying complexity—vast archipelagoes of islands that seem to splinter off from every mile of an epic coastline that itself is pierced by uncountable fjords, bays, and snaking inlets. "After two generations of incessant effort on the part of Danish explorers," the New-York Tribune reported not long after Koch returned, "Greenland, the most northern country of the world, has been thoroughly explored, its mountains and waterways defined, and its contours mapped out."[14]

This might have given Americans the impression that the far, frozen island was perfectly known. But that was hardly true. Greenland's weather, for instance, created in part by the massive stretch of ice in the center, was barely understood. Georgi's measurements of temperatures and winds—along with his balloon ascents when conditions permitted—were intended to help shed light on how these factors influenced the weather of Europe. The work was thought to have other practical applications, too. By the time Eismitte was established, the question of whether an airport or refueling station should be built in Greenland had become a matter of serious debate on both sides of the Atlantic. Aircraft were not yet dependable enough, in Wegener's view, to use them to supply his expedition in 1930 (in fact, the first flight across the Greenland ice sheet didn't occur until 1931). But as one polar historian noted, a realization had dawned around that time that "commercial aircraft, if they could be safely navigated across the Arctic . . . could greatly shorten the flying time between the great centers of population in the eastern and western hemispheres."[15] This possibility led other scientific expeditions besides Wegener's to investigate Greenland's aviation potential, including possible landing sites on the coasts.[16]

There were larger unknowns here, though. In 1930 the science of Greenland's ice sheet remained mostly a mystery. How did it move and fracture and melt? What lay below? How deep was it, and how old? How did it build up, and how might it wear down? Some of these questions might eventually be answered by Sorge's seismic work with explosives at Eismitte, which was scheduled for the late spring or summer. For now, the task for Sorge, the station's glaciologist, was to dig a deep hole. Wegener had perceived the value of probing the ice sheet as far back as his 1913 Greenland expedition, when he and his colleagues had drilled into the ice with augurs, and in the midst of their crossing had dug the "big hole," as J. P. Koch had described it. They had done so to measure the average temperature of the ice sheet and to view what glaciologists would eventually come to describe as the ice sheet's *stratigraphy*—the layers of snowfall that build up and densify over time to create the bulk of its ice.

Sorge began to excavate what would eventually be called "the shaft"—his primary job that winter—in mid-October 1930. If you were to enter Eismitte at that time, you would go down the short flight of steps into the ice sheet behind the central castle tower and find yourself in a tight, tunnel-like passageway. To the immediate right was the small subglacial vault for Georgi's barometer and meteorological equipment. If you took a few steps farther, you faced three doors. To the right was an entrance to the men's living quarters, which included their bunks and kitchen; to the left was a storeroom for food, fuel, and equipment. Straight ahead, behind a wooden trapdoor, was the entrance to Sorge's shaft. Seeing as they had no rope ladder, Sorge decided to construct it not as a hole but as a set of steps at an angle of 45 degrees. "Of course that means more work," Georgi wrote in his journal at the time, "because it involves more moving of material. But, on the other hand, it is not dangerous, as we shall have proper steps to walk on and a shaft thus built can never fall in."[17] Sorge would dig blocks of the firn with a shovel and saw; then he would haul them up the stairs. By mid-January 1931, his staircase had reached twenty-seven feet deep; by early March, it was thirty-three feet deep. When Sorge reached thirty-six feet down he decided to stop building steps and continued the shaft as a perpendicular hole that he could access with handholds on the icy wall. The shaft ended at forty-eight feet, but another, smaller hole went down an additional four feet. In all, he reached fifty-two feet below the ice sheet surface.

The ice shaft terrified Georgi. He feared slipping on the steps and breaking a bone and being left incapacitated and abandoned at the pitch-dark bottom. He went there as rarely as possible. Sorge seemed untroubled by his daily descent, however; he worked assiduously, and daily, digging by pale lamplight and also taking readings from the fourteen thermometers he had inserted deep into the walls. He was especially meticulous in sawing out excavation blocks and carrying them up to the living quarters, where he could measure them. In particular, he wanted to see how the depth of the snow related to its density. Thus he would saw the blocks to a precise width, depth, and height, so as to calculate volume; then he would weigh them on a scale. Density equals mass divided by volume. The farther down one went, the denser the snow. That didn't seem a surprise. But he also could measure significant differences—"perfectly regular fluctuations"—between the density of snow from each summer and winter, even if he couldn't see any horizontal stripes on the ice block itself.

With this insight, Sorge calculated that in going down fifty-two feet, he had reached snow that fell on the ice sheet twenty-one years before. In essence, some of the deepest blocks he had cut were a record of temperature changes from snow that had fallen around the time that Wegener had crossed the ice sheet so long ago, back when he and J. P. Koch were urging their snow-blind ponies forward. And here it was, preserved in perpetuity.

* * *

With three men now living at Eismitte, life changed in significant ways. Georgi kept to his ritual of recording the weather data three times a day; Sorge remained committed to his afternoon ceremony, the descent into his deep shaft, where he would slice out ice blocks and gauge his thermometers. Georgi made a regular breakfast of oatmeal, and Sorge took care of afternoon dinners, which usually involved canned meat and vegetables but on rare celebratory occasions (a birthday, for instance) involved a slice of whale meat, simmered in water and butter, which had been sawed from a frozen, forty-pound hunk that Sorge had brought to the central station. On Sundays, each man got an apple or an orange that was frozen hard like a billiard ball. It was the culinary highlight of their week.

Sorge gave Fritz Loewe his own bunk and carved out another for himself. The men constantly debated whether the ceiling of their room at Eismitte—furred with hoarfrost now, and inching lower and lower as thousands of pounds of snow piled up on the surface above them—could collapse without warning. For safety, or at least for peace of mind, Georgi built a support column out of ice blocks in the middle of the room to hold up the roof.[18] And then life continued as it had before: meals, measurements—and above all conversation, often conducted in the dim blue light to conserve fuel. The atmosphere was cold but warmly intellectual. The men passed around books to share (a favorite author was Thomas Mann) and on Sundays they discussed philosophy, "optimism and pessimism, religion, the growth of human consciousness, war and peace among the nations, and the probable changes in the world outside."[19] They also discussed Wegener and Villumsen, and how they might have fared after they left Eismitte to return home. Loewe, the newest guest, turned out to be a delight: affable, arch, polymathic. He spoke German, English, French, and Danish, and could read Italian, Norwegian, and Swedish; he would regale the men with stories about his experiences climbing in the Alps or of the daring airborne meteorological missions that he had flown over Iran. He had won the Iron Cross, first class—a high award for bravery—during the Great War.[20] As Sorge and Georgi went about their science work on most days, Loewe, essentially an invalid, would stay in his bunk and help them by doing complex calculations in his head. Also, he would entertain them with witty, grandiloquent speeches that—he predicted—would be made in their honor after they returned to Europe. "In his opinion, 'world fame awaits,'" Georgi noted skeptically in his journal.[21]

It was not clear to Georgi whether Loewe would even make it through the winter. The man was in agony. After his frozen toes had thawed, a necrosis began to spread through the tissue, darkening the skin and wafting the unbearable odor of gangrene into the cool air of the vault. All the men agreed the decaying flesh should be removed quickly. If they didn't act, it seemed possible that Loewe would lose his feet, and might succumb to deadly blood poisoning. "We have nothing," Georgi noted, "no book on frost-bite and its treatment, no surgical instruments, no bandages; all this—a whole chest of it—is lying somewhere on the road here."[22] But Georgi did have a pocket knife, which he sharpened "as thin and fine as possible." He also had metal-cutting shears to snip the bones. On November 9, he stayed awake most of the night going over the surgery in his head. And on the morning of November 10, at eleven A.M., using a small amount of iodine and a mild antiseptic called chinosol, he cut off all the toes on Loewe's right foot. Sorge held the leg down. Several days later Georgi cut off three toes on Loewe's left foot. In the hours afterward, Georgi noted he had difficulty getting the smell of putrefaction out of his mind. Loewe's feet healed slowly, and painfully, but in a few weeks, they showed a marked improvement. In a few months he could begin to hobble around.

Living inside the ice sheet from the fall of 1930 to the spring of 1931, the men often discussed the past. But increasingly they existed in a suspended state of the frozen present. *Who was making the measurements today? What was on the menu for supper? How were Loewe's toes faring?* In a few years' time, Sorge would continue his glaciological research in the Arctic and, later still, train Nazi soldiers in cold weather survival techniques.[23] He would be dead by 1946. Loewe, the only Jewish scientist of the expedition, would flee Hitler's Third Reich in 1934 for England with his wife and two young daughters, and then in 1937 move to Melbourne, Australia, where he would carve out a life as an esteemed meteorologist. He would forever walk with a shuffling gait and ride a bicycle everywhere he went. As an old man lecturing to students, he would recount at length the history of Greenland's exploration—from Nansen, to Peary, to Rasmussen, and then to Wegener—and sometimes quote an entry from Wegener's personal journal: "Nobody has accomplished anything great in life who did not start out with the resolve, 'I will do it or die.'"[24]

Georgi would spend the rest of his long life dealing with the events that would unfold over the next few months.[25]

These futures were inconceivable. When the men thought of what might come next, they lay in their frozen bunks and focused mainly on any kind of marginal improvement for the near term. An extra piece of icy fruit during the week. The reappearance of the sun in the late winter. The relief mission from the west coast that was due to arrive in the spring. Outside Eismitte, the temperature fell to –83 degrees Fahrenheit at one point, leading Georgi to note: "We are here without doubt in the coldest place on earth inhabited by human beings."[26] Often, writing in his journal, he seemed convinced that the winter would never end.

In Europe and America, the Wegener expedition had been a front-page story almost from the moment it began. During the summer of 1930, newspapers chronicled the fateful delays owing to the late breakup of the sea ice in west Greenland. Later, they followed Wegener's subsequent difficulties in getting his supplies up the glacier and onto the ice sheet. After that, readers could learn about the German team's high hopes for the propeller-driven sleds, along with details of their plans to study the meteorology of the Arctic at three different Greenland science stations. When Wegener set off with his dog teams in late September to bring cargo to Eismitte—"Dr. Loewe and I are pressing forward to Dr. Georgi and Dr. Sorge at the central station and hope to reach them and be able to return to the coast before Winter sets in in earnest"—his cable about this new development was published around the world.[27] But after that, anyone curious about the station in the middle of the ice sheet encountered silence or speculation. In late November, journalists and scientists began to ask whether the lateness of Wegener's return was cause for worry. On November 26, 1930, the London *Times* noted "a feeling of anxiety." For nearly two months, no one involved with the German expedition had heard anything of Wegener or the mid-ice station.[28] And because a wireless radio had never been delivered to Eismitte, the blackout continued through the winter.

At the west coast station, it was understood that a rescue party would need to go to Eismitte as early in the spring as possible—by then, the central station would be in dire need of food and fuel. All along, the belief at the west coast station had been that Wegener, Loewe, and Villumsen had decided to stay the winter at Eismitte and that the five men had survived on the mid-ice camp's meager supplies. "The longer we waited and considered all the possibilities," Karl Weiken explained, "the more probable, nay the more certain, did it seem to us."[29] And so in February and March, a huge effort began in the west to outfit dog teams and amass enough dried fish for the dogs to eat during the journey. On April 23, seven sledges and eighty-one dogs set out for the middle ice. "We were so full of hope in those days," Hugo Jülg, a schoolteacher who spent the winter at the western station, would later recall. [30] The propeller-driven sledges, beset with mechanical problems, had been abandoned on the road to Eismitte in the autumn—one was discovered twenty-five miles from the western edge of the ice sheet; another was thirty-two miles in. Now, the German mechanics dug the machines out and repaired them so that they could make the journey along with the dog sledge teams. In fact, the motor-sleds passed the dogs and arrived first at the central station, late in the day on May 7. The dog sledge teams arrived a few hours after.

Karl Weiken, who had come with one of the dog teams from the west station, would later recall "a sinister stillness" as he approached the camp.

He rushed up to a tent and shouted to anyone who might hear: "What's the matter?"

He received no answer. Then finally Fritz Loewe came out of the tent limping, with a full beard.

"Wegener and Rasmus left for the west on the first of November," Loewe told Weiken. "So they are dead."[31]

In the hours that followed, men sat about the camp, dazed and grim. It was obvious to them now; they had spent the winter in blissful ignorance. The mid-ice team had not known the fate of Wegener and Villumsen. But neither had the west coast scientists.

The relief team had at last brought along the wireless transmitter, so the news of Wegener's death was sent to Greenland's west coast, and then to Europe, the following day at noon. The next task was to find out what happened to the leader and his companion. At mile 118—almost exactly halfway between Eismitte and the west coast station, at a place where Wegener's skis had been placed upright in the snow—a search party led by Weiken and Sorge dug deep in the snow and found Wegener's body. It had been carefully sewn up between two sleeping-bag covers by Villumsen. Wegener's eyes were open, the men later noted, and the expression on his face was calm and peaceful. The search party seemed to concur that Wegener had died in his tent, probably of a heart attack from overexertion, and perhaps from skiing alongside Villumsen's dogsled in −50 degree Fahrenheit weather for most of the day. Almost certainly his heavy smoking, and possibly a prior heart condition, had exacerbated his vulnerability.[32] Villumsen had no doubt taken Wegener's journals with him so he could deliver them to the west station. But what had happened to *him*? The search party located sites where he had camped after leaving Wegener behind, and then Sorge, accompanied by three Greenlanders, began searching a surrounding region of the ice sheet marked by long, undulating ridges and "the whole panorama of ten thousand snowdrifts." The search continued for several days. Soon the ice appeared to Sorge like a hall of mirrors, where evidence of a campsite or a body was both everywhere and nowhere. He dug into snowdrifts and looked behind hummocks, believing he had noticed something. Then he would see it was illusory; there was no trace of Villumsen.

Sorge dispatched the Greenlanders back to the west coast station as their supplies ran low. But he continued searching, alone and obsessively, zigzagging with his dog team for eight more days over hundreds of miles. When he finally decided to give up, he did so by accepting the fact that neither Villumsen nor the journals would be found by him. The Greenlander may have died on the ice, Sorge surmised, in which case his body would be buried more and more deeply in snow with each passing storm. Or he may have reached the outer part of the ice cap "in which case his body under favorable conditions may eventually be revealed by the yearly melting of ice."[33]

As the rest of the world followed the search in weekly news reports, eulogies for Wegener poured in from foreign leaders and from old Greenland hands such as Knud Rasmussen and Peter Freuchen. "To a polar traveler there is only one finger's breadth between life and death. Without luck as a helping factor, nature is too strong," Freuchen wrote in tribute to his old friend, at precisely the same time Sorge was searching the ice fields for Villumsen. It was important to see that Wegener's scientific plans have succeeded, Freuchen added, "and he alone has failed." How Wegener and Villumsen died, Freuchen correctly predicted, would never be completely understood.[34] In addition, the journals that explained what might have happened were likely gone forever.

Else Wegener, Alfred's widow, asked that her husband's remains be buried in Greenland, and in keeping with her wishes a high iron cross, about twenty feet in height, was erected over the location on the ice sheet where his body was found. All involved knew that in several years' time the grave and its marker would disappear. Snowdrifts would rise up and up; the pressure would meanwhile build so that the snow surrounding Wegener would turn to firn. And then the surrounding firn would in turn harden to ice. Ultimately, the scientist would be subsumed into the subject of his studies.

At Eismitte, work began anew for the summer. Loewe had departed. He had been transported to the west coast by motor-sledge so a physician could examine his injured feet as soon as possible. And while Sorge searched for bodies in the west, Georgi was again alone in the center of Greenland. He continued his weather and atmospheric balloon tests, and was forced to take up the record keeping in

Sorge's frightening ice shaft. High-strung by nature, the combined effect of loneliness and grief seemed to push him to the edge of a breakdown. "It is a long, long time to live here alone like this," Georgi noted in his journal at the time. "It is a thing no one can really imagine."[35] Between factual entries on his daily activities he would sometimes pen raw disquisitions on suffering and compunction. "Never again on the inland ice," he vowed to himself privately.

The cause of his anguish was clear to see. He knew that his note to Wegener—the note where he threatened to walk to the west coast if fuel were not delivered by October 20—must have been a factor in what had happened.[36] "Wegener's death affects me more deeply than that of any other human being would, you excepted," he wrote to his wife at the time.[37] But was that because of Wegener's stature and the circumstances of the tragedy, or because it cast doubt on Georgi's judgment and besmirched his reputation? In any event, Georgi was consumed by the fires of guilt. One day at Eismitte he came across a copy of *Hamlet*, frozen in a chunk of ice in a garbage bin at the bottom of the main staircase. It must have been Loewe's, Georgi realized, and was dropped by mistake, perhaps when Loewe's possessions were brought up during his departure. Good things to read were scarce here and prized. Georgi thawed the book, dried it leaf by leaf, and then began reading.

Sorge returned in late July. At that point, it was time to blow the explosives for the seismic tests. He set up a charge a mile and a half away from Eismitte, involving about 160 pounds of TNT, buried in deep snow. A film crew recorded for posterity the preparation and the massive detonation—snow and ice blown high in a huge explosion which Sorge was able to feel but not see, since at the time he was in a darkened room next to the living quarters in Eismitte, recording waves from the blast on photographic paper. The experiment struck him as successful. "I developed the film and found waves recorded [and] was practically certain [these] were waves reflected from the bottom of the ice," he later noted. By looking at the time it took for the "reflected wave" to bounce off the bedrock under the ice and back to the surface, he calculated the depth of the ice sheet. His rough estimation was that the ice was between 8,200 and 8,850 feet thick.

This led to his conclusion that Greenland, ringed by coastal mountains, was evidently lower in its center and resembled "a soup plate filled with ice."[38]

In later years, Sorge's estimation of the ice thickness would be seen as an imperfect measurement. It was the first of its kind, though, and allowed Kurt Wölcken, a seismologist and one of Sorge's colleagues on the Wegener expedition, to make a general assessment about the extent and volume of Greenland's ice. The island, Wölcken posited, "must carry a mass of ice amounting to a million cubic miles," an area equal to all the mass of Europe, or about forty times all the water in the North and Baltic Seas combined. If the Greenland ice were to melt, he continued, "the oceans all over the world would rise more than 25 feet and extensive low-lying tracts of country all over the world would be submerged."[39]

Wölcken had no sense of what the future held, of course. But he was correct in raising the specter of terrible floods and submergence. In time, scientists would conclude that his figure of twenty-five feet of sea level rise was inexact. But not wildly so. The German scientist had overestimated the oceanic impact of a world without Greenland's ice by only about eight inches.[40]

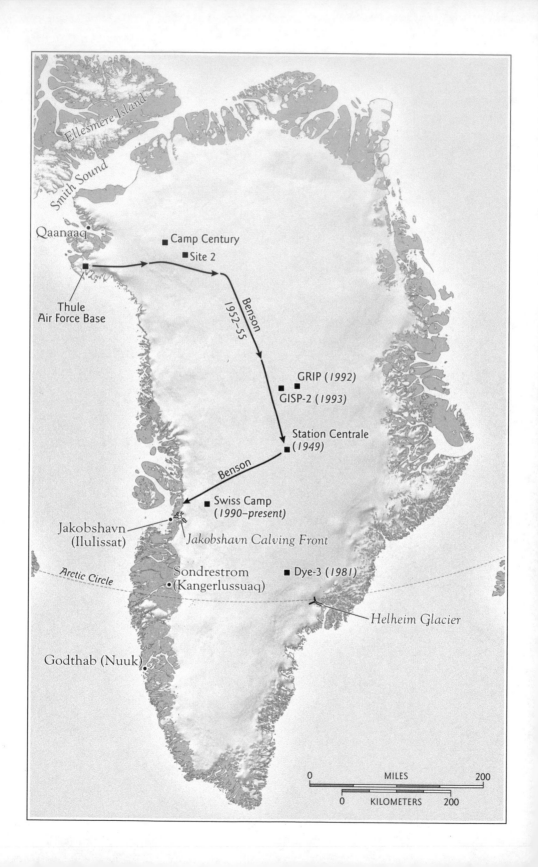

Ellesmere Island

Smith Sound

Qaanaaq

Camp Century

Site 2

Thule
Air Force Base

1952–55

Benson

GRIP (1992)

GISP-2 (1993)

Station Centrale
(1949)

Benson

Swiss Camp
(1990–present)

Jakobshavn
(Ilulissat)

Jakobshavn Calving Front

Arctic Circle

Sondrestrom
(Kangerlussuaq)

Dye-3 (1981)

Helheim Glacier

Godthab (Nuuk)

0	MILES	200

0	KILOMETERS	200

Part II
INVESTIGATIONS

(1949–2018)

Paul-Émile Victor on the way to Station Centrale
(© Fonds Paul-Émile Victor/Zeppelin)

9

Machine Age

There were twenty-two men in all, most of them exhausted from moving forward at three or four miles per hour as they made their way up the long incline toward their destination. They were traveling east. When they began their journey a few weeks before, the men had found themselves mucking through vast fields of slush that tested their equipment and their patience; after that, they had come to icy stretches cut with deep white ridges—S-shaped, carved by the wind, known by the Russian word *sastrugi*—that made for a rattling, uncomfortable ride. But things had improved lately. On recent afternoons, they had reached colder terrain. The upper altitudes. They motored effortlessly over regions of soft snow, hours and hours of nothing but the featureless white world, stretching out empty on all sides. There was only the sound of the engines. Their convoy was powered by five beat-up tractors known as "weasels"—sturdy, squat, tank-like vehicles that had been painted a brilliant orange to stand out from the snow. At last, on the evening of July 17, 1949, they reached the center of Greenland's ice sheet.

It was nearly midnight, a fact that would have been meaningful anywhere but the high Arctic, which was still awash in summer sunlight. The tractors were hauling seven aluminum sleds loaded with gear, along with two small trailers that housed mobile laboratories. Coming from the western edge of the ice sheet, the men had traveled to an altitude of 9,900 feet and a location that their compass readings indicated nearly matched the coordinates of Wegener's mid-ice station.[1] After stopping, they got out and gathered around the vehicle in the front of the caravan. Then their leader, a handsome, broad-shouldered Frenchman named Paul-Émile Victor, asked a colleague to lend him a pair of binoculars. Victor walked forward a few steps and began scanning the area.

He realized almost immediately that the act was futile; no one had been here for eighteen years, since Sorge and Georgi had abandoned their camp in the wake of Wegener's death. Eismitte would now be buried under more than forty feet of snow and firn. In looking about, Victor saw the same thing Georgi had seen when he first arrived here, so long ago: A pale world that stretched uninterrupted to the horizons. He brought the field glasses down from his eyes.

"So?" a colleague asked. "Nothing?"

"No," Victor answered after a pause. "Nothing, of course. Nothing at all."[2]

His team then began setting up a makeshift camp. Victor collapsed in his tent for the night. And the next morning, just after he awoke, the work of building another ice sheet station—this one christened Station Centrale—began in earnest. A year before, Victor had approved a plan for the camp. His men began measuring where the corridors and main rooms would be. Then they began to dig into the snow, taking occasional breaks for soup, coffee, and cigarettes. Everyone working a shovel that morning understood their intentions. At the center of the ice, they were to pick up the frayed ends of Wegener's work.

Almost all of the legendary explorers of the region had by now passed away. Robert Peary was the first to go, in 1920—dead at the age of sixty-three from pernicious anemia, a blood disease related to his body's inability to absorb vitamin B12. In his later years, Peary, though often ill, divided his time between Washington, D.C., and a large house off the Maine coast, where he worked at a desk chair backed by an enormous, hanging polar-bear hide. To the end of his days he ferociously defended his legacy—insisting, above all, that his claim to have reached the North Pole on his final Arctic expedition was legitimate.[3] He had long since cut off contact with his Inuit mistress, Ally, and his two children by her. He hadn't been to Greenland in more than a decade.

Peary's lifetime rival, Fridtjof Nansen, died of a heart attack ten years later, in 1930, at age sixty-nine. The Norwegian's early years of exploration were followed by a charmed later career as a diplomat, writer, and humanitarian. Though he continued his scientific work in oceanography, after World War I Nansen mainly involved himself with the League of Nations, the international peace organization. There, his work focused on helping to repatriate eastern European refugees from the war, and he eventually won the Nobel Prize for his peace-making efforts. Before he died, Nansen also grew close to Knud Rasmussen—the men would see each other on ceremonial occasions, and regularly exchanged letters.[4] In speeches he gave to audiences, Nansen would often relate stories that Rasmussen had told him about his travels, and he would use those stories along with his own to encourage younger audiences to take risks and plot adventures. He would also urge them to seek out solitude and willingly endure suffering. Only by doing so could they understand the world's deeper truths. "Most people are satisfied too soon," Nansen would say, "and that is the reason why there is so little wisdom in the world."[5]

Following the first five Thule expeditions, Rasmussen had continued to do ethnographic research and write books. He spent most of his later years in Denmark, working in a comfortable house by the water in Hundested, a quiet town about an hour and a half northwest of Copenhagen. He made yearly summer visits to west Greenland, however, and in 1933 made an extended visit to east Greenland, where he was making a feature film—*The Wedding of Palo*—about the native culture. During the course of his visit, Rasmussen became violently ill, possibly from food poisoning related to the fermented foods he loved. His illness was then worsened by a bout of the flu. Evacuated by ship to Copenhagen, upon arriving Rasmussen "courageously attempted to walk down the gangway, although on the verge of collapse."[6] Taken immediately by taxi to a hospital, he passed away a few weeks later at age fifty-four, a death that triggered a shock wave of mourning in both Greenland and Denmark.

Rasmussen's friend Peter Freuchen later reflected that the death elicited "sorrow from all parts of the world."[7] Freuchen's health remained robust, however, and his life at times resembled that of a wayfaring character in a Scandinavian saga. There was the wooden foot, the long beard, the hulking six-foot-four frame, and the world-weary knowledge evident in the folds and deep creases that began to spread around his eyes. After years in the Arctic, Freuchen pursued a successful life of writing in Denmark until Hitler's rise in the mid-1930s prompted him to speak out aggressively against fascism in newspaper columns and at international conferences. Freuchen had Jewish ancestry. He watched as his books about Arctic exploration were banned in Germany, yet he kept writing and protesting, and he willingly hid political refugees in his home. He was temporarily arrested in Denmark in 1942, which was at that point controlled by Germany. But he shaved off his beard and escaped in 1944 to the neutral country of Sweden, and from there to New York City, where he settled down, married, grew back his beard, and began again to enjoy public life.[8] In his case, fate seemed mischievous. Moving from the solitude of Thule to being a man-about-town in one of the most crowded places on earth seemed to suit him, and many of his books about Greenland, now available in English, began finding a new audience. For a brief moment, he even became a household name in America, when he appeared on the televised game show *The $64,000 Question.*

Making small talk during an early round, the game show host had asked: "Have you ever gone whale hunting, Peter?"

"Oh sure," Freuchen answered in his deep Danish accent, seeming surprised that anyone would think he hadn't. Now balding, limping, and beaded with sweat from the studio lights, he appeared unsuited for prime-time television. But when questions for the quiz began, mostly regarding oceans and polar exploration, he responded with flawless answers. He knew *everything*, from the direction of the current that flows through the straits of Gibraltar to the characteristics of the eyetooth of a walrus.[9] Ultimately, he won the jackpot.

Regardless of where he was living, Freuchen kept abreast of what was happening in Greenland. But he rarely visited anymore. Much of the work done on the island during the 1930s involved coastal exploration—geographical studies as well as archaeological digs that sought to understand the early settlements of the Inuit and the origins of settlers who had apparently come to Greenland hundreds (or even thousands) of years earlier.[10] Paul-Émile Victor, then in his late twenties—a factory owner's son who had spent a good part of his teenage years paging through books and magazines about exotic places and wondering if he, too, could become an explorer—was one of the few visitors venturing deep onto the ice sheet during this era. Inspired by Nansen and Rasmussen, whom he described as the last of the "traditional" explorers, he crossed Greenland for the first time in 1936, accompanied by two French colleagues—Robert Gessain, an anthropologist, and Michel Perez, a geologist—as well as a Danish archaeologist named Eigil Knuth.[11] Victor started from Jakobshavn on the west coast on May 17, with three sledges and thirty-three dogs. During much of June, he would later relate, the weather was "deplorable," and "snow fell practically without interruption."[12] The men ran low on food; twelve dogs had to be killed. They reached the eastern coast safely on July 5, covering 420 miles in forty-five days. But the crossing of the ice sheet had made a permanent impression on Victor, and would help to explain his later approach to polar expeditions. As much as he respected the traditions of the Inuit, the dogs, sleds, and weather made the going miserable. There had to be a better way.

Victor's life was already unusual; it soon became extraordinary. As Europe began its descent toward total war in the late 1930s, all research on Greenland's ice ceased. Victor was stationed in Stockholm, officially working at the French embassy as an intelligence officer, as Germany's army invaded Poland in September 1939, signaling the official start of the war.[13] When France fell to Germany in June 1940, Victor received orders to return home, and embarked on an odyssey through Europe. He moved stealthily through Russia, Turkey, Greece, Portugal, and Spain (a roundabout course dictated in part by the need to avoid the most perilous war zones) before returning to France, now controlled by a government beholden to Hitler. He was at that point thirty-three years old. For a number of reasons—most crucially, his belief that in the United States he would have far greater research opportunities—he decided to leave Europe.[14] Traveling by way of Morocco and then Martinique, he arrived in the United States in July 1941.[15] For the next few months, he bounced around Washington and New York—corresponding intensively with his family in France and wondering what part he should play in the war effort and French resistance. In the meantime, he staked out a living by writing articles and lecturing about Greenland. He often concluded his evenings in New York at the apartment of his expatriate friend Antoine de Saint-Exupéry, the French writer and aviator. At the time, Saint-Exupéry was working on a fanciful, illustrated book about a little prince.

The desire to help with the allied effort against Hitler became a compulsion for Victor. He enlisted in the U.S. Air Force as a private in 1942, and the following year was assigned work in Colorado, where he was asked to test and develop survival equipment for troops deployed to inhospitable regions. His experience crossing the Greenland ice in the mid-1930s suited him perfectly for the work. A few months later, he was promoted to lieutenant and was asked to lead search-and-rescue missions in Alaska. Now a licensed pilot as well as an American citizen, he prepared by spending a month learning parachute techniques at a school run by the U.S. Forest Service in Missoula, Montana.

In retrospect, he would be able to see that the components of his future polar expeditions were now falling into place.[16] And all of it—military planes, parachute drops, and an expanded sense of range and possibility—was part of a larger shift that would not be widely understood for many decades. The world of Nansen and Peary and Rasmussen was receding; technology would soon alter everything about how scientists worked in the Arctic and Antarctica. Victor could envision it already.

By 1944 he had become particularly enthusiastic about a new type of U.S. military vehicle that was built to traverse ice, snow, and slush.[17] "I loved these Weasels, like children I had nursed," he later wrote of the sturdy, amphibious tractors, built by the Studebaker Company of Indiana, which the military's cold weather troops relied upon. "Then, after the war, for five long months, I looked in vain all over Europe for them."[18]

Victor had returned to France by then, and was newly married. He was about to give up his search when amidst the postwar chaos he discovered eight hundred of these tractors abandoned in a surplus equipment camp in the Fontainebleau forest, forty miles outside of Paris. He made some inquiries and bought ten weasels on the spot. By his reasoning, they were exactly what he would need when—not if—he returned to Greenland.

With a background in the social sciences rather than the physical sciences, Victor might not have seemed an ideal candidate for the job of Wegener's successor. In 1935 and 1936, around the time he had crossed the ice sheet by dogsled, he spent several seasons in Greenland living with the Inuit on the island's remote eastern coast; for a year he spoke almost nothing except for the local dialect.[19] Eventually, he wrote a popular book of anthropology—My Eskimo Life, as the English version was called—that comprised a diary of his time with the east Greenlanders. What linked him to Wegener, more than anything, was a fascination with Greenland and a breadth of ambition.

In Paris in 1947, not long after discovering the weasels, he founded an organization he named Expéditions Polaires Françaises, or EPF, for which he scraped together funding from a host of private benefactors and the French government. His energy and relentlessness—he later said he made fifteen hundred visits to French authorities to convince them to help fund his idea, and even demanded personal meetings with Charles de Gaulle—dovetailed with a talent for describing to potential donors the grand adventures he planned.[20] Much like Nansen and Peary, Victor understood the value of public relations and the need to pique the public interest by way of lectures, books, and drawings. (Among his many talents, he happened to be a witty caricaturist.) And he often sought to elevate the drama of a polar journey. The feeling of crossing the Greenland ice sheet, he once wrote, was "extraordinary, appalling"—like being "thrown into another world, into an element unknown, where nothing is at the scale of man, an inhuman world of white chaos and sparkling serenity where man has no part, into lifeless, eternal, immovable space."[21] His ideas of what to do with EPF were actually less poetic. He meant to establish a research and engineering corps that would employ leading scientists from a variety of disciplines and examine both the Arctic and Antarctic regions—sometimes in the same year. Victor's true intent was the painstaking collection of data, and the pursuit of discovery. And almost immediately after forming EPF he began making preparations for the trip to the middle of Greenland.

In a world trying to put itself back together after World War II, uncovering the secrets in Greenland's ice would have likely seemed well down the list of pressing problems. Europe was at that point broken physically as well as socially, a landscape of hollowed-out cities, ruined economies, and ruptured communications systems. The fact that Victor could cobble together enough funding in 1947 and 1948 for his Arctic expeditions, appealing to his benefactors' sense of curiosity and patriotism, astonished his colleagues. But it seemed equally true that he succeeded *because* of his country's social and economic distress, not in spite of it. "Precisely at that moment," he later explained, "France needed a 'lift.' Even such a strange thing such as a successful polar expedition would help in its way to revive our national prestige."[22]

His ideas, moreover, carried the mark of an innovator. To Americans, Victor's name and work would never become synonymous with adventure as they did to the French. But in Greenland he helped bring to glaciology new technologies and a genius for logistical planning.[23] He saw, before almost anyone else, that the cold regions of the earth could be—indeed, should be—enduring places of scientific interest. "Whereas earlier expeditions to the Arctic or the Antarctic had been sent out to solve particular problems," the ice sheet historian Børge Fristrup later explained, "and their first task had been to get hold of the necessary equipment either by purchasing or by borrowing, Victor decided to build up a permanent organization that would have supplies of the necessary equipment and would have a reserve of engineers with specialized experience of journeys under Arctic conditions."[24] The same equipment could be used for both the Northern and Southern hemispheres, thus saving costs and sharing expertise. And while it was not Victor's stated intention, his approach supplanted the idea that progress in the Arctic would depend on another Nansen, Peary, or Rasmussen coming along.[25] With Victor's expeditions, Fristrup pointed out, "members of an expedition no longer needed to be outstandingly robust physically to make a journey across the ice sheet."[26] The new French methods—big teams, sophisticated equipment, a staff with broad expertise—suggested the notion of heroism would be different in the modern era.

The goal now was to create a fine-tuned, machine-powered system of scientific polar inquiry.

"On 14 May 1948," Victor wrote, "some six months after preparations were begun, the 25 members of the expedition, with their 90 tons of equipment, sailed from Rouen aboard the Norwegian freighter *Force*, which had been chartered for the purpose."[27] They landed in west Greenland on June 1, near a glacier known as Eqip Sermia, a place his colleagues jokingly renamed Port Victor.[28]

Eqip is legendary—even today—for the regularity with which it calves icebergs into the adjacent bay. And in fact not long after Victor arrived, one of his landing boats was capsized by a thirty-foot wave that rolled out after a massive chunk of ice crashed into the water from the crumbling glacier. The group shrugged off the setback and eventually succeeded in unloading 180,000 pounds of supplies on the rocky coast. Victor would later list the cargo as "dehydrated food and surgical instruments, typewriter ribbons and snow vehicles, glue and seismographs, mustard and sextants, a few bottles of champagne for birthdays, spare lenses for spectacles, and almost anything else you can name."[29] The crates were marked by stripes in order of importance: red (essential); blue (useful); green (desirable). The goal now was to move everything up to the ice sheet in stages. Right away, Victor's men laid out a rough, six-mile road up the rocky coastline toward the edge of the ice—to a place they called Camp II. The weasel tractors they had brought with them did most of the hauling.

Getting to Camp II was difficult. But the next step—Camp III—involved surmounting a five-hundred-foot cliff that separated the rocky Greenlandic coast from the ice sheet. To move the supplies over the cliff, the EPF team had brought twelve thousand feet of steel rope, which they used to string up a *téléphérique*—a cableway up the mountain, much like those at a ski resort. In fact, Victor began to communicate with engineers in Europe at this point, seeking advice from experts who built similar cableways at Chamonix in the French Alps. With the long-distance help, the men succeeded in relaying forty-three tons of crates to the edge of the ice by the elevated cable, as the weasel tractors made a roundabout run over several more miles of rocks and gullies to rendezvous at the next station. The workdays were brutal, lasting on average fourteen hours, though sometimes the men worked forty hours straight.[30] In photographs and films made during these weeks, the French team looks exhausted, besieged not only by physical exertion but by a warm spell of weather that brought on what Victor described as "billions" of mosquitoes. During this period, he would later write: "We could have learned that a shipload of pretty girls was coming to visit us, and it wouldn't have made any difference. All we wanted was to sleep and eat and sleep again and eat again."

Forty-six days after the French group's arrival in Greenland, Camp III was finally established at the edge of the ice, at about two thousand feet above sea level.[31] There was some debate about going farther. But progress at that point was impeded by the typical summer melt, brought on by warm weather, and the usual hazards at the edge of the ice sheet: huge lakes and crevasses in the ice, as well as deep holes filled with meltwater.[32] So Victor decided to pause at Camp III. And in late August he concluded they should all return to France, with the intention of coming back the following spring to bring their gear to the center of the ice and establish Station Centrale. The team departed Greenland on September 22, 1948.[33] If they could come back early the following year, Victor had reasoned, the melt season would not have started and progress in getting to the center of the ice would be greatly accelerated. They left the weasels behind in Greenland.

They returned to Camp III in the spring of 1949. Yet as the team set out to reach the center of the ice sheet, they again encountered slush fields and dangerous crevasses, which resulted in time-consuming detours or their tractors ripping a tread, leading to a forced rest for mechanical repairs. "During many of those hours," Victor would recall of that leg of the journey, "we were bumping inside our weasels as they jumped from hummock to hummock, plunged into rushing rivers, or got stuck on a crumbling crevasse snow bridge." As the men moved closer to the center of the ice, though, the ride grew smoother. "Morale was good," Victor wrote of the journey. "We stopped only every 10 miles to build snow cairns, play soccer with an empty tin can or fight with snowballs, and record our scientific data. At each stop we detonated dynamite to make a seismic sounding."[34] Victor wanted to create a more detailed map so as to understand Greenland's entire bedrock landscape—in effect, to see what the island would look like without any ice at all.

On the day in 1949 when he finally arrived at the center, Victor confided that "the dream I had cherished for 15 years was indeed coming true." For the entire journey over the ice he had thought of Wegener, had obsessed on Wegener.[35] In the wake of the German's death nearly two decades prior, Sorge and Georgi's results had been published and widely discussed, yet in the research community they seemed to have raised as many questions as answers. Victor later wrote: "They brought back observations that gave the scientific world a faint idea of what the icecap is: a huge reservoir of cold which affects the weather of the whole North Atlantic."[36] At the same time, Sorge's work in his shaft—his methodical cutting of ice blocks, his hauling them up to the surface to measure their density, and his conclusion that he could calculate the age of the ice at the bottom of the shaft—had shown scientists something intriguing and mysterious. It appeared to be the case, Victor remarked, that Greenland's ice was "a tremendous recording machine of times and climates past." He had come to investigate that, too.

As the French team began their work at the center of Greenland's ice in the summer of 1949 they resembled in some ways a repertory company restaging the drama of the Wegener expedition. On the ship's journey from France to Greenland, Michel Bouché, a young meteorologist hired by Victor who would later write an account of the winter of 1949, had read through the memoirs of Sorge and Georgi, and even scrutinized the details of a photograph of the two scientists sitting at their work table in their ice cavern at Eismitte. He marveled that in such miserable circumstances they still adhered to their work schedules, like monks. (At the end of the winter, when the rescue team had come to Eismitte, Georgi's watch was only running an hour late, a fact that amazed the young Frenchman.) The differences between his own team and the German expedition were nevertheless clear to Bouché. Unlike the Germans, they had the weasels; they had functioning radios; and perhaps most important, they had French pilots who could fly over from Iceland upon request and make an airdrop of fuel, equipment, or even fresh radishes. In the late summer, in fact, as the men began building their camp upon the buried site of Eismitte, they received a dozen airdrops with seventy tons of food and supplies. If they were lucky, the pilot would add to the airdrop a cask of *pinard*—slang for cheap red table wine.[37]

By the middle week of August the team at Station Centrale had assembled an insulated, prefabricated hut, measuring twenty-six feet long and sixteen feet wide, with walls that measured six-and-a-half feet in height. The hut had a peaked roof, making it eight-and-a-half feet tall in the center, which partially alleviated the sense of claustrophobia. To protect it from the wind and storms, the men excavated an area and nestled the building nine feet below the snow line—knowing that as the winter wore on, it would be covered by drifts, and buried deeper and deeper. This hut housed their kitchen, kerosene stove, living room, bunks, and dining room. But the camp extended far beyond the hut: Four hundred and fifty feet of under-ice corridors led off from the men's living quarters and connected to storerooms, scientific labs, and surface exits. Nearby, technicians installed generators that provided the camp with electricity for about ten hours per day.

Victor spent some of his time mulling over which of his eight men should stay for the winter. He himself would be managing operations from Paris, where he was also planning an EPF expedition to Antarctica. By late August, he had settled on a wintering team and was ready to meet a ship on the west coast of Greenland for passage home. At Victor's moment of departure, a storm kicked up and snowflakes whipped against their faces. "We formed groups, exchanged a few words on weather conditions on the trail, shook hands, and the vehicles drove away in the whiteout," Bouché recalled. "The eight of us remained for some time, winking in the wind, soon tired of not seeing anything more."[38]

Winter began at Station Centrale with the start of September. Apart from all the scientific gear, there were ample supplies—fuel and stoves and books and medicine; crates of coffee, powdered milk, potatoes, meat, rice, and canned tuna. They kept most of the food stacked in shelves they had carved from the firn walls in the narrow corridors leading off from the hut. Their bunk rooms and kitchen were often kept close to room temperature (around 65 degrees Fahrenheit), even as the air outside went as low as −89 degrees during the depths of winter.[39] And yet, day-to-day life was difficult and the work unceasing, with constant measurements of wind, temperature, and atmospheric pressure, done with radio balloons. "We knew this wintering would not be what one imagines, what we imagined almost irresistibly," Bouché would write. He later reflected on how he had underestimated the difficulties. He suffered headaches and shortness of breath from the ten-thousand-foot altitude. Fumes from the stove choked the indoor air. Broken equipment required constant repairs. The kitchen needed cleaning and scrubbing over and over again, and still it never seemed clean.

Most of all, there were "the storms, the cold, the night," which often made venturing outside an ordeal, and which left them resigned to pass much of their time in the small hut with each other.[40] But sleep was difficult, too. "Since October," Bouché observed, "we have been suffering from insomnia. It hounds each of us periodically, and likes to pick on whoever has to get up early."[41] Tossing in his narrow bunk, Bouché would eventually arise and shuffle into the dirty kitchen and light a lamp. The sun would not be coming up; there was no expectation of dawn. He would make a cup of Nescafé, sit down at a table covered in a checkered tablecloth, and start the day's first cigarette. Then he would get ready to begin work. In addition to his weather measurements, the young Frenchman had taken on the task of secretary, gathering every detail for his diaries about day-to-day life at Station Centrale.[42] Sometimes in his spare time Bouché would think of Georgi, Sorge, and Fritz Loewe, sharing books of philosophy together in their cold crypt, on this very spot, two decades before. With eight men and a higher degree of comfort, the mood at Station Centrale, he noted with a slight note of dejection, was busier, more professional, less intimate.

"One is tempted to conclude that we have voluntarily deprived ourselves of a unique human experience," Bouché worried.[43] As the months of under-ice living dragged on, and as the sense of isolation increased, the sense of adventure dimmed. Bouché recalled it as "the crushing monotony of identical days organized by strict schedules." Yet an adherence to routine carried the men forward, and a colorless work regimen was brightened by occasional moments of pleasure— long conversations about life and work on the ice, and birthday celebrations with wine or rum. The men had brought along a phonograph to play records in the hut. Listening to Bach's *Brandenburg Concertos*, songs by the chanteuse Édith Piaf, and spirituals sung by Paul Robeson and Marian Anderson, they forgot their plight for a few hours at a time. Yet it was the station's radio receiver more than anything else that kept them from succumbing to the isolation. Victor's offices in

Paris would check in with the station daily, and on holidays such as New Year's the men enjoyed an extended time talking with their families in Paris. Once a week, moreover, they would converse with Lauge Koch, the Danish geologist who had traveled with Rasmussen years before and was now spending the winter seven hundred miles away, on Ella Island in east Greenland.

One evening they heard from Koch's team the story of how some of the Inuit with them had left their camp on Ella Island and embarked on a three-hundred-kilometer sledging trek down the Greenlandic coast, without food or tents. "Crammed around the table with a smoking lamp and the nightly dishes piling up, we talked at length of this life they lead—the coast, the race of sleds in the snow," Bouché noted. The French team debated for hours the differences in the life of the Inuit and life at Station Centrale. Amazed by the daring of the natives, the scientists agreed that the lives of the Inuit barely resembled their own. Venturing three hundred *meters* from Station Centrale in a storm of blowing snow could seem as dangerous as going three hundred kilometers, which was why the scientists often ventured out wearing goggles and gripping a rope that tethered them to an entranceway. The wind would sometimes lift them off their feet. "Here," Bouché wrote, "the unknown begins at the door."[44] It did not seem lost on him that his team's job was to look up, into the stratosphere, and to look down, into the depths of the ice, and that if the old means of exploration was roving far and wide, the new means—at least here, now, in Station Centrale—was to stay put, stay healthy, stay focused, and see whether the exploration of a single location might reveal a world of new knowledge.

By early June 1950, the men at Station Centrale, still in frequent radio contact with Victor and his colleagues, became aware that a relief team was on its way. Actually, they began making bets on what time it would arrive. Victor got there on June 30, 1950, almost ten months after he had left the eight men behind to winter over. When he pulled up, he noticed that the camp—no longer appearing as a collection of tents and huts, but rather as a landscape of snow hills and ventilation pipes and metal towers—"seemed deserted." He honked his horn and got out of his weasel. Still nothing. He honked again.

"Then a head appeared, flush with the snow surface," he recalled. "We did not move. It disappeared. And then a man jumped up and came running and stumbling. He was Lucien Bertrand, the radioman. He could not say a word. He fell into my arms, and we kissed each other as brothers."[45] Victor's actual impression was that Bertrand and the other men looked tired, dirty, pale, and thin. But he could see that they seemed elated, too. Winter had gone largely as planned, with no mortal difficulties. No one had perished; no one had been injured.

The dark mood at the station began to lift. For the next few weeks, several dozen French researchers arrived at the camp to conduct studies on ice and atmosphere. "The Central Icecap Station," Bouché noted at that point, "very much looks like a trodden anthill." Victor would later add, astutely: "Never had the central icecap seen so many living creatures in its hundreds of thousands of years."[46] There were arrivals and departures of groups studying the ice sheet's geodesy, gravity, and underlying geology. The camp was now busy twenty-four hours a day, making for "tumultuous evenings" of long dinners where the soup was repeatedly thinned down for extra visitors. Chunks of snow, melting on the stove for drinking water, sizzled at all hours, and still there was never enough.

One lingering question was whether Victor's men could accomplish much beyond their own survival—that is, whether their systematic approach to researching the ice sheet would bring rewards. After the work of the summer of 1950 was complete, another French team came to Station Centrale to winter over to spring of 1951. And after that, Victor presented to the world a number of new insights into Greenland's ice. On the most basic level, his group had made temperature and atmospheric measurements, so that meteorologists had a clearer sense of conditions in the air above the ice cap in mid-winter. The French team had also looked downward. Going nearly five hundred feet into the sheet with a hollow drill, they recovered thin cylinders of ice just under two inches in diameter—cylinders that would later be described by glaciologists as some of the earliest "deep ice cores" ever made. The holes the drillers made in extracting the ice allowed them to measure temperature and density inside the ice cap, and to conclude that Greenland's ice, deep down, seemed to retain its colder temperatures from the past, even as the world had obviously warmed considerably over the last few thousand years. The effort to dig down was aided by another tool the French team had brought for the summer: A one-ton drilling "grab" that looked like an enormous version of a game you might play on the boardwalk of a beach town, where a scoop reaches down to pluck toys from a pile below. The grab was meant to allow the team to dig into the snow and ice and create a narrow hole, about three feet in diameter and a hundred feet deep, that the men could rappel down into with a rope. Victor was eager to try and later reported that the dark, narrow well gave him the uncanny feeling of being inside a factory's tall chimney. When he pointed his flashlight on the sparkling, icy walls, he decided the descent was worth it: "Here before my eyes," he said of the layers of the ice cap's compacted snow, "was the record of much more than a century of its history."[47]

The science historian Janet Martin-Nielsen would later note that the French drilling team extracted a thin ice core "of poor quality [that] proved difficult to analyze." In other words, the success of the endeavor was as a "proof of concept" rather than as useful science, in that the drilling experiment seemed to point toward a future where scientists might be able to extract better-quality cores and look at the old ice to gather meaningful insights—assuming they could figure out how to interpret the ice.[48] Meanwhile, the seismic soundings with explosives were more useful and had produced a map of the bedrock in Greenland's south-central region. These soundings indicated that parts of the island now buried beneath miles of ice were actually below sea level, likely because the massive weight of the sheet was pushing down the underlying rock. And the seismic readings helped Victor glimpse a bigger picture of how the ice sheet behaves. He could see that large amounts of Greenland's ice seemed to flow from the center toward a drain on the west coast—in other words, the ice flowed slowly from the central region of the island, and then toward the sea, by way of big glaciers like the one at Jakobshavn, where the ice broke off into giant icebergs and floated away. What also struck him was that some of the biggest glaciers—Jakobshavn, for certain— seemed to have carved out canyons below their bottoms that were twenty to thirty miles wide and as much as twelve hundred feet below the sea level.

Many years later, this particular discovery would be augmented by the work of a team of scientists in California—mostly French, as it would turn out—who took a closer look at the extraordinary depths under the rivers of ice that flowed out from the central ice sheet and asked whether these troughs somehow made Greenland's ice more vulnerable to melting and destabilizing forces.[49] For now, though, it was barely clear whether these insights were pieces of a larger puzzle or a curious array of independent facts. That would have to be settled later. Victor knew he couldn't keep Station Centrale open indefinitely to find out. By the end of 1951, he was feeling pressure to raise more money—an expedition to Antarctica was a large burden on his organization's resources—and would soon lack the funding to keep all his teams in the field.[50]

Just around that time, however, Victor was becoming aware of a secret American project under construction. He had heard about it through various polar contacts and old friends within the U.S. military. The project was sited on the northwest coast of Greenland, and involved building a defense outpost so sprawling that it would eventually become one of the largest U.S. air bases in the world. What was coming to the island, Victor surmised, might be a military endeavor, but for some scientists it could also bring an abundance of money, machines, and opportunity. It might be precisely the kind of opportunity he needed to continue his fieldwork. He began discussing getting involved in the project with American military officials. He wanted in.

Operation Blue Jay (Getty Images)

10

The Americans

The secret project that Paul-Émile Victor heard about as his work in the center of Greenland's ice sheet was winding down was known as Thule Air Base. Construction had begun in spring of 1951 and been given the code name Operation Blue Jay. The U.S. Department of Defense unveiled the new base to the public the following year, in September 1952, just before it became fully operational. Touted by the press at its curtain-raising as "an engineering miracle" that served as a "guardian that looks steadily over the top of the world down into Russia," Thule was an outpost conceived in paranoia and dressed gaudily in propaganda.[1] The Americans had built it with astonishing scale and speed—a frenetic, expensive effort that was planned to cost at least $250 million (about $2.5 billion in today's dollars) and was comparable, in logistical aspects, to the Allied landing at Normandy or the Manhattan Project behind the first atomic bomb.[2] In June and July 1951, about 120 ships left Norfolk, Virginia, and landed near Thule, hauling an enormous array of plows, dump trucks, rock crushers, pavers, diggers, and cranes. The ships carried enough lumber, prefabricated building panels, sheet metal, and pipes to construct a small American city—which was, in effect, the goal.[3]

About eighty-five hundred men came ashore. Some were members of the Army Corps of Engineers, but most had been hired discreetly in the suburbs of St. Paul, Minnesota, which the army concluded was a good cold-weather location in which to recruit. These Thule workers—carpenters, plumbers, ironworkers, electricians—were informed only that they were "going overseas and to a very cold climate"; even when they landed, some had no idea they were in north Greenland.[4] The men were told the pace of work would be intense, which was indeed true: Operation Blue Jay ran twenty-four hours a day—a morning shift and an evening shift—for the entire summer. "They were putting up, basically, a building every day," one scientist who visited at the time recalls.[5] And when the gloom of winter descended, the work schedule slowed but did not stop. After just thirteen months of effort, Thule had a ten-thousand-foot airstrip, plane hangars, gymnasiums, barracks, an officers' club, a post office, library, churches, a baseball diamond, and a small network of roads.[6] It was designed to house about ten thousand personnel. A tavern served beer but no hard liquor; known as Guffey's, it was packed in the evenings as well as in the mornings, just after the night shift got off work.[7]

To understand why the Americans had come to Thule in 1951, it helps to look back a decade, to the early days of World War II—and even more precisely, to April 1941. That month, the Danish ambassador to the United States signed the "Defense of Greenland" contract with the U.S. Secretary of State.[8] By asserting that defending the island against attack "by a non-American power" was now "of vital concern to the United States of America and also to the Kingdom of Denmark," the deal gave the United States virtually unlimited authority to protect Greenland from Germany's incursions. On the day of the signing, American troops and engineers arrived to begin building a stealth network of airstrips, weather stations, and communications outposts.[9] The American code name for Greenland was "Bluie." Five stations were constructed on the east coast and nine were built on the west coast. Each was identified by its location and number. Bluie West Eight, for instance, was a new airstrip by a fjord that the Danes called Sondrestrom, on Greenland's southwest coast.[10] Bluie West Six was a small radio and weather station at Thule, on the northwest coast, where Rasmussen and Freuchen's old trading post was located. Bluie East Two was a radio station on the rugged east coast, near a place called Angmagssalik, which was where Paul-Émile Victor had spent much of his time in the mid-1930s doing ethnographic research.

Back home, news accounts offered scant details of what the United States was doing at these locations, or why five thousand soldiers were being stationed there. "It is not a big show like Russia or Italy," one United States Air Force colonel would later explain, referring to his experiences with Arctic warfare. "There are no vast armies, no major campaigns, no epic battles or a million men."[11] Nevertheless, the U.S. investments were far larger than many Americans and Europeans realized, reflecting the fact that Greenland was given a high value by military strategists. One reason was that a mine at the southern tip of the island was a vital source of a rare mineral called *cryolite*, which was useful for processing aluminum. Just as crucial, though, was the idea that meteorological stations situated along the island's coasts could help predict the coming weather in Europe and provide a crucial tactical advantage for convoys and maneuvers, such as the D-Day landing, where the timing for the invasion derived in part from north Atlantic meteorology reports.[12] Having control of the island provided other advantages, too. U.S. military commanders considered Greenland "a logical stopover point" for ferrying fighter planes and bombers going from the United States to Great Britain.[13] In an era when military airplanes could not be built for intercontinental travel, in other words, a pit stop in a place like this would be essential.

The Bluie stations operated—quietly, for the most part—for the duration of the war. But when an armistice brought peace to Europe

in May 1945, America's interest in Greenland broadened into a desire for total control. The Danish political historian Nikolaj Petersen
notes that in June 1946, the U.S. military concluded that air bases
built in Greenland during the war were still considered "essential" to
its national security.[14] And in fact, such a realization made some U.S.
officials worried: What would happen if Denmark—no longer requiring American help to defend its Greenland colony—became unwill-
ing to let the United States continue operating those stations? "It
might be a good idea to take prompt action toward securing [Green-
land] from Denmark," U.S. Secretary of War Robert Patterson wrote
to Undersecretary of State Dean Acheson. Patterson suggested purchasing Greenland from Denmark—an offer tendered by the United
States to the Danish foreign minister a few months later, in December
1946. The United States proposed paying $100 million for the world's
biggest island. Danish officials, however, showed no apparent interest
in either negotiating or selling.[15]

What concerned the American generals most was the U.S.S.R.—or
more precisely, the notion that the most direct line of attack from
Moscow to Washington was a transpolar route that ran through
Greenland. "Greenland loomed large in the American psyche in the
late 1940s," the American science historian Ronald Doel would
write, "because the U.S. felt newly vulnerable to the Soviet Union,
especially in the north, a part of the world it knew little about."[16] As
it turned out, after the war Denmark allowed the United States to
maintain its new air bases in Greenland. But at the highest levels of
the U.S. military, discussions then turned to the question of whether
this was enough to avert a grave new threat. The Soviet Union exploded its first atomic bomb in August 1949. Some U.S. generals
began making the argument that it would be necessary to build bases
for heavy military bombers even *farther* north than the existing airstrips in Greenland—to construct a true Arctic air base, in other
words, which was situated much closer to Russia's main cities. Northern Canada and Alaska were discussed as possible locations. So was
northern Greenland. In 1950, the air force brass turned the question
over to a highly decorated officer named Bernt Balchen, who had
years of experience working as a military pilot in the Arctic regions.[17]

Balchen had spent extended periods of time in Greenland during the war, and he understood the island's geography and military value.[18] He had also once had a serendipitous encounter with Knud Rasmussen. One day in the late 1920s, Balchen had been in New York City and met Rasmussen, then visiting on a lecture tour. They had a long conversation about Greenland in a New York speakeasy. During the course of the evening, Rasmussen mentioned to Balchen that Thule had an excellent harbor and a flat swath of land nearby that would make for an excellent runway.[19] Balchen recalled the conversation some years later, in the summer of 1942, when he decided to fly to the north Greenland coast and make his own aerial survey.[20] "Everything Rasmussen said is correct," he would later report. "Thule has all the potentials for a great air base, extensive gravel flats for long runways, plenty of room for hangars and barracks, and, in North Star Bay, a deep-water harbor, open two to three months a year, that could handle any cargo ship."[21]

And so, asked in 1950 whether Thule would be workable for a new Arctic air base, Balchen thought again of his discussion with Rasmussen years before and considered Thule's advantages. At the end of that year, he submitted a study to his bosses that recommended northwest Greenland as an ideal strategic location.[22] The idea of Thule—a new Thule, one with only a vanishingly slight historical connection to Rasmussen and Freuchen, and none at all to Robert Peary, whose legacy originally led Rasmussen to set up a trading business—had little to do with exploration or science, at least at the start. For now, the point of the new base was to stake out a massive advantage in terms of bombs and planes, and to wrest victory in the coming Cold War.

The construction and activity at Thule in 1951 and 1952 took place just over a hill from where Rasmussen and Freuchen once traded fox pelts with their friends and waited, sometimes for years, for a ship to arrive with coffee, mail, and news from Europe. That kind of isolation was now a thing of the past. The local Inuit population, which had been living in the area for centuries, mostly in small, hand-built houses of stone and turf, watched the arriving armada and the subsequent construction with fascination. During the initial months of the Thule project, some of the natives living nearby traded with American soldiers and workers, and were delighted to receive gifts of surplus food and supplies. Jean Malaurie, a French scientist living with the Polar Inuit in 1951, looked on and watched the culture clash with concern. He was not witnessing direct mistreatment of the Inuit on the part of the Americans; as he later understood it, he was witnessing the deepening erosion of their culture. One day, he crossed paths near the base with a local man named Qisuk, an acquaintance who "was pushing a sledge piled high with cases of tinned corned beef, ham, marmalade, and bundles of magazines." A large group of children following behind Qisuk were smoking Lucky Strike cigarettes, which had been given to them at the Thule store. *We don't have to work anymore*, the natives announced, partly in jest. This obviously wasn't true. But the activity and noise around Thule was apparently driving away native wildlife that the Inuit depended on for hunting. "Men who lived by the harpoon," Malaurie would later conclude, had "found themselves in the atomic age."[23]

At the urging of Danish administrators, U.S. commanders began to enforce a strict cultural separation between the base and the local Inuit. And a year later, in late 1953, the Americans began moving the local Inuit community to a newly built and more sequestered settlement known as Qaanaaq, which was located sixty-five miles to the north.[24] In the blinkered view of the American military, the displacement of the locals was a minor inconvenience that paled next to their more important challenges. The Inuit would have to adapt.[25] In a world where global stability depended on meeting force with force, too much depended on Thule's smooth operations to allow for delays.

Still, for all the Americans' engineering wizardry—for all their willingness to roughly push obstacles, material or human, aside—it was clear they had a flaw in their plan to fortify Thule.[26] The U.S. Army and Air Force lacked one vital capability when it came to the work at the top of the world. Russia had defeated Germany on the Eastern Front of World War II thanks in part to its soldiers' ability to fight, or at least endure, in cold weather; the American forces, by contrast, had a less sophisticated understanding of how to persevere in a region of near-constant winter.[27] U.S. military strategists were not wholly oblivious to their weakness: Paul-Émile Victor's work testing equipment in Colorado during the war, for instance, and his later assignments running rescue missions in Alaska, were early attempts to hone cold-weather strategies.[28] After the construction of Thule, moreover, Victor—who was seen by that point as more knowledgeable about the ice sheet than anyone in the American armed forces—was hired as a consultant to figure out the safest and most accessible place to bring tractors and men from Thule onto the ice sheet, which began about fifteen miles from the base. He was also entrusted to plan a crossing of the northern ice sheet in summer 1952, by weasel, on a path that resembled the ones Peary and Rasmussen had taken many years before.

Victor's contributions weren't nearly enough to satisfy Cold War strategists, however. At the start of the 1950s, the U.S. military discerned the need for a much larger domestic program to research the Arctic. The questions its leaders confronted as they began the occupation of northwest Greenland were endless: Could you land a plane safely on compacted snow or floating ice? What was the best way to build a runway on a bed of permafrost? How do snow crystals change with time and subzero temperatures, and how might those changes affect, say, ski-equipped airplanes or tractor treads? In addition, they began weighing more conceptual questions: How could structures or tunnels be built under the ice sheet, which slowly flows and deforms, so they last years and don't end up buried, crushed, and useless, like Eismitte?

The advent of the Cold War and the birth of modern glaciology happened in tandem and happened here. Many decades later, a team of academics in the United States and Denmark would scour declassified documents from the period to give a definitive assessment of the close links. "Control and power in the Arctic regions required new scientific knowledge," one of these studies, from a team at Denmark's Aarhus University, concluded. "Science therefore became an integral part of [the] military operation."[29] It wasn't just any kind of science, however. One army analyst noted at the time that there existed within the U.S. military a deep knowledge of how metals and other substances behaved in various conditions. But there was "a scientific ignorance of many of the physical properties of snow and ice."[30] For precisely that reason, the U.S. military founded several organizations—obscure to the American public, even at the time—around the mission of trying to understand the deeper nature of two of the most plentiful and familiar substances on earth.

The most important was a new division within the Army Corps of Engineers that established its laboratories in a three-story industrial building in Wilmette, Illinois, about a mile from Northwestern University. The lab had six cold chambers that ranged in temperature from –65 to –23 degrees Fahrenheit; in the coldest areas, the parka-clad scientists could store ice and snow samples from around the world, which they would use for testing and analysis in the "warm" chambers. They would compress the ice, smash the ice, melt the ice, and use an array of sophisticated tools to measure results. The researchers had on hand samples from frozen lakes in Canada and glaciers in Alaska, as well as firn blocks from the under layers of Greenland's ice sheet. They even had ice from far-flung outposts known as "drift stations"—installations set up in the Arctic Ocean on huge ice floes many square miles in size, which were manned for anywhere from a few months to a few years for weather observations and research.[31] The lab building in Wilmette was known as the Army Corps' Snow, Ice, and Permafrost Research Establishment. Almost always, though, it was known by its acronym—SIPRE.[32]

A dapper and sometimes irritable former professor named Henri Bader worked as the chief scientist at SIPRE. A native of Switzerland—he would often thump his chest and tell people he had just met: "I am a Swiss export!"—Bader had come to the United States to teach at Rutgers University after an unsuccessful stint as a geologist during the war, when he'd worked for several mining companies in South America. Bader was a man of medium height with a goatee and thinning, combed-back hair. He smoked heavily, read science fiction with a passion, and carried an intimidating air that bordered on imperiousness. A product of the European academy, he was fluent in

German, French, English, and Spanish, and had a deep knowledge of physics and chemistry. His early research trips in the United States involved visits to the Mendenhall Glacier in Alaska, where he dug out ice crystals to bring back to his lab for study.

Bader's scientific instinct was to scrutinize small things, rather than large. Or to put it another way, he had an inclination to look intensively at the fundamental properties of ice and snow, such as its strength and molecular structure, rather than focus, as many of his colleagues did, on the broader icescape—for example, on the back-and-forth movements (known as *advances* and *recessions*) of large glaciers.[33] Years before, while in college in Switzerland, he had written a dissertation on snow crystals that would become legendary amongst the tiny contingent of scientists who studied the subject at the time and would still be consulted as an authoritative work seventy-five years later. Bader had tracked what happened to snowflakes as he watched them for hours and then days under a microscope in a cold room as they aged and melted. The work might have seemed esoteric, but snow—how the shape of grains smoothed and changed over time, or how microscopic pockets of air or water vapor filled the pores between flakes and dissipated as the snow compacted—turned out to be an enormously complex material.[34] Bader correctly believed that to better predict the conditions for deadly avalanches, especially in Switzerland, where such disasters had sometimes killed dozens of people, it was necessary to understand the microscopic fabric of the snow pack and how it evolves over time. What was small and hidden and granular, in other words, could lead to effects that were large and significant and sometimes terrifying.

In the mid-1950s, one Chicago newspaper called Bader "the biggest ice man in the United States."[35] It was almost certainly true. But it was likewise correct that Bader was a giant in an obscure and seemingly irrelevant field. In mainland Europe, the formal study of glaciers was now more than a century old, dating back to the work of Agassiz and Rink and Nansen; in Great Britain, the study of ice by the middle of the twentieth century was attracting the attention of several brilliant physicists who were curious to understand in mathematical terms how glaciers flowed. In the United States, meanwhile, glaciology was an arcane practice as well as an arcane word—a little-known extension of geology that one journalist noted was "a rare branch of learning which attracts only the most hardy intellectuals."[36] Mainly, the early glaciologists in the United States were either focused on measuring the glaciers of Alaska and the Pacific Northwest or were intent on trying to work out a theoretical framework for how and why these large masses of ice move.[37] But in sum, there were very few of them. "Glaciology was something that if you wanted to do it, nobody understood *why* you wanted to do it," recalls one scientist of the era.[38]

By an accident of history, Bader found himself to be in the right scientific field, in the right country, at the right time. Managing the work at SIPRE, he could merge an interest in the science of ice and snow with an interest in practical applications that captivated the U.S. Army and U.S. Air Force. What's more, he could tap immense financial reserves for travel, equipment, and research, thanks to the military's huge Cold War budgets and the massive new air base at Thule. "He convinced the Army Corps of Engineers that their standard approach was just not adequate," a colleague recalled. Bader would explain to generals and colonels that the army's idea of "going at it"—building bases, or landing planes, or mounting radar defenses for Thule—by merely using a practical, cut-and-try engineering approach was going to fail. To build without truly understanding their frozen environment was folly. "He said to them, you have to understand the fundamental physics, the behavior and processes that go on in snowpacks and ice. Without the basic science you're not going to get anywhere. And they bought it."[39]

With a big budget, Bader could hire a big staff. And in a field without much of a past in America, it made sense to look to the young, to the curious, to those working on the fringes. The early ranks of SIPRE glaciologists comprised a mix of men (and not a single woman) who tended to work at the juncture of mountaineering, geography, and geology.[40] Some of the recruits were intent on using their research work at SIPRE to supplement their PhD studies.[41] But some shared a spiritual kinship with the first balloonists or supersonic test pilots— individuals united by curiosity and a love of the outdoors, and who appeared to have little fear and few inhibitions. Others were merely outdoorsmen who enjoyed looking at the strata of rock along the walls in abandoned mineshafts in the Rockies, or who had a hobby of recording changes (sometimes by using photogrammetry, the science of measuring objects and natural changes over time by precision pho-tography) in far-flung ice fields in Alaska. Still others came from the U.S. Geological Survey, where they had spent years climbing peaks and mapping desolate regions of the west. By the standards of the modern era, they took risks on their journeys that were often extreme and sometimes insane: Weeks in the backcountry of Alaska or Cali-fornia, for instance, with no physicians nearby, scant air support, poor communications equipment, and slender rations. But to the pilots who dropped them off in remote coves or isolated mountain valleys, they would say: *Come back and pick us up a month from today.*

That might have been what appealed to them about the jobs Henri Bader had on offer. The perils of doing field research on ice floes in the Arctic Ocean, or in Greenland or Antarctica, or the specter of long months bent over a microscope in a subzero laboratory in the Chicago suburbs, were outweighed by the promise of adventure and possibility. "I mean, this was a first-class government research lab working on the cutting edge of a new subject," recalled Wilford Weeks, a glaciologist who worked with SIPRE after an earlier career in geology had brought him through some of the most remote back-country locations in the United States and Canada. "Laboratories like it are hard to find. The subject was wide open. Everything was new. Anything you'd think of doing, no one had ever done before."[42]

Bader suffered a painful fifteen-foot fall into a deep crevasse while he was in Alaska chipping out ice crystals in the fall of 1951.[43] And eventually, a bad back limited his ability to do outdoor work, espe-cially when it came to hand-digging deep pits in snow and firn—a requisite part of the fieldwork in those days for a glaciologist working in the Arctic. Yet such limitations didn't hamper his career. He had a knack for management, and he used his intimidating presence to se-cure for his staff at SIPRE ample funding and freedom from bureau-cratic interference. "Leave them alone," Bader would say in heavily

accented English to army officers, reflecting a belief that a good research program on ice and snow was not to be hurried or managed too
closely. Despite his prickly nature, his protective instincts won him
the deep devotion of his employees. At SIPRE he had the habit of
walking from office to office, or from laboratory to laboratory, dropping in unannounced, and demanding that each researcher answer
three simple questions:

> What are you working on?
> Why are you working on it?
> What do you expect to get out of it?[44]

One of the scientists who went to work for Bader was a tall, rugged
twenty-three-year-old from Minneapolis named Carl Benson. The
grandson of Swedish immigrants, Benson had joined the navy after
high school, in 1945; when the war ended he took advantage of the
GI Bill and enrolled in the University of Minnesota to study geology.
To help fund his college studies, he also took a variety of jobs during
school breaks. He worked in logging camps around the Northwest,
and in railroad gangs in Oregon and Washington, laying down track.
Benson loved being outdoors, and he especially loved winter. When
the opportunity arose in 1950, just after his college graduation, to
join a U.S. Geological Survey project in Alaska's Brooks Range, he
jumped at the chance. For several months, his team crossed streams
and mapped large areas of Alaska's backcountry. After that, Benson
returned to Minnesota to begin a master's degree in geology. He took
an assistant position with SIPRE during that time, which turned into
a full-time research job in 1951. That was how he found himself on a
military transport plane with Henri Bader, flying to the new Thule
base in Greenland, in the spring of 1952.

The military had asked Bader to look at several possible sites for
radar installations near Thule.[45] Benson recalls the sprawling Thule
construction site, then less than a year old, as a place of chaos and
wonder, a project so massive and crowded with contractors and equipment that the total economic costs struck him as beyond compute.
He had been asked to come along because he had recently gained
expertise studying the accumulation of snowdrifts in the Sierra Ne-

vada Mountains—but Benson recalls that Bader's wife liked the idea
of him going along, "because I was big and strong and could take care
of him."[46] The men were planning on digging deep rectangular snow
pits at several locations, work Bader tended to avoid because of his
bad back. The point of their digging these pits was twofold: By studying yearly accumulation layers of snow, they could give the military
insights as to how fast big radar equipment might get buried in annual
snowfalls; at the same time, as they dug down, Bader and Benson
could measure annual ice sheet temperatures and look for evidence as
to whether a particular site was vulnerable to melting in the warmer
summer months. Some of the possible sites for Thule radar stations
could be places where meltwater tended to pool into slushy lakes in
June and July, or where streams flowed through. To build a defense
installation on unstable ice like that could be disastrous.

For several weeks, Bader and Benson took day trips from Thule to
locations nearby. They traveled on a C-47 transport that alighted on
the ice sheet on skis. The men would disembark, and Benson would
dig by shovel a rectangular pit nine or ten feet deep. The work was
strenuous but not as arduous as it might sound: The top layers in these
areas were soft snow; and beneath that, the layers were composed of
firn—airy, coarser snow from seasons past "that cut like Styrofoam,"
as Benson recalls.[47] (To reach solid ice the men would have had to dig
at least a hundred feet deeper.) Together, Benson and Bader would
make measurements of the yearly layers, which they could usually
discern by sight, although they also used a more exact process of measuring how the density of the snow related to different seasons, similar
to what Sorge used at Eismitte. All told, the work took them four or
five hours, and the pilots waited patiently. At the end of the day the
group would fly back to Thule for dinner.

Usually the pit-digging work went smoothly. But on one trip, Benson stayed behind at Thule while Bader flew to an area on the ice
sheet a few hundred miles east of Thule that army officials had named
Site 2. After a few hours of digging, it turned out the plane, a model
known as an SA-16, could not gain enough lift to take off from the
ice sheet. "Poor Bader," Benson recalls.[48] The sophisticated European
professor was stuck at Site 2 with the pilot for the next thirty-two

days, with only a few tents and the grounded aircraft. Back at Thule, Benson helped organize some food and fuel supplies, as well as construction materials for a hut, which were then air-dropped to Bader by a search-and-rescue plane.

Something important came out of that mishap, though. To get the plane at Site 2 back up in the air, the pilot stripped everything off to lighten it—extra fuel tanks, seats, cargo. He then left the discards on the ice sheet. Not long after, a team was organized at Thule to travel by weasel to Site 2 and retrieve the stripped-off parts. Benson went out with the group to help, but also to do his own research—after a day of driving, he would dig snow pits all night long, when everyone was asleep, to measure the subsurface temperatures of the ice sheet and gauge the snow layers and accumulation. And at that point he got the idea that maybe he could do the same thing all over Greenland. That summer at Thule, he had run into Paul-Émile Victor, who was organizing a northern ice sheet crossing at the request of the U.S. Army. The French team's preference for using weasels had made an impression on Benson. And he could now see for himself how easy it was to get around on the ice by tractor.

Benson recalls: "The goal I had at that point, and I wanted to try, and which Bader gave the backing to, was to see if we couldn't come up with a temperature distribution for the whole ice sheet, and the picture of snow accumulation for the whole ice sheet."[49] To the army and air force, such a project made good sense. SIPRE had been established to help the military transport goods and men across the ice; more knowledge about snow and temperatures in different locations would therefore be useful. From the point of view of Benson, though, the idea addressed a scientific curiosity to better understand how the ice sheet *worked*. For instance, such a project should be able to show where Greenland gained the most snow during the winter and lost the most water in the warmer months of snowmelt. And by doing so, the study might afford some insights into a question that had been of interest to scientists dating back to the days of Fridtjof Nansen and Alfred Wegener: Simply put, was the ice sheet growing or shrinking?[50] If Greenland was accumulating about as much snow every winter as it was losing in warmer months, when meltwater ran off the edges and icebergs calved from glaciers at the periphery, its mass could be said to be "in equilibrium." On the other hand, if Greenland was losing more ice each year than it was gaining through snowfalls—a process of subtraction that defined the retreat of ice sheets at the end of previous ice ages—it was likely out of balance and growing smaller. That also meant its frozen water was adding to the world's oceans and sea levels.

"Bader went down to Washington and came back and said there was support," Benson recalls of his idea, noting that his boss urged him at that point to map out the project. Eventually Benson devised a plan for a "traverse" around Greenland's ice sheet during the summers of 1953, 1954, and 1955. During those summer tours, a small team of three or five other men accompanied him. They used weasel tractors to cruise around the ice sheet, stopping every ten to twenty-five miles to dig a pit, ranging in depth from nine feet to eighteen feet, to make measurements. The men slept in the weasels or in boxy, canvas-topped army campers they towed along. Because much of their food was frozen, Benson recalls that "if [we] wanted something thawed for the next day, we would take it in the sleeping bag with us at night."[51]

The work went like this: Find a site; dig a pit; collect data; drive on. And repeat the process over and over again, for eleven hundred grueling miles. The longest trek occurred during 1955, when the men worked and traveled for 120 consecutive days without rest. Along the way, Benson's team would receive airdrops of fuel and food similar to the ones pioneered a few years before by Paul-Émile Victor. During these air deliveries, heavy-duty barrels were usually "free dropped" without parachutes into the snow from an air force plane flying about fifteen or twenty feet above the ice sheet. Afterward Benson's group would spend most of the day locating, collecting, and unpacking the barrels, which sometimes included home-baked cookies from Benson's wife in the United States and bundles of letters from home. Then they would go back to their trek. Benson recalls he was never bored and rarely fatigued. Most of all, he notes, "I've never had anything approaching that kind of logistical support." America's Cold War policy put few limits on military-related spending. "It was an infinite budget."[52]

On a map of Greenland, Benson's traverse over those summers forms a clear line that moves east from Thule, at Greenland's northwest corner, and toward the center of the ice sheet; then it turns south for hundreds of miles toward the location of Eismitte and Station Centrale. From there, it arrows west in a direct line from the central ice sheet toward the coast, ending at the western edge of the ice. From the data Benson collected along the way, he ultimately made the most accurate picture of how temperatures and the rate of snowfall varied across different regions of the Greenland ice sheet. He also sketched a picture of how some regions were subject to different amounts of surface melting each year and discovered that some areas—notably the central part of the ice sheet, at the highest altitude—did not melt at all, even in summer. In the end, Benson recalls, "the conclusion we had was the ice sheet was close to equilibrium."[53] In other words, he estimated that the yearly melt and iceberg losses were probably balanced—that is, replenished—by the amount of snow that was falling. He couldn't be certain. But in the mid-1950s the Greenland ice sheet looked to be stable.

We tend to think of scientific field research as conducted by teams in which colleagues each lend their complementary talents to a project. But in the Arctic—so large, so difficult to access, so new to modern science, so difficult to understand—certain questions have lingered for more than a century, and various projects have remained in progress for decades. Scientists who are retired or deceased therefore serve in some aspects as colleagues to those still living, having left their data and ideas for others to use and build upon. To an unusual degree, problems in the Arctic are worked on not just at a particular moment in time, but over generations.

When Wegener and Koch crossed the ice sheet in 1913, during the horrendous trip on which their ponies died one by one, they were doing the first long-distance comparisons of annual layers across the Greenland ice sheet. That was why they had dug rectangular pits—to look at snow accumulations. "I read enough German that I could follow Wegener's notes with his traverse with J. P. Koch," Benson recalls. He was keenly interested in comparing Wegener's old calculations to the ones he was making where their paths intersected—and he discovered that the accumulation measurements from 1913 and 1955 were exactly the same. As important to Benson was how his traverse in 1955 crossed the location of Station Centrale, where there also existed detailed records of snowfall and weather. It had been four years since Paul-Émile Victor's team had abandoned its research camp. Benson knew when he was getting close to the site of the old station— "Our navigator said, 'We ought to be here'"—but the visibility that day was extremely poor, with fog and snow making it hard to see out the windows of the weasel. "But we stopped anyway," Benson recalls. And when they got out and looked around they immediately glimpsed a radio antenna sticking up from the snow nearby. They called in to Thule to tell them— actually, they sent a Morse code message at eight o'clock that evening, which was their customary time—and as it happened Paul-Émile Victor was at the air base. "They told me he cried," Benson recalls, "because he didn't think we'd ever find it."[54] When the weather cleared, Benson and his colleagues began digging down and found an entrance into the old camp's corridors. And in the partially crushed ruins of the frozen station, amidst a mess of strewn clothing and canned goods, he discovered a discarded pair of boots from one of the French scientists. He brought the boots home and eventually gave them to a museum in Alaska, thinking that they represented something meaningful in the history of science, the willingness of some researchers to do what is exceedingly difficult in one of the most unfriendly locations on earth.

In the mid-1950s little of this mattered to the public, though, or to the press. Outside the Army Corps of Engineers, the work of SIPRE was almost entirely unknown, and even among scientists—as the age of the transistor and the atomic bomb began—the Greenland ice sheet could seem largely superfluous. No one could see that the scientists were putting down markers that would be of immense value in a future era. Wegener and Sorge and Victor had done that. Benson was doing that, too. Sixty years after his traverse, in fact, several Army Corps scientists would replicate his journeys over the Greenland ice sheet to measure temperatures, snowfall, and melting. Using GPS, they located their sites within fifteen feet of where Benson had been, and moving around by snowmobile they dug pit after pit. In their excavations, they hoped to find an old candy wrapper or food tin from Benson's traverse. They had little luck in that regard. But they did establish that parts of the ice sheet were up to six degrees warmer than during Benson's day, due in large part to increasing amounts of summer meltwater dripping down through the snow and dispersing heat. Except for the center of Greenland, everywhere the researchers traveled they discovered evidence that the ice sheet was melting more, and at higher elevations, than what Benson had recorded decades before.[55]

Sixty years means everything to human beings but very little to geologists, who tend to think in spans of thousands or millions of years. For the features of earth that command their attention— continental plates, for instance, or ice sheets—alterations can be effected so slowly and so imperceptibly that they go beyond the meager human ability to observe them visually. For that reason they've invented sophisticated tools to measure tiny gradations in magnetism, minuscule ripples in gravity, hairsbreadth upticks in elevation. Comparing Benson's old traverse with the updated study, however, suggested that things were changing in obvious and profound ways. On an ice sheet, six degrees in sixty years is *fast*. So fast, in fact, that it meant changes in Greenland, in the years after Benson's research, were no longer in the realm of geological time. The ice was being transformed in human time, too.

The approach to Camp Century (U.S. Army)

11

Drilling

The question that tantalized Henri Bader beyond all others was what secrets the ice sheet contained in the depths below the pits that Carl Benson had shoveled out. Alfred Wegener and Paul-Émile Victor had harbored a similar curiosity about the mysteries of the deep ice, but neither had come close to finding an answer. Bader, as the head of the Army Corps' SIPRE division, held fast to the belief that drilling into Greenland's center, deeper than 150 feet or so, where under tremendous pressure the firn begins to squeeze together and turn to ice, could yield something miraculous.[1] To his thinking, the ice sheet contained a frozen archive of long-ago events and temperatures—it was encrypted, in some yet-to-be-deciphered way, with a code to the past.

Discovering the temperature from a time before human societies maintained records, or before human civilization existed at all, is not a trivial bit of knowledge. "In the 1950s we had very limited information on temperatures from the past," recalls one of Bader's younger hires at SIPRE, a geologist named Chester Langway. "Very limited. And some of it was questionable. Old records from ships' captains. Farmers with broken thermometers. It went back maybe 150 years, and with spotty numbers."[2] Langway and Bader knew, though, that if they could use the ice to somehow look further back, the questions that could be answered would be extraordinary. For instance, had a change in climate extinguished a particular species of dinosaur or mammoth? Had it prompted vast ice sheets that once covered most of northern North America and Europe to shatter into icebergs? Had it vanished the Greenland Norse from their settlements in southwestern Greenland in the 1400s? Accurate insights into ancient climates would open up a world of possibilities, not least of which was an ability to see into the causes of events that otherwise remained hazy matters of conjecture.

As he looked to Greenland, Bader was aware that scientists were already tapping other parts of the natural world for clues. By the midpoint of the twentieth century, it was becoming increasingly clear that shreds of archived history were locked within every cleft of the earth—in caves, under lake beds, and in traces of plants buried in the loam beneath bustling cities. The earth's oceans seemed especially promising. As far back as the late 1800s, scientists had been aware that mineralized deposits from tiny sea creatures covered the bottoms of the seas. As the oceanographer John Imbrie noted, "The organic oozes had been formed over a very long period of time by the slow rain of skeletons on the sea floor."[3] Some of these layers on the ocean bottom were riddled with certain types of microscopic plankton known as *foraminifera* that were prevalent in warmer waters and gave an indication of warmer temperatures during a particular time period. Other layers seemed to lack these particular "forams," as they were usually called, and were instead richer in skeletons that flourished in colder waters, and thus colder temperatures—perhaps making them an indicator of a previous ice age. In the late 1940s, Swedish ocean-ographers learned how to efficiently plumb down with hollow tubes and extract cylinders from the ooze; then they began using their knowledge of tiny fossilized remnants to start estimating climate changes of the past. Some researchers imagined these marine cores, striped with layers built up over time, were worth their weight in gold. Hans Pettersson, the Swedish scientist leading a global effort to scour the ocean bottom, predicted it would yield an "unrivalled archive" of the planet's past. By probing into the seafloor "by means of tubular corers," he predicted in 1949, "sediment columns can be raised which contain the records from tectonic, volcanic and climatic catastrophes of a remote past."[4]

The study of tree rings, a branch of science that came to be known as *dendrochronology*, was coming into its own, too. Tree rings had at-tracted scientific interest in the early 1800s, at which point some sci-entists observed that a severe winter in Europe beginning in 1708 seemed to have led to very narrow growth rings.[5] But it became an established research area in the 1930s, thanks largely to the work of Andrew E. Douglass, a scientist trained in astronomy who began looking at tree rings in the American Southwest, usually by scrutiniz-ing samples of buried logs or old beams he found in ancient pueblo structures. He saw that the wood cross sections, which he and his as-sociates traced as far back as the year 700, could be an excellent means of gaining insights into long-ago periods of drought and weather variation. What seemed particularly intriguing was how his studies offered more than just data about the past. They offered a way to *interpret* the past. By combining a new understanding of climate history with what was known about human history, Douglass re-marked, it was now possible in Arizona, Colorado, and New Mexico "to correlate the increases of rainfall that permitted these villages to expand and the drought years that placed upon them the heavy hand of starvation."[6] Understanding our past climate, in other words, was a key to understanding our civilization. It might likewise shed light on the potential for climate changes in the future.

Even in the early days of paleoclimatology, as the study of ancient climates came to be known, it was clear that nature's archives weren't all the same. Each would have distinct virtues and shortcomings. Ocean cores, for instance, could go back far in time—likely many millions of years—and might thus offer information from very distant geological epochs. Yet these records could be difficult to date with precision, and they lacked the resolution to say much about a specific year or short period. On the other hand, tree rings seemed to promise great specificity as to what happened in a particular year, such as the chilly winter of 1708 in Europe, or the ravaging droughts that occurred in the twelfth century in the American West. But they didn't seem to go back much further than five or ten thousand years.[7] To Henri Bader, the ice sheet promised to capture a year-by-year resolution in its layers, meaning that if one could figure out how to read precise temperatures in these layers, one would find (as Bader put it) "a treasure trove." Just as important, the layers were *depositional*: Everything in earth's atmosphere had been deposited there along with the snow that turned to ice. In theory, that meant that an ice core from deep in the ice sheet would contain telltale vestiges from, say, the start of the industrial revolution, and include evidence of how atmospheric gases and pollution intensified over time. An ice core might likewise contain traces of ash that blanketed the earth after the volcanic explosions of Krakatoa in Indonesia (in 1883), Laki in Iceland (1783), or maybe even Vesuvius, near Pompeii (79). And judging by how thick the center of the ice in Greenland appeared to be, the record might go back much, much further. On the ice sheet, Bader predicted, "every snowfall, including everything that fell with it, is, so to say, separately and safely filed for future reference by being buried under later snowfalls."[8]

The ice sheet, moreover, contained something that could not be found in ocean ooze or tree rings. There were bubbles of air trapped inside. Bader was not alone in this realization—during the winter of 1930–31, for instance, some members of the Wegener expedition who did not go to Eismitte camped on the western edge of the ice sheet in what they called the "winter hut." Under their floor they dug a vertical pit, sixty-five feet deep, and scaled down on a rope ladder to see what the ice could tell them. Ideally, a scientist looking inside the

Greenland ice sheet would prefer to sample an area where it is cold at all times—in the center of the island, for instance, where Ernst Sorge was digging his own shaft at the time—rather than at the periphery, where springtime melting and glacial movements could play havoc with the accumulation of layers. Still, one resident of the winter hut, Hugo Jülg, would note that the pit on the western edge seemed intriguing. He and his colleagues could make a series of temperature measurements in the shaft and measure the stratigraphy—that is, the layers—of the ice. Also, they could see "bubbles of air enclosed in the ice" which they analyzed by measuring the air pressure inside.[9] Two decades later, in the late 1940s and early 1950s, Bader worked on bubbles in some of the early ice cores drilled in Alaska. "He could see the bubbles were under pressure," his SIPRE colleague Carl Benson recalls. "And he wrote a little paper in the *Journal of Glaciology* called 'The Significance of Air Bubbles in Glacier Ice.' Now, the bubble is recording the atmosphere at the time the bubble is sealed off. And in other words, these little bubbles in the ice have a history of what the climate was like at the time. He knew this. We knew this, but it was a question of: How do you measure it?"[10]

Bader did not expect to find answers quickly. Inspired by the science fiction books he loved to read, he was thinking that the results from what he called "deep cores"—as opposed to "shallow" cores that went down a few dozen feet—might only be realized in the distant future. "He understood the significance of this *before* we even knew what we'd get out of it," Benson says. He carried a vision that in the future glaciologists would have cores of ice stored away in special freezers—libraries of ice, in effect—and would be able to take them out for sampling and investigation. "They would put them on a light table and examine them, do chemistry and electrical and optical experiments on them." And whatever methods were someday developed, by testing the tiny pockets of air trapped in the bubbles, and examining the trace chemicals trapped in the ice, the distant past might become all the more clear to those living in the present.

Bader was known to say: "Snowflakes fall to earth and leave a message."[11]

*   *   *

There was reason to think that using old ice to find temperatures of the past was at least possible.

In 1931, Ernst Sorge had found that the density of snow on the ice sheet was different in summer and winter, which allowed him to count layers—and, therefore, years—with reasonable accuracy. Several decades later, as Carl Benson traveled around the ice sheet and stopped every few days to dig a snow pit, he could usually count the annual layers of snow accumulation by sight, just by looking at the strata—the thick stripes—on the wall of the pit. But to verify his conclusions Benson also called on the expertise of a Caltech professor named Samuel Epstein, who had helped develop a new method to chemically analyze snow layers.[12] It so happens that not all the snow that falls on Greenland is precisely the same. To be sure, all of it is frozen water—that is, frozen H_2O. But Epstein understood that the ice contains H_2O molecules with tiny differences in their mass. In some snowfalls, there are more "heavy" oxygen atoms (known as ^{18}O) that have two extra neutrons in their nuclei. These extra neutrons make them weigh slightly more than typical, "lighter" oxygen atoms (which are known as ^{16}O). The greater abundance of ^{18}O in some snowfalls is caused by temperature variations on stormy days and can be immensely useful for telling temperatures in the past. Epstein surmised that snow deposited in summer had more heavy oxygen isotopes than what is deposited in winter. That meant Benson could sample snow layers in the pit and send those samples to Epstein for testing, so he could be sure he was correct when he counted back through the years. "We found the oxygen isotope ratios were giving a beautiful representation of the annual units," Benson later reported.[13] And from this small demonstration, chemistry seemed a promising way to accurately count layers and perhaps find a pathway to more precisely understand ancient temperatures.[14]

But the problem was not only about how to interpret traces in old ice. It was how to get a deep ice core in the first place. Drilling a hole in ice at a remote location—a hole that might be thousands of feet deep—is not an easy proposition, especially if your intention is also to pull up a pristine cylinder for scientific research. By the early 1950s, there had only been four long ice cores extracted in all the world. Two were done by Paul-Émile Victor's team in Greenland—the longer of the two went down about 450 feet. Another core measured about three hundred feet in length and was drilled by Bader at the Taku Glacier near Juneau, Alaska. A fourth core was also about three hundred feet in length and had been pulled by a British-Norwegian-Swedish Antarctic expedition on a coastal part of the frozen southern continent, near an area known as Queen Maud Land. None of these cores offered any breakthroughs in terms of explaining past climates, and some contained only fragments usable for research. What's more, none of the projects yet made clear what equipment was best for drilling a deep ice core, or what diameter a core should be.

Henri Bader was already tapping the U.S. military to fund his science projects on snow and ice. But in the mid-1950s he saw an additional way to boost the idea of ice drilling. At that point, preparations were beginning for a global collaboration known as the International Geophysical Year, or IGY. Eventually, sixty-seven countries took part in the IGY, which was scheduled to run from July 1957 until December 1958, and which involved a dizzying array of scientific experiments on land, on ice, in the oceans, and in space. The most historic of the IGY projects involved rocketry and astronomical observations (and included Sputnik, the world's first satellite). But a good part of the work involved multinational expeditions to Antarctica and the Arctic Ocean; in fact, IGY expeditions were the first to estimate the average depth of ice in Antarctica and prove that under all that ice was one continent.[15]

Bader argued strenuously that deep core drilling should be included in the American itinerary—"If the research succeeded," he insisted, "it would be a major contribution by the U.S. to the IGY."[16] Eventually, the national committee planning the IGY agenda agreed, and Bader's group was directed to try to drill a deep core in two locations in Antarctica. Bader believed that before that work could begin, a fair amount of preparation would need to be done, so he looked toward Greenland to try out machinery and drilling techniques. Then (as now) the island was seen as a staging ground for the more severe challenges—

especially with regard to transportation and infrastructure—around Antarctica. Because of Thule, Greenland was far more accessible than in the past, and ski-equipped planes made it possible to get almost anywhere on the ice sheet in a matter of hours, weather permitting. What's more, you could now drive right on to the Greenland ice cap by weasel or tractor. After building Thule, the Army Corps of Engineers, acting on surveys done by Paul-Émile Victor, had constructed several roads, ranging from about ten to twenty miles in length, which connected the air base to the ice sheet. Engineers had then constructed massive gravel ramps *onto* the ice sheet and taken the further step of filling in big crevasses on the ice sheet's periphery and smoothing them over with snowplows.[17] Knud Rasmussen had once envisioned Thule as a station that would make it easier for science expeditions to roam about the north of Greenland; by creating a small economic hub, he thought it could hasten the world's understanding of the island. He was both right and wrong. Decades later, it wasn't trading that made his scientific vision come to fruition but the Americans' relentless militarization of the ice sheet.

Just as the International Geophysical Year was about to begin, Peter Freuchen died at Elmendorf Air Force Base in Alaska. He had traveled there from New York to take part in the filming of a network television special in connection with IGY activities; he intended to fly with a group of aging polar explorers from Alaska to the North Pole. "The giant bewhiskered explorer," as one newspaper put it, had insisted on carrying his own suitcases up a staircase at his hotel. He collapsed and died on the top step.[18] Freuchen was seventy-one. According to his wishes, his body was cremated and his ashes were brought to Thule, the place that he had named forty years before. The ashes were then sprinkled from an airplane flying above the old trading station, now defunct, where he and Rasmussen had lived so long ago.

Bader decided to test deep drilling at Site 2, situated on the ice sheet about 220 miles east of Thule. It was the same place where he had been stuck a few years before, when he came there to dig a snow pit and then found his plane couldn't take off again. But Site 2 was different now: it had a new radar station and, under Bader's direction, a hundred-foot-deep research pit, which had been dug into the ice sheet in the summer of 1954—with chainsaws, shovels, and picks.[19] Not far from the pit, army engineers had also begun to dig long tunnels so as to measure how quickly dug-out areas were deformed over time by the weight of falling snow and the movement of the ice sheet. Usually eighteen people were stationed at the camp, a group that included American military personnel along with a few SIPRE engineers and scientists. The U.S. military had essentially picked a point on the map—quite literally the barren middle of nowhere—and decided to see if it could, with the force of technology, money, and will power, make it useful.

Although there are a few existing photographs, it is difficult to conjure what day-to-day life was like in an outpost so remote and forsaken that it existed for only a few spectral years, and was thereafter crushed by decades of accumulating snow. If you were staying at Site 2 in the early 1950s, you had access to electricity, hot water showers, a mess hall, and barracks. You had a Ping-Pong table and recreation area. You were not living in a normal army camp, however. Almost all the living areas were located within prefabricated housing units that had been assembled *inside* giant corrugated metal pipes that measured eighteen feet in diameter. The Army Corps of Engineers had trucked the corrugated steel across the ice sheet to Site 2 and then assembled it, believing the pipes to be a good means of testing how to prolong the life of an Arctic military installation, especially as yearly snowdrifts buried it.[20] Site 2, at an altitude of seven thousand feet, was therefore an experiment in ice sheet living. At the same time, it was also an experiment in human resiliency, with residents having little to do on bitter days except venture out of "the tubes" to make measurements on local ice sheet conditions. Weather interruptions often made radio contact with Thule impossible. Even with occasional visitors, ample rations, warm bunks, and frequent movie nights, the loneliness of Site 2 caused some men to break, especially in the dead of winter when storms meant no mail or flights could come in. One journalist described Site 2 as "among the most lonely and isolated habitations on the face of the earth."[21]

Chester Langway arrived at Site 2 in the summer of 1956. He was hired by Bader a few days after finishing his master's degree in geology at Boston University—in fact, Langway graduated on May 15 of that year, and on June 1 he was ordered to take a flight from McGuire Air Force Base in New Jersey to Thule in Greenland. He had not intended to study ice or work in the Arctic; he had not even been aware that ice drilling existed, or that it was deemed an important part of the upcoming International Geophysical Year. Langway recalls: "On that first trip we went out from Thule to Site 2 on what they called a 'swing'"—Greenland jargon for an ice sheet convoy—"on a tractor that was about six or seven feet tall, with 58-inch tracks. These were huge machines that pulled six or eight sleds and fuel for the tractors." The tractors moved slowly—two or three miles per hour—and the journey took several days. When Langway arrived, he and his team were assigned to live and work a few miles from the tubes of Site 2, in insulated Jamesway huts that resembled huge half-barrels nestled in the snow. "The army fed us and sheltered us and built us a camp," says Langway, "and that was the route to success."[22]

Langway's primary job was to catalog and analyze the ice cores, rather than to oversee the drilling process, which was the responsibility of the engineers and technicians. Not long after the SIPRE crew arrived, they cut a deep trench with plows, roofed part of it with wood for protection from storms, and began setting up the ice drill, which resembled a small derrick and was so tall it couldn't be covered by the roof but instead needed to poke into the open air above the snow line. The men had decided to extract a core by using a Failing-1500 rock driller adapted with special hollow drill bits that SIPRE engineers surmised would work for cutting through ice.[23] The entire process was arduous. "At Site 2," Langway recalls, "all you did was work, eat, and sleep—seven days a week until you finished." His team stayed three months. By the end, they had reached a depth of nearly 1,000 feet. But half of the ice core they brought up was broken and unusable, and the work—constantly connecting and disconnecting drill pipes and equipment every time they brought the bit to the surface—was slow and frustrating.

They returned the following year, in June 1957. This time things went better. The engineers were now pulling ice cores that looked usable and intact. And when Henri Bader visited as the summer season was ending, he seemed pleased. He took Langway aside to ask him some questions.

"Chet, it's getting dark now, it's getting late, and we're down now almost four hundred meters," Bader said. "That's the deepest ever. And the core is pretty good. Do you want to go to Antarctica?"

Langway hadn't been thinking of much else but Greenland for the past year and a half.

"When?" he responded.

"October," Bader said. That was his way of saying he was considering bringing Langway into the IGY drilling projects.

"October of what year?" Langway asked.

"Next month," Bader said.[24]

When his shock receded, Langway said no. He knew that part of the deal of working as a researcher at SIPRE was that you lived a life of perpetual winter. Summers were spent in the field in the Arctic, then you endured Chicago's cold seasons. And when buds appeared on the trees, it was time again to go back to the Arctic. But if Langway were to sign on to the IGY project in Antarctica, it would mean nine more months of constant drilling, constant work, constant cold. And what he most wanted was to spend time in the SIPRE lab in Illinois, trying to figure out what to do with the ice from Site 2. By the end of drilling, the SIPRE team had reached a depth of about 1,350 feet. The cylinders they recovered varied in length from six inches to one and a half feet and were just shy of four inches in diameter. Each had been carefully cataloged and numbered as it came to the surface, so that Langway could trace a continuous record of the past.

Many years later, researchers would establish protocols for analyzing ice cores: There would be precise instructions as to how to slice and saw them, both crossways and lengthwise, and how to measure out wedges and crescents and cubes so that certain cuts could be dedicated to specific methods of physical, chemical, and optical analysis. At this early moment in time, however, there was no protocol. At Site 2, Langway had a makeshift laboratory in one of the Jamesway huts where he conducted preliminary tests on the core. By counting the visible layers, he was now fairly sure that he had reached back about nine hundred years, meaning the deepest parts of the core were made from snow that fell at the time that the colonies of the Green-land Norse were still thriving. He had measured the core's tempera-ture and density; he had looked at the stratigraphy through a light table, which highlighted the seasonal changes in snowfall. He had focused on understanding how the deeper cores were affected by pres-sure from bubbles of air inside, which seemed to make the ice brittle when it came to the surface. But in the end, he concluded he could do far more back home and decided to saw all the pieces of core in half, lengthwise, and bring a large portion of it, packed in dry ice, to a cold lab in Illinois.[25]

A few months after returning to the United States he wrote a re-port about the Site 2 drilling project that was published in the *Journal of Glaciology*. He listed all the tests he planned to conduct next on the core—for more precise dating; for determining the pressure, shape, and distribution of air bubbles; for understanding the ice's permeabil-ity and porosity; for considering isotopic quantities of oxygen, deute-rium, and tritium inside. He even noted he would make an effort to look for bacteria trapped within. He seemed willing to throw every-thing he could think of at the ice, scientifically speaking, with the hope it could somehow prove demystifying. And Langway suggested that SIPRE was open to still other suggestions, since no one knew for sure how to decode an ice core anyway. "Interested persons," he wrote at the conclusion of his article, "are cordially invited to submit pro-posals for further studies."[26]

The two IGY drillings in Antarctica did not reach the same depths as the hole at Site 2. But these projects convinced the scientific estab-lishment that drilling deep cores was a promising endeavor, even if the ice couldn't yet be properly interpreted.[27] The question for the glaciologists was what to try next. Chester Langway would recall that by the end of 1958 the discussions at SIPRE turned to the idea of developing an ice-coring system "capable of reaching bedrock depths."[28] Bader was setting his sights on extracting a top-to-bottom core from one of the world's two major ice sheets, with the intention of seeing how far back in time he could go. His plan was to drill first in Greenland, rather than Antarctica. Again there were two main reasons. Thule provided tremendous advantages for travel and infrastructure. And because the Cold War continued to escalate—especially in the months after Sputnik was put in orbit—part of the military's enormous Arctic budget remained at Bader's disposal.

Bader hitched his plan to drill a hole through the Greenland ice sheet to a very different idea concocted by the U.S. military's Cold War strategists—one which involved building a massive new base under Greenland's ice. This was fairly typical. Much of the work Bader's organization had already done in Greenland had been in conjunction with the U.S. military's construction plans. In 1957, for instance, as Chet Langway and his team were drilling at Site 2, the Army Corps of Engineers was conducting a program, also near Site 2, that involved building a small under-ice camp named Fist Clench. Rather than being housed within giant corrugated pipes, Camp Fist Clench was situated in deep, wide trenches that had been cut into the top layer of the snow and firn. More specifically, army engineers had dug four parallel trenches that they then bisected with a central trench. These deep cuts, twenty feet deep and eighteen feet wide, were then covered over with metal arches.[29]

It was an experiment. To cut the trenches atop the ice sheet, Bader suggested that the Army Corps of Engineers use a machine called the Peter snow miller—a tractor-like snowblower that could clear a wide, crisp path through even the deepest snowfall, and which served as a workhorse for road crews in Switzerland, Bader's country of birth. His associates made tests at Fist Clench to show that the Peter miller was extremely adept at cutting deep trenches. And after the trenches were covered over by metal arches, engineers used the Peter miller to blow surface snow back over the shiny roofs.[30] This "Peter snow," as those at SIPRE called it, would quickly bind into a dense, durable roofing material—so durable and strong that Bader's researchers determined the Camp Fist Clench arches could even be removed, if necessary, so that some of the test trenches there resembled narrow church naves. Inside these long white alleyways, lit by electric lighting, it was blindingly bright, with ice crystals glittering on the walls and roof.

The spaces did not remain empty long, though. Within the completed Fist Clench trenches, the Army Corps of Engineers assembled prefabricated huts for barracks, mess halls, and storage spaces—leaving room under and around these huts for cold air circulation. This was part of the experiment, too. In sum, the engineers seemed intent on demonstrating whether there was an efficient way for the army to build Eismitte after Eismitte. Several decades before, Sorge and Georgi had realized the firn of the ice sheet could protect humans from bitter cold and ravaging winds. The military now surmised that with snow and ice it could camouflage soldiers and armaments from the prying eyes of an enemy's aerial observations.

All of this served as preparation for what came next.

In 1961, a British physicist named John Nye flew from Suffolk, England, to Washington, D.C.; and from Washington, Nye took a military flight to Thule Air Base in Greenland. Amongst glaciologists, Nye was famous for a number of academic papers he wrote in the 1950s that explained how glaciers and ice sheets flow—his observations showed that large bodies of ice behave like a plastic that "deforms" to a new shape under gravity and various stresses. Nye was visiting Greenland as a guest of Bader, to see what life was like under an ice sheet. From the Thule base he traveled about fourteen miles to the edge of the ice sheet, where a ramp to the ice was located; from there, he and a group of soldiers boarded a "wanigan"—a heated, insulated trailer hitched to a massive Caterpillar tractor, which would be part of a convoy known as a "heavy swing." Their destination was a point on the ice sheet 138 miles from Thule.[31] Here, the lessons from Site 2 and Fist Clench had been directed to a new under-ice station, known as Camp Century, which the army had built between June 1959 and October 1960.[32]

On a heavy swing, you slept, you chatted, you ate steaks grilled in the kitchen. Though it was possible to fly from Thule to Camp Century and land on a runway of groomed snow, often such flights were limited to army officers or to clergy who flew in briefly on Sunday to conduct religious services.[33] So instead, in the wanigan, you looked out the windows in boredom and in wonder for three or four days—or sometimes six or seven days, when bad weather intruded—as the panorama of ice stretched on and on. And on. The U.S. Army had built an ice road to the camp, lined with flags, emergency huts, and emergency supplies, and thanks to a new type of electrical crevasse detector, fitted to the front of each tractor, the journey had become fairly safe. Any resemblances to the travails of earlier explorers like Wegener or Rasmussen had mostly vanished. "It took three days," Nye would recall of the heavy swing, putting the speed at about two miles an hour, or slightly slower than the walking pace of a human being. The tractors and wanigans "rumbled over the snow and ice, and sleeping in them was very like being in an express train going at about eighty miles an hour with that sort of noise."[34] The sound came from the crunch of the wanigan's sledge runners on the ice sheet. When the tractor finally reached Camp Century, it descended down an entrance ramp—not unlike a car pulling into an underground parking garage—and motored into the camp's interior. What Nye saw was a place similar in concept to Fist Clench, a hidden catacomb of snow and ice, but far, far larger.

A high, wide, long tunnel—more than a thousand feet long, big enough to serve as a road for the tractors, and known as "main street"—ran the length of the camp. Perpendicular to main street, about two dozen long trenches had been cut in parallel to thirty-foot depths by the Peter miller. All the trenches were covered with metal arches and topped with snow. Within the side tunnels, the army installed prefabricated huts; all told, they held the capacity to house about two hundred and fifty soldiers and scientists. Camp Century contained an officers' club, a mess hall, showers, toilets, barracks, a church, and a store; you could drink beer or whisky on Saturday nights, read one of the books in the four-thousand-volume library, or watch movies. A haircutter named Jordon, the self-proclaimed "best barber on the ice cap," charged enlisted men fifty cents for a chop and officers one dollar.[35] The ambient temperature under the ice was well below zero, but the huts stayed at a pleasant 65 degrees Fahrenheit. To avoid having to tow massive amounts of fuel from Thule to the camp, however, engineers eventually installed a small, experimental nuclear reactor in an under-ice trench to provide electricity. The reactor cost $5.7 million and weighed 310 tons. It was towed on sled runners—gently, in the late summer of 1960—by Caterpillar tractor over the ice sheet.

The same year Nye traveled to Camp Century, the American television news anchor Walter Cronkite visited, too. Cronkite endured a similar ritual: the multiday haul from the edge of the ice sheet by wanigan; the wide-eyed surprise at the scale and audacity of the camp; and the glimpse—now shared with a television audience of millions—of a pint-sized nuclear reactor and the army's strange and exhaustive efforts to make life under the ice sheet comfortable. The military brass was forthright about the fact that Camp Century was doomed. At best it would last ten years, they acknowledged, at which point the overburden of snow would push down on the roof, push in on the walls, and thus destroy it. For the time being, though, officials strained to present the camp as a project built in the spirit of national defense and scientific inquiry—"as part of man's continuing efforts to master the secrets of survival in the arctic."[36] In fact, when Cronkite asked Tom Evans, Camp Century's commanding officer, about his objectives, Evans rattled off three: "The first one is to test out the number of promising new concepts of polar construction. And the second one is to provide a really practical field test of this new nuclear plant. And, finally, we're building Camp Century to provide a good base, here, in the interior of Greenland, where the scientists can carry on their R & D activities."[37]

At that point in time, some researchers and soldiers working at Century were aware this was not entirely true, and that one goal of the camp was unrelated to exploring the viability of ice sheet habitation. In a horseshoe-shaped, under-ice trench about a quarter mile from the main camp, and not open to visitors like Nye and Cronkite, an Army Corps engineer named Gunars "Chuck" Abele was experimenting by moving massive hunks of pig iron on a flatbed rail car—thousands of pounds of raw metal meant to approximate the weight of an intermediate range ballistic missile (IRBM). "We found there was no problem doing that," recalls Austin Kovacs, an engineer working nearby at Camp Century. "The firn was more than capable of supporting the load."[38] Wayne Tobiasson, an engineer working at Camp Century at the time, remarks: "That was something going on there that made no sense to us. And there was never an explanation except: Can we build tunnels and runways in Greenland?"[39]

Several decades after Camp Century was abandoned, a study by the Danish Institute of International Affairs uncovered a 1962 memo that detailed a U.S. military proposal for what was known as the Iceworm system: A nuclear arsenal of six hundred IRBMs, targeted toward the Soviet Union, which could be moved around under the Greenland ice sheet by rail. A blueprint for a project that was never built, Iceworm involved "thousands of miles of cut and cover tunnels twenty-eight feet beneath the ice" where "Iceman" warheads "would be in constant motion." The installation was expected to cover an area of the ice sheet that historian Erik D. Weiss approximated as the size of the state of Alabama.[40] In light of how the hollowed-out tunnels at Century were subject to the powerful and crushing dynamics of the Greenland ice sheet, the idea would come to seem preposterous in later years. Almost as soon as Camp Century was built, in fact, an entire crew of men had to be assigned to wander the camp corridors to shave and trim the encroaching snow from the tunnels. They removed as much as forty tons of firn from the walls every week. And yet, in the few but feverish years of Cold War strategizing, a contingent within the military not only thought Iceworm was possible, they thought it ingenious.

The irony was that Henri Bader and his colleagues were actually using Camp Century for an inspired research project. It just happened to be a research project that the U.S. Army didn't care as much about. In Trench 12, the Camp Century drilling crew, led by an engineer named Lyle Hansen, had set up a new rig. "The army allowed us to freeload with them," Chet Langway, who was in charge of cataloging and analyzing the ice cores that Hansen's drill brought up, recalls. And since the army was maintaining the appearance that the camp was for scientific research rather than for nuclear missile research, officials at Camp Century welcomed the prospect of showing visitors what the drillers were doing. Nye himself visited the early stages of the drilling project, as did Cronkite. "We were sort of a cover, if you will," Langway says, even though his team's goal—to reach bedrock—was deeply serious.[41]

The drilling group made some test holes, with mixed results, in 1961 and 1962. Then the effort to go from top to bottom began in earnest in October 1963. Bader estimated the distance was about a mile down. He expected the drilling team would reach near to bedrock—by his guess, ice that fell as snow seventy-five thousand years before—in four months.[42]

Drilling rig, Trench 12 (Herb Ueda/American Institute of Physics)

12

Jesus Ice

The crew that gathered every morning in the Camp Century drilling trench during the summer of 1963 worked in a frozen underground cavern between walls of firn that were spaced a narrow distance apart at the top and stepped back every few feet so as to make the room wider at the bottom. It gave their work space the appearance of being located inside an ivory pyramid. A thirty-foot high drilling tower stood in the center of the room and reached from floor to ceiling. Hung with icicles, wooly with hoarfrost, the room was cold enough ("not comfortable, but bearable," one driller recalls) that it required those inside to don thick army-issue polar jackets and billed hats with wool earflaps. The men stored barrels of fuel near the derrick—not because they needed to burn it, but because they had come to understand that it was necessary to fill the deep hole they were making with fluid that held a density equal to ice. That way, the hole wouldn't close up on them. The ice sheet tended to seal itself like that. If you were drilling deep and weren't careful the ice would grip your bit, freeze it in place, and not let go. Then all your work would be for naught. You'd snip the cable; drilling season would be done; and you'd need to come back to Camp Century the following year with new boring machinery. The only drawback to filling the hole with fluid—a potent mixture of diesel oil and trichloroethylene—was its smell. Visitors who ventured back toward Trench 12 could find the drilling crew just by sniffing.[1]

Drilling rigs that are customized to recover ice cores are fantastically complicated contraptions. To work properly, these elongated machines must go a mile or two down a narrow hole, digging into the ice inch by inch. During this process, a length of core—a cylinder of ice anywhere from three to ten feet—must safely be carved out of the ice sheet, gripped, severed, and pulled to the surface by a winch. Then the drill must go back down and carve deeper, inch by inch. The coring process must be repeated (cut, grip, sever) and another length of ice must then be raised and retrieved. For the Camp Century drilling, Henri Bader suggested that his engineering chief, Lyle Hansen, create a new kind of drill. Rather than saw circles through the ice, why not use a hollow-tipped "thermal" bit that *melted* the ice as it went down? In other words, a hot ring of metal could burrow into the ice and produce long cylinders that could then be brought up. Going along with Bader's wishes, Hansen's team built precisely this type of machine.

Ever since the drilling project at Site 2 a few years before, it was understood that keeping the ice in rigorous order would be just as crucial as a good drill. In other words, if a team lost track of the sequence in which the cores came out of the ice—or if the depth from whence the cores were pried from the ice cap became mixed up—the scientists could lose track of climate history and jeopardize their entire experiment. For that reason, on most summer days during the early 1960s, the cores that reached the surface in Camp Century's drilling trench were carefully bagged and logged and stored in card-board tubes on racks against the wall. Before they were put away, however, Chet Langway, the scientist in charge of scrutinizing the ice, would usually look them over closely on a light table. Cores that came from closer to the surface exhibited seasonal stripes, and some-times pockets of frozen dust, suggesting remnants of an ancient volca-nic eruption or dust storm. But as the drill reached farther down, the cores were less obviously marked with annual layers. What's more, Langway could see that some cores came to the surface hazy and loaded with bubbles, resembling cylinders of frozen milk, whereas deeper ice emerged clear like glass—only to become hazy a few weeks later as gases that had been under tremendous pressure deep in the ice sheet coalesced back into bubbles. Some of the cloudy, bubbly ice could be as fragile as crystal stemware. Minutes after retrieving it from the drill's core barrel, Langway could see it fracture, and hear it crackle, as the air inside "relaxed" in reaction to the pressure changes at the surface.[2]

Herb Ueda was usually the technician in charge of the day-to-day drilling work. He would typically fly in to Camp Century every April and stay until September. How he had ended up here surprised him as much as anyone. At his birth in 1929, Ueda's father decided to name his son for Herbert Hoover, taking the suggestion of a local man who thought it wise to name a child after an eminent politician. "Then the Great Depression hit six months later," Ueda would recall.[3] By his own assessment, his family was dirt poor, and Ueda often worked as a laborer alongside his parents in the crop fields and orchards of the American Northwest—work he would sometimes think about when he drilled in Greenland, especially as a reminder of how much he preferred the cold to those brutal childhood summers of hundred-degree heat. After the attack on Pearl Harbor in December 1941, Ueda and his family were forced by the American government to move from the Tacoma, Washington, area to Idaho, to an internment camp for Japanese-Americans. For three years, his family lived in what was essentially a concentration camp, ringed with barbed wire, with about nine thousand other Japanese-Americans.[4] Ueda never-theless finished high school, got drafted, and served in the U.S. Army, beginning in 1951. He spent a year stationed in Germany, and upon returning home pursued a college degree in mechanical engineering at the University of Illinois. He was twenty-nine years old when he graduated, aimless and living in Chicago, when in the course of interviewing for jobs he got a call from "some kind of a snow and ice lab" in Wilmette, Illinois. So he went to Wilmette and talked with Lyle Hansen, Bader's top engineer at SIPRE. Hansen discovered from their conversation that Ueda had worked harvesting sugar beets in Idaho. Hansen had a background in farming, too; he knew how difficult such work could be. So he offered Ueda a job. It was August 1958. The next summer, Ueda flew to Thule and learned how to drill holes in ice.[5]

Ueda was much less focused on what the cores might say about the history of the earth than how to get them out of the ice sheet. He soon knew every quirk and problem of the drilling rig. At Camp Century, his habits tended toward workaholism—six days a week, ten hours a day, he was drilling in Trench 12. (Sundays at the camp were for recuperating and, for many, sleeping off a hangover from Saturday night.) During one summer at Camp Century, Ueda stayed below the ice sheet for a month without once going to the surface, perplexing colleagues who couldn't imagine how someone could endure for so long without fresh air, sun, or a glimpse of the outside world, even if it happened to be a featureless white icescape. Still, by the end of 1964 he and his fellow drillers had reached 1,755 feet down and recovered an ice core for about 96 percent of that length.[6] It was slow and difficult work, and Ueda was increasingly frustrated by the thermal drill. On average, it could only melt through the ice sheet at about one inch per minute. Worse, its constant breakdowns and need for maintenance meant that for much of his day in the trench, Ueda made repairs. To his boss Lyle Hansen, Ueda said, "We've got to start considering something else."[7] At this rate, the hole that Bader had predicted would only take a few months would take five more years.

The Army Corps of Engineers decided to try a machine that had been built to extract oil in the American South. In 1926, a Russian immigrant living in California named Armais Arutunoff had invented an electric submersible pump for oil drilling, which meant it could operate in a hole filled with liquid. A friend of the mogul Frank Phillips of Phillips Petroleum, Arutunoff eventually built around the invention a successful drilling equipment company in Oklahoma that he named REDA—an acronym for Russian Electrical Dynamo of Arutunoff. In 1964, on a field trip to Oklahoma, several Army Corps engineers discovered a REDA rig they thought might work for Camp Century. "They found it abandoned, in some cornfield somewhere," Ueda recalled.[8] "The owner offered to sell it to us for $10,000, so we bought it, and we modified it to work in ice." This "electrodrill," now tweaked to drill a core, was shipped by air to Camp Century in the spring of 1965. The drillers set it up in Trench 12, inserted it into the existing hole, unwound the cable, and lowered the bit to the current depth of 1,755 feet. Then they revved it up.

"It was like drilling at the end of a long noodle," recalls Tony Gow, an Army Corps of Engineers scientist who was involved with a number of early drilling projects.[9] The electrodrill was an ungainly machine—eighty-three feet long and weighing 2,650 pounds, not including the drilling tower and eight thousand feet of thick cable that provided the drill's stability and power. With flexible segments comprising barrels and pumps and motors, the drill rig resembled a thin, long, multi-stage rocket. At the tip, the electrodrill had a hollow, circular cutting bit studded with diamonds that rotated at a rate of 225 revolutions per minute. The teeth of this bit cut an ice core from the sheet which could be captured in a pipe-like sleeve just behind it. One of the significant challenges for drillers involves ice chips that get produced as a by-product of cutting a cylinder out of the ice cap; these chips can clog a hole and hinder progress as well as the purity of the ice core. Ueda and Hansen found they could remove chips by dissolving them in a solution of glycol—essentially antifreeze—they pumped into a tank on the drill. And in time, they found the glycol and the new machine worked surprisingly well. "We were getting cores twenty feet long with this drill," Ueda recalls of the summer of 1965. "And so you can cover a lot of depth like that. On a good day we could do more than a hundred feet." On average, the new drill from Oklahoma seemed to be moving at about five or six times faster than the old thermal drill.

By the end of summer, Ueda's progress was especially encouraging in light of the fact that Camp Century was beginning to collapse around him. Inside the trenches, heat from the buildings, humans, and machinery was softening and destabilizing the floors and walls. Main street—the wide trench that ran through the center of camp—was bedded with what Langway recalls as filthy white quicksand. At the same time, snow falling on the surface forty feet above was piling up and pushing down on the ceilings.[10] To live in Camp Century, residents had always needed to master their fear that the steadily compressing roof would lead to a catastrophic collapse; it was the same fear that had bothered Sorge and Georgi at Eismitte, which led Georgi to build a supporting column.[11] But things were now getting worse. As many as fifty men were on duty and tasked with shaving and trimming the walls and ceilings—usually with chain saws—to maintain the camp's viability. And it was undoubtedly a losing battle. The season before Ueda tried the electrodrill, army engineers had decided it was necessary to decommission and remove the reactor, which meant the base was now functioning only on diesel-generated power. As the summer of 1965 came to an end, Ueda had not yet reached bedrock. But a total evacuation from Camp Century seemed inevitable, and very near.

It was anyone's guess what the drillers would find at the bottom of the ice sheet. In 1962, Henri Bader had laid out a few thoughts about the glacier bed. The drillers might encounter solid rock or *till*—that is, small stones and dirt produced by the powerful grinding movement of the ice sheet as it moved over the bedrock—or a pocket of natural gas. Bader also noted that a Russian glaciologist, Igor Alekseyevich Zotikov, had developed a bizarre hypothesis that "heat flux" from the earth's interior would produce at the glacier bed "stores of energy" from compressed air—so much energy, in fact, that it could "turn the giant turbines of a great electric power station for many thousands of years."[12] Whatever the result, it was clear the drillers were on the verge of tapping into a place on earth that had not been disturbed for tens of thousands of years.

The team returned to Trench 12 and started up the electrodrill in the late spring of 1966. Their coring work was still the same: cut, grip, sever; pull the core up for capture and analysis; repeat. On July 4, 1966, they hit bedrock at 4,450 feet. There was no gush of gas or explosion of air. A photo exists from the day that Ueda reached bottom: Wearing army fatigues and an insulated hat, he stands beside a long cylinder of ice and rock that has been slid from a drilling sleeve onto a trough for observation. He looks fairly amazed and also relieved. Ueda would later recall that it was the most satisfying minute of his career. "You can imagine—it took us how many years?" he says, before concluding the number was six.[13] To celebrate the accomplishment, some of the men at Century took a small chip of ice from a core that approximately dated to the birth of Christ and toasted the occasion by putting it in a glass of Drambuie. But Ueda and his fellow drillers weren't quite finished. Once they extracted the final core of ice, they used the drill to cut into the bedrock and dirt below, pulling up about a hundred feet of it. When they concluded they had about all they could want, they finally packed up.

Camp Century was by now pretty much finished, too. The summer of 1966 marked its final season as an army base. With the nuclear reactor gone, the army's main concern was bringing the tractors, trucks, and wanigans back to Thule. Herb Ueda and his team took the electrodrill; they had plans to ship it to Antarctica and use it to extract a core at a camp called Byrd Station, beginning that fall.[14] But almost everything else was left in the Camp Century trenches: prefabricated huts that served as dorms and mess halls, tables, chairs, sinks, mattresses, bunks, urinals, the billiards table. Waste products from the camp—human sewage, diesel fuel, toxic chemicals such as PCBs, and radioactive coolant from the reactor—were left behind, too.

The working assumption was that everything would soon be crushed by the overburden of snow, anyway. And after that, it would be locked forever into the ice sheet.[15]

With the drilling complete, Chet Langway, the ranking scientist, was ready to leave. He took a toothpick dispenser from the wall of the

Camp Century mess hall as a memento, but Langway also took more than a thousand ice cores. In time, they would prove to be the only thing of lingering value that came out of the military's strange and expensive Camp Century experiment. He used army transport planes to ship the ice to a freezer in New Hampshire, which was where he was now working. The organization under Henri Bader that he had originally joined—the Snow Ice Permafrost Research Establishment, or SIPRE—had merged with another division within the Army Corps of Engineers to become the Cold Regions Research and Engineering Laboratory, or CRREL; in the process, its offices had moved from Wilmette, Illinois, to a larger building in Hanover, New Hampshire, not far from Dartmouth College.[16] At around the time of the merger, Bader had also left to take a professorship at the University of Miami. He lived with his wife on an upper floor of an apartment building by the water, which meant the responsibility for interpreting the cores fell almost entirely to Langway. What to do? Langway recalls that he went around the country looking for help in interpreting the trace gases and shreds of evidence in the Camp Century cores. He also went around the world.

In 1962, while Langway was attending a conference in Austria, he struck up a conversation with a genial Swiss physicist named Hans Oeschger. When Langway mentioned the trapped bubbles of air in the old ice of Greenland, Oeschger suggested that he might be able to measure and date carbon isotopes in the air bubbles, which no doubt contained traces of carbon dioxide, or CO_2. This was a variation on the established process of radiocarbon dating, which was already being used to date fossils and ancient wood back to about forty thousand years. Langway kept up with Oeschger, and in the decade following, Oeschger's lab, focusing on Antarctic ice, figured out how to use the ancient air for dating as well as for something even more important: They could match the concentrations of carbon dioxide and methane within the bubbles to the approximate year that they were trapped. That meant they could discern in their laboratory what the composition of the atmosphere was like during a long-ago epoch.

Langway found another partner who could reconstruct ancient climates by studying the traces within the ice. In 1964, a Danish scientist named Willi Dansgaard had come to Camp Century with some colleagues from Copenhagen to conduct a chemistry study on the ice sheet. Dansgaard never actually visited the Camp Century drilling trench during his visit. Nor did he get to meet Langway or Herb Ueda at that time. He was informed by one of the camp's military officers that he was unauthorized to visit the coring experiment, and in his diary Dansgaard wrote: "What a shame . . . What the Americans are going to do with the ice core is unknown." Later, back in Denmark, musing about the drilling experiment again, he concluded that the Camp Century ice "would be a scientific gold mine for anyone who got access to it."[17] In 1966, when he heard of the coring's completion, he decided to write Chet Langway a letter that proposed his doing an analysis of the ice. One of Dansgaard's students would later say, "That letter is the birth certificate of ice core climate research."[18]

Ice scientists are detectives at heart. Dansgaard was by that point one of the pioneers of measuring oxygen isotopes. These are the naturally occurring variations that reflect whether an oxygen atom has six or eight neutrons in its nucleus.[19] The differences are expressed by comparing the prevalence in a water sample of the heavier and rarer isotope (^{18}O) to the lighter and more common isotope (^{16}O). Dansgaard began some of this work in 1952, when he collected rainwater in his yard with a beer bottle and a funnel. What he then began to understand was that warm weather storms produce moisture with a higher percentage of "heavy" ^{18}O than cold weather storms. He made a further leap and soon concluded that the temperature of a cloud helps determine the amount of ^{18}O in the snow or rain it produces. In essence:

Higher temperature = a higher concentration of ^{18}O in H_2O
Lower temperature = a lower concentration of ^{18}O in H_2O

The significance of this connection to Greenland ice cores might not seem immediately obvious, but Dansgaard surmised this made it possible to connect the oxygen makeup in the water of old ice with climate. In other words, if he had a sample from a deep ice core that could be dated to an approximate year, he could likely measure the concentrations of ^{18}O in the ice. Then he could look at the results and discern the temperature of the surface air on the day the snow-flakes fell to earth, even if it was ten or fifteen thousand years ago.[20]

The tool he used to do this was known as a mass spectrometer. Dansgaard prepared a sample of ice by processing it with carbon dioxide in a sealed container and then feeding part of the mixture into a small vacuum chamber. The instrument—the "mass spec," as they called it in the lab—then bombarded the sample with electricity so as to charge its oxygen molecules; once charged, the sample could then be separated into the heavier and lighter components by passing it through a magnetic field. The physics were complex but the outcome was simple: Within the machine, the heavy and light oxygen isotopes from the ice sample could be detected and their concentrations measured. The result was vaguely akin to producing a rainbow that allows someone to see for the first time the individual colors within the light spectrum.

"I offered to measure the whole ice core from top to bottom," Dansgaard recalled of his 1966 offer in the letter to Langway.[21] Langway readily agreed, and Dansgaard and several associates flew from Copenhagen to the new CRREL laboratory in New Hampshire. The men cut seventy-five hundred samples of the Camp Century ice core and brought them back to Denmark, where Dansgaard had technicians working long hours in his mass spectrometer lab. Out of that big trove of ice, he formulated his first study. And on October 17, 1969, Dansgaard's team and Langway published the results in a paper in the journal *Science*, entitled "One Thousand Centuries of Climatic Record from Camp Century on the Greenland Ice Sheet." Dansgaard created a graph tracing the oxygen isotopes—and, in effect, the climate—back approximately one hundred thousand years. Langway recalls, "When Willi made that, he shocked the world. Because one of the most difficult things to look at is the temperatures of the past. How do you get that information? You can't get it by carbon-dating rocks. It doesn't work. But it can with gases in ice, if you've got a tag on their age."[22]

In the *Science* article, Dansgaard wrote, "It appears that ice-core data provide far greater, and more direct, climatological detail than any hitherto known method."[23] It was nevertheless clear to him that his study wasn't perfect. Many parts of the ice core were hard to read, and it seemed to be the case that chaotic changes in temperature characterized the earth's climate at various points during the period that stretched ten to fifteen thousand years before the present era. This would have been about the time that the earth was emerging from the last ice age. The period of wild, swinging indicators could have been some noise in the climate signal, errant pulses of information that need not be taken too literally, for they might have originated in ice that had flowed and folded over bumps in the Greenland bedrock.[24] Then again it might suggest something else: That climate could change quickly and drastically.

In some respects, field experiments to retrieve deep ice cores soon became like ice sheet expeditions without the dogs or the roving or the hunger. Rather than pursuing the vision of a charismatic leader, though, you were exploring an idea—seeking clues, in effect, to construct an understanding of the vanished earth. It was now becoming clear that the mysteries of Greenland were writ small as well as large. They were buried in microscopic chemical traces, scattered around and within the ice sheet, which would only yield to a combination of technology, persistence, and imagination. Ernst Sorge's realization from nearly a half century before had been striking: Looking down while flying over the ice, he saw an apparently empty landscape *which yet conceals a thousand secrets*. The era of concealment, however, was now ending.

Drilling once was never enough. And as the age of coring began in the wake of Camp Century, Langway and Dansgaard began to plot what kind of field experiments they needed to undertake in the coming years. There were some new and complicating factors. In the 1950s and early 1960s, Henri Bader had piggybacked on the U.S. military's Arctic programs to fund his research efforts, but by the late 1960s, the United States became engaged with the Soviet Union in a different manner—from a distance, rather than up close, due to the development of longer-range ballistic missiles and bombers. Thule's air base was reduced both in size and scope, and dollars for the American military began flowing toward other geographic regions.[25]

For Langway, Dansgaard, and Oeschger—now called by their colleagues "the three musketeers"—this meant that drilling deep holes in ice was no longer just an engineering and scientific problem. It was an economic problem, too. The army's declining interest in ice meant new projects had to be funded through universities and national science agencies. As Langway recalls: A flight to the center of the ice sheet that would have been free to an ice driller in the mid-1960s, thanks to the U.S. military, now cost six thousand or eight thousand dollars an hour. And another wrinkle made the potential costs even more formidable. If anyone wanted to drill deep again, they would need to build a new machine that might cost hundreds of thousands or even millions of dollars. The big electrodrill Herb Ueda had used at Camp Century was gone. It had gotten permanently stuck at the bottom of a 7,100-foot-deep hole he was drilling at Byrd Station in Antarctica. Given up as unrecoverable, its cable was snipped.

Langway, Dansgaard, and Oeschger decided to form an international research team they named GISP, for Greenland Ice Sheet Project, in the spring of 1970. They scraped together funding from sources in Denmark and Switzerland, and in the United States from the National Science Foundation, an independent federal agency.[26] The change officially meant that glaciological research was no longer a military endeavor, but something done for the cause of scientific discovery. Langway, who became the chief scientist of GISP, recalls that the small community of glaciologists and ice drillers tended to think of the early 1970s as the start of the next era of researching the world's ancient climates. But one of the first goals of GISP was simply practical rather than conceptual: Where was an ideal place in Greenland to remove the next deep core from the ice sheet? "Camp Century wasn't built to drill the ice core," Langway says. "We took *advantage* of that, because it was there."[27] He and his new colleagues therefore wanted to find a more optimal place—a place where they could get a core that had a clearer and longer record of temperature changes. Almost certainly, they knew, the real prize would be in the center of Greenland, somewhere along what is known as the ice divide, the place where the altitude of the ice sheet is highest, where the ice sheet never melts in summer, and where one side of the ice sheet flows steadily to the west and the other side to the east. Along this divide, which happens to fall near where Eismitte and Station Centrale were situated, the layers of firn and ice would presumably be undisturbed by the steady glacial movements of the ice sheet, leaving behind a clear record of perhaps 125,000 years of snowfalls. In theory, it would be the perfect ice, the ultimate core.

For the better part of the 1970s, the GISP team looked all over Greenland to find the right spot to drill down to bedrock again. At sites with names like Crete, Milcent, and North Site, they flew in on ski-equipped planes, set up temporary camps, and drilled test cores. These weren't *places*; they were just dots on a map of Greenland's ice cap. And the cores that came up, ranging in length from three hundred to fifteen hundred feet, were meant to give them a sense of the underlying conditions in the ice—in effect, to see if drilling deeper was worth it. In the meantime, the Danes decided to build a heavier drill. Smaller and more nimble than the Camp Century machine, this Danish deep drill was named Istuk—a neologism that came from combining the Danish word for ice (*is*) with the Greenlandic word for drill or spear (*tuk*). Measuring about thirty-eight feet long and weighing about four hundred pounds, Istuk was attached to a cable by way of a thirty-three-foot-high tower. The drill sliced through the ice with three sharp knives arranged in a circular pattern. What also separated it from its predecessors was its electronics. The machine had a microprocessor inside. In the trench, a driller could issue orders to the drill by computer, and no matter how deep, the drill would listen.[28]

As the limits on funding became clearer, the GISP team in the late 1970s was essentially told where to drill by the National Science Foundation: A radar station on Greenland's south central ice sheet that was a remnant from the early Cold War days. It was known as Dye-3. In addition to a radar dome it had a runway and a few insulated buildings. But mainly what it had going for it was location—a quick flight from an airstrip on Greenland's southwest coast known as Sondrestrom Fjord.[29] Also, it was seen as a low-risk choice: The project's funders thought Dye-3 would be a good place to test Istuk, the new Danish drill, before it was used in a remote location, like somewhere along the central ice divide.

Dansgaard and Langway worried about the new site. Dye-3 happened to be located in a place where melting during the summers might have altered the ancient climatological record. Worse, it appeared from radar soundings that the underlying bedrock was mountainous, suggesting a "flow pattern" from the ice sheet that might have wreaked havoc on the steady accumulation of layers. The scientists were looking at the ice sheet in a different way than anyone had before; what was apparent on the surface was not necessarily relevant to them. What was below, what was inside the ice, mattered more. "I pointed out that from a scientific point of view, Dye 3 was not an ideal drill site," Dansgaard recalled. "But I got the answer that to begin with, it was Dye 3 or nothing."[30]

And so it was Dye-3. In the summer of 1978, the Danes and Americans and Swiss converged on the site, along with several Icelandic and Japanese scientists, to set Istuk up. The engineers cut trenches and built a connected complex of under-ice trenches that they covered with wood for protection from the elements. These were for drilling, ice core analysis, and storage. The plan was to drill around the clock, all summer long. "When everything functioned optimally," Dansgaard noted, "45 people were in action. Three drill teams, three persons each, worked in turn around the clock, and two teams worked in the science trench from morning to late evening."[31] Ski-equipped planes ferried scientists and technicians in and out on a regular schedule. The process was technologically sophisticated, exhausting, and expensive. But unlike the expeditions of Greenland's distant past, they knew exactly what they were looking for.

Sitting underneath the Greenland ice sheet in the summer of 1980, deep into the quiet of the Dye-3 night shift, staring at the primitive computer console in the drilling room, Jørgen Peder Steffensen—J.P. to his friends and coworkers—wondered how he had ended up there. A week before, he had been a senior at the University of Copenhagen, intending to do graduate work in astrophysics. He'd planned on working as a projectionist at a local art house cinema for the summer. But then on a Friday afternoon in early July he had gotten a phone call from a professor he had heard of but didn't know. Willi Dansgaard asked the young man if he'd like to come to Greenland for six or seven weeks, to take part in a project he was working on in the middle of the ice sheet. He had heard that Steffensen was smart, handy, and technically minded—"that he liked to take his hands out of his pockets."[32]

Steffensen had never given a minute of consideration to working as a glaciologist or climatologist. But after a moment of thought he said that he would try to cancel his summer plans and Dansgaard should call him back in half an hour. When the phone rang again, Steffensen told him he would go, and Dansgaard said: "Okay, tomorrow, you go to the institute. I'll be there. And I'll give you a ticket to Greenland. Bring your passport, because you have to be cleared through the U.S. Air Force base at Sondrestrom and at Dye-3. Then I'll issue you some polar gear."

"It was only the second time in my life I was to fly," Steffensen recalls. "I had to leave Tuesday. So I had three days warning. Wednesday at noon I was at Dye-3. And I didn't know *anybody*. I was briefed in the kitchen, and was told I was going to help them with cutting the ice cores and collecting samples. I came down the staircase to this underground vault where they were doing the drilling, and as I was passing the console the guy in the chair said, 'No, you're not going to process ice cores. You're going to drill.' So he put me in the chair of the driller. And I got a crash course of two and a half hours. And then I discovered later that Niels Gundestrup, who was sitting at the console, just that morning had received a call from his highly pregnant wife saying that either he returned tomorrow or there's going to be a divorce. So within a week of not knowing anything, I became a driller."

Unlike Dansgaard—a genius in the laboratory, but so wanting at mechanical tasks that the drillers made a concerted effort to keep him away from the rig and the practical aspects of ice coring—Steffensen was well suited to the engineering demands of the job. He also loved the collegial, "flat" hierarchy of the drilling camp; scientists who would later enjoy great fame in glaciology—Langway, Dansgaard, Henrik Clausen, Claus Hammer, and others—were just Chet, Willi, Hank, and Claus. On Saturday evenings, despite the fact that they were in the middle of the Greenland ice sheet, they all put on neckties and had a dinner party. Most of all, Steffensen found the idea of extracting history from the ice sheet an experience of constant revelation. "I had fantastic teachers—Hank Clausen was one of them," Steffensen recalls. "He worked in the science trench, next to the drilling room. And he had this fantastic historical overview." Every time Steffensen pulled a core up, Clausen would come down and discuss the relevance. "Hank would say, 'You know, this fell as snow when Marcus Aurelius was going into Germany.' And then we'd go back in time, the deeper we'd get. And then it was the time of Augustus. And then it was the time of Caesar."

To help date the cores, the scientists had begun using an electrical conductivity test. By moving two electrodes along the ice, they could run a current through a core segment and note the response to the voltage, which varies from season to season due to dust particles, and which can jump in sections where the ice is suffused with acidic volcanic residue. "There's this eruption that shows up in the electrical conductivity test that we call the Caesar volcano, because it's pretty close to 42 B.C.," says Steffensen. "Plinius the Elder wrote that when Caesar was killed the gods were so ashamed of what Rome did that they hid the sun behind a red veil for an entire year. Very poetic. But it was exactly how you would describe sulfuric aerosol after a major volcanic eruption. So all that was there for us to see. And then we continued drilling. Down to the Punic Wars. Farther down. To Hannibal. And farther down. And Hank always had these points in history to relate. For a young student with an open mind it was just awe-inspiring, because you're standing there, with your hands on the core, and it came from that exact time."

When the season ended, the drilling had reached about 2,950 feet— and Steffensen decided he was hooked for life.

The following summer, in 1981, Steffensen returned to Dye-3 and was teamed with another student, a woman a year younger than him, in the drilling trench. "Only the two of us," Steffensen says. "So, night after night. And sleeping during the day." Dansgaard had an arbitrary rule that reflected the hidebound traditions of his field: No women could come to the drilling camp. But he had relented in this case, seeing as the young student, Dorthe Dahl-Jensen, happened to be serious about becoming a glaciologist. What's more, she happened to be brilliant.

The two night-shift drillers fell for each other. It was hard to see ahead, but eventually, the duo would spend the rest of their careers drilling into the Greenland ice. Steffensen would become the logistical expert, and Dahl-Jensen would become one of the world's most eminent glaciologists. Also, for someone who was initially told she couldn't even visit the ice-coring camp because she was a woman, destiny imposed a kind of justice. In time, Dahl-Jensen took over Dansgaard's job at the University of Copenhagen.

The crew at Dye-3 reached bottom on August 11, 1981, at 6,683 feet. *The New York Times* published a front-page story about the strange drilling experiment a few days before the team hit bedrock. The writer, Walter Sullivan, had visited the underground tunnels and drill site, and he explained to readers how the researchers in the science trench had already discovered evidence in the ice cores. "Hidden in the deep layers of ice are samples of the earth's ancient atmosphere," he wrote, "clues to volcanic and climatic factors that led to past ice ages and could set the stage for a new one."[33] Sullivan noted that the Dye-3 team had also started looking at the ancient bubbles of trapped air to see how the earth's atmosphere had changed over time.

It was obvious that it would take time to understand the traces within the ice. But just as he had done for the Camp Century ice, Willi Dansgaard shipped pieces cut from the cores—a continuous chronological sequence—to his laboratory in Copenhagen. He and his colleagues had built a new mass spectrometer that could automatically analyze about 260 ice samples every night.[34] At this rapid pace, his lab proceeded to make tens of thousands of measurements on the oxygen isotopes in the ice cores, and the results again tracked

Greenland's wavering climate over the course of about a hundred thousand years. It was immediately noticeable that the strange and dramatic temperature variations that had shown up in the Camp Century cores a decade earlier appeared here, too. There seemed to have been wild swings in climate between ten thousand and thirteen thousand years ago, the cores were saying—swings that seemed to have no relation to gradual trends of warming or cooling that were assumed to characterize climate shifts.

Dansgaard and one of his Danish colleagues, Sigfus Johnsen, began discussing whether they should try to get a more precise picture of this time period. Though the Danes had already measured the entire ice core, they still had a trove of the ice in their freezers in Copenhagen. They could therefore go back and examine some eras in closer detail. This time, they decided to try using a different measure: a hydrogen isotope called *deuterium*, which responds much like oxygen isotopes to changes in ancient ocean conditions and temperatures. Dansgaard and Johnsen didn't have a machine for this kind of work in Copenhagen. But the men eventually teamed up with a young American post-doctoral researcher, James White, who was working in Paris in 1985. At the time, White was working with a pioneering French geochemist named Jean Jouzel, who had the world's best spectrometer system to measure hydrogen isotopes.[35]

These Dye-3 samples comprised small vials containing melted ice. Each was labeled carefully with the depth of the core that sample had come from. When the vials were shipped to Paris for White to analyze, a few broke in transit. So White visited Dansgaard and Johnsen in Copenhagen, cut replacement samples from the ice cores in their freezer, and stuffed the vials of twelve-thousand-year-old ice into his carry-on luggage when he returned to Paris. Not long after, Sigfus Johnsen decided he wanted to bring White even more sections from the Dye-3 core, so he took a detour during a family vacation to meet White. The fact that a French laboratory was being used to analyze Danish ice samples was politically fraught, so the meeting took on a stealthy air. "Sigfus and I met on a park bench in front of Notre Dame," White recalls. "He had a bag with the ice samples in them. He put it on the ground. And I sat down next to him. He shoved the bag over with his foot. I picked it up. It wasn't as if we were trying to be like spies, but that's what it was."

White returned to his work in the lab. And when the results from the Paris spectrometer were ultimately compiled, they looked as strange as the earlier results. "It's important to recognize that even though we saw these big abrupt climate changes in the Camp Century ice core," White says, "a lot of people thought that was some artifact in the ice." In other words, they thought it was a false signal.[36] "Then, when we saw the Dye-3 ice came in with the same changes, I know that Willi and Sigfus were convinced that these were really climate signals." White recalls that most people he spoke with at that point were still not convinced. "When I first started to talk about this," White says, "I remember people just staring at me, like, *Come on*."

It was the speed of the climate changes that elicited disbelief. At the time, there was growing agreement that ice ages and warm periods resulted from small variations in how solar energy falls on the earth. Predictable cycles in our planet's orbit and tilt ("wiggles" that occur at regular intervals of about 20,000, 41,000, and 100,000 years) alter the way sunlight reaches the arctic regions. Such cycles—named Milankovitch Cycles, after the Serbian astronomer Milutin Milankovitch, who calculated these patterns—seem to work very slowly to affect climate. And yet, they can explain how, as the earth's exposure to the sun varies over long periods of time, polar regions can grow colder or warmer, ice sheets can expand or shrink, and a number of feedback loops can kick in to then make the climate shifts even more pronounced. Some of these Milankovitch cycles had shown up already in the Dye-3 and Camp Century cores, and were emerging in an even longer core that was being drilled during the 1980s in a remote Soviet science camp in Antarctica known as Vostok Station. The Vostok cores, as they're known, which eventually went back 400,000 years, happened to be the ice that White was measuring on weekdays in Paris while he worked on Willi Dansgaard's Dye-3 cores on the weekends.[37]

So the general sense was that barring any catastrophic event—a massive asteroid impact or immense volcanic eruption that shrouded the sky in gases and particulates—the world's climate tended to change, based in part on the planet's exposure to the sun. But it shouldn't jump drastically from one level to another, which is what Dye-3's cores were indicating. "You're talking here about 10 degrees Celsius change"—about 18 degrees Fahrenheit—"in less than a human lifetime," White says. "That was the period we had narrowed it down to at that time. And people just shook their heads. They said, 'You're crazy.'" But White was almost certain he was not.[38]

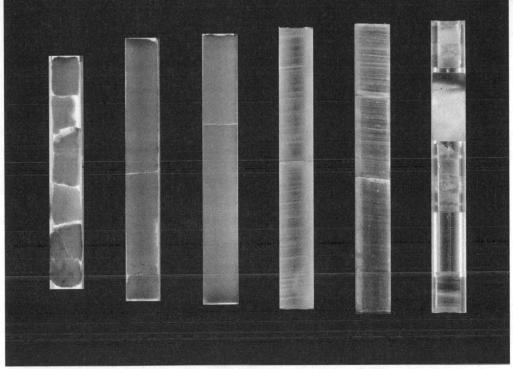

Ice cores from GISP-2 (© Peggy Weil/"88 Cores")

13

Deeper

While researchers in the United States and Europe debated the meaning of the Dye-3 ice, a number of climate scientists were privately discussing a related matter: Whether a powerful force—a force different from the changes in earth's orbit and tilt that led to the coming and going of past ice ages—could alter the environment in a dramatic way. As he was doing measurements on the Dye-3 cores in Paris, James White recalls that "those of us who were in that small community knew that carbon dioxide levels were climbing and knew that it was going to change climate. In my defense, I was in my twenties. I naïvely thought society would latch onto this and do something about it."[1]

The notion that carbon dioxide could trap energy close to earth's surface and thereby warm the planet actually dated back more than a hundred years—to the research (in 1856) of an American named Eunice Foote; to the lab experiments (in 1859) of a British scientist named John Tyndall; and to the calculations (in 1896) of a Swedish chemist named Svante Arrhenius.[2] In 1938, a British engineer named Guy Callendar evaluated thousands of weather measurements and put forward proof that the world's industrial emissions were already warming the earth. Callendar went before the Royal Meteorological Society in London to explain that "man has added about 150,000 million tons of carbon dioxide to the air during the past half century," and that the effect had been to increase temperatures "at the rate of 0.003 degrees Celsius per year at the present time." He added that the warming trend was not automatically dangerous; as a benefit, it would likely block the onset of the next ice age. "The return of the deadly glaciers," he noted optimistically, "should be delayed indefinitely."[3]

By the midpoint of the century, it began to seem conceivable that rising CO_2 levels might do more than just put off another ice age. Emissions from cars and industry might also melt much of the ice that sat upon the bedrock in Greenland and Antarctica. A handful of scientific evaluations—notably, one by a physicist named Gilbert Plass, in 1956—led to articles in publications like *Scientific American* that sketched out a calamitous scenario suggesting stupendous rises in sea levels and a future polar region of slush and mud. Against a backdrop of crumbling glaciers, a 1958 television show even asserted that "man may be unwittingly changing his climate due to the waste products of his civilization."[4] Around this time, a story buried deep in the Sunday edition of *The New York Times* explained how altering the amount of CO_2 in the atmosphere by only a small amount could change the planet's surface temperatures. And what appeared to be a slight jump in temperatures would nevertheless have extraordinary effects: "If the average temperature should fall only a few degrees centigrade, glaciers would cover a large part of the earth's surface," the article warned. "Similarly, a rise in the average temperature of only 4°C would convert the polar regions into tropical deserts and jungles, with tigers roaming about and gaudy parrots squawking in the trees."[5]

These alarms were mainly being rung on the scientific periphery. None seemed to penetrate far enough into the mainstream to influence public thinking or policy discussions. Still, in the academic literature, a few researchers continued to ask whether carbon emissions could become a worrisome threat. In a paper published in 1967, two scientists working at a government funded laboratory in Washington, Syukuro Manabe and Richard Wetherald, explained how they had created a crude but effective model to simulate earth's atmosphere and see what would happen if atmospheric CO_2 levels in the future doubled from their concentrations from before the industrial revolution. This, the first computer modeling study of climate change—unknown to the general public, but later voted by earth scientists as the most important climate paper ever published—showed that global temperatures would go up on average by 2 degrees Celsius, or about 3.5 degrees Fahrenheit.[6] When Manabe improved his computer model in 1975, the increase actually looked to be as high as 4 degrees Celsius, or slightly more than 7 degrees Fahrenheit. Whether or not this meant parrots would be squawking in the Arctic, the ice sheets would surely be at risk, and so would the expansive cover of floating sea ice around the North Pole. Around the time Manabe's papers were published, in fact, a Russian climatologist named Mikhail Budyko calculated that rising CO_2 levels would drastically raise Arctic temperatures and melt the sea ice cover, perhaps as early as the year 2050. In one of his studies, Budyko also pointed to the tenuous balance between a livable and non-livable climate, and seemed to think that man's industrial activities could tip the future toward "climate catastrophe [where] the existence of higher forms of organic life on our planet may be exterminated."[7]

To go back to the scientific and government reports of this era is to have the uneasy sense of looking at a forensic photograph, akin to one taken at a disorderly crime scene, where details of immense importance have been repeatedly overlooked.[8] In 1977, the National Academy of Sciences, the organization of the country's most accomplished scientific thinkers and researchers, issued its first long report on possible climate changes and their relationship to fossil fuel emissions. It was entitled *Energy and Climate*. In the text, a respected meteorologist named J. Murray Mitchell wrote that

> the climate of the earth is now known beyond any doubt to have been in a more or less continual state of flux. Changeability is evidently a characteristic of climate on all resolvable time scales of variation, from that of eons down to those of millennia and centuries. The lesson of history seems to be that climatic variability must be recognized and dealt with as a fundamental quality of climate, and that it would be potentially perilous for modern civilization to assume that the climate of future decades and centuries will be free of similar variability.

As it happened, Mitchell had a keen interest in ice cores and what they could tell us about the past—on the day that the drillers at Dye-3 reached the bottom of the ice sheet, he was in Greenland and snapped a photograph of Willi Dansgaard, Chet Langway, and Hans Oeschger looking victorious and elated.[9] But in his report he noted that the immense complexity of the climate system made it difficult to predict precisely what would happen next, based on past evidence of ice age cycles and on rates of industrial emissions. Without question those rates were increasing, and atmospheric concentrations of CO_2 were growing.[10] Mitchell was far ahead of almost all other American scientists in believing that CO_2 would soon create a distinguishable warming trend. "Today, we find ourselves uneasy because through our use of energy we may be significantly disturbing the natural climate system," he noted. "This uneasiness is justified."[11] Yet there was still too little data, and too little insight, about what happens when natural and human forces combine. To take just one essential question, would warming happen slowly over millennia, or would it happen quickly, over centuries or even decades?

The scientific community tends to respond to unanswered questions about research results with a call for more research. The early and mid-1980s were characterized by a growing body of scientific analysis on the potential of carbon dioxide to alter climate, along with the stirrings of a global political movement to craft policy so as to avert catastrophe.[12] But even with new evidence from Dye-3, there were still profound uncertainties about the science of ice and the future of the polar regions. The notion that earth's climate could change abruptly from one state to another, to take just one example, required validation that could only be demonstrated by another ice core, ideally one done in the pristine center of Greenland, along the ice divide. A heightened interest amongst some academic geologists and oceanographers led to another government report in 1986 that put forward a detailed blueprint for why and how the United States should invest in extracting another ice core. Declaring that this was now "a major scientific priority," the study explained that "human modification of climate, either purposeful or inadvertent (for example, by CO_2-induced warming) may cause major changes in the global ice volume and sea level."[13] It now seemed clear that a better understanding of past climate changes that might be decoded from ancient ice could translate into a better understanding of climate changes in the future.

It the cores went back far enough, moreover, they might not only settle questions of whether sudden spikes and drops in climate had actually occurred. They might demonstrate correlations between CO_2 levels and temperatures. And they might even contain evidence from a time more than a hundred thousand years before, when temperatures in the Arctic were believed to be a few degrees warmer than the present day. As such, they could explain whether Greenland's ice sheet, and the earth's ice in general, could survive what might be coming.

During this period, a group of eminent American scientists urged the U.S. National Science Foundation to fund another Greenland ice core project as soon as possible. Various rounds of conferences and false starts followed, characterized by disagreements between American and European researchers over equipment, methods, leadership roles, and funding. In January 1987, during a small meeting at a Boston hotel, a group led by an American geochemist named Wally Broecker worked out a plan with a European group that included Willi Dansgaard, the Danish scientist.[14] Rather than team up on a new ice core, the decision that day was to split up amicably. The Americans and the Europeans would each drill their own ice cores near Greenland's ice divide, and each group would keep their ice for investigation. "We decided to separate the ice cores by ten ice sheet thicknesses, which is about thirty kilometers," recalls Paul Mayewski, a scientist then at the University of New Hampshire who attended the Boston meeting.[15] The agreement was to start planning immediately.

Two years later, in May 1989, Mayewski waited at the center of the Greenland ice sheet and watched as a succession of ski-equipped planes landed on a runway of freshly groomed snow. The cargo loads kept coming—tents, huts, food, fuel, scientific equipment. And then, when the camp was up and running, Mayewski watched as a plane landed with the most crucial shipment: an ice drill, which had been built over the past year in Alaska. "I remember taking a minute just to stand and stare at it," he would later recall.[16] Part of the reason was that it was so massive—with a steel tower of over a hundred feet, bigger than any ice drill ever made.[17] Yet its arrival also marked a moment of relief, since work could finally move forward.

The new camp was called GISP-2, which stood for Greenland Ice Sheet Project-2; it had been designed to house around fifty scientists, drilling engineers, and staffers. Mayewski had been appointed the project's principal investigator, which in academic jargon meant he was in charge of organizing and managing the research. He intended to run GISP-2 through the summer until early August. Then everyone would fly home and return the following May to work. The pattern would continue after that, summer after summer, until the objective—drilling to the bottom of the ice sheet, yet again—was complete.

The presumption was that the Americans and Europeans would drill at the same rate. When both camps were set up, though, it turned out that the Europeans—equipped with their Istuk drill and several decades of experience in deep drilling—took the lead. The European camp was called GRIP, which stood for Greenland Ice Core Project. Willi Dansgaard was at GRIP, and so were J. P. Steffensen and Dorthe Dahl-Jensen, who were now married. A number of French and Swiss glaciologists had joined in, too, along with scientists from several other European countries.[18]

Some of the delays for the Americans were a result of technical problems with the big drill. But it was also the case that the first generation of American drillers and ice scientists, most of whom had worked with the Army Corps of Engineers in Greenland and Antarctica—Herb Ueda and Lyle Hansen, for instance—were no longer part of the effort. Chet Langway had decided to retire from fieldwork, and Henri Bader was now living the life of a retiree in Florida. The learning curve for the Americans was steep. On the other hand, having so many young researchers in the U.S. camp had an upside: The energy level was high, and a younger generation of ice scientists and engineers—including, for the first time, a group of accomplished women—could join the ranks of what had been an almost exclusively male and military-oriented profession.

Traveling the distance between GISP-2 and GRIP—or vice versa—wasn't much trouble. By hitching a few sleds to a few snowmobiles, a crowd could pile on and get towed the thirty-kilometer distance, which was equal to about eighteen and a half miles. Mayewski had marked a route between the camps with flags, but just in case a storm blew in, he declared that all travelers had to tow survival gear. James White, who knew many of the Europeans from his time looking at ice cores in Paris, and who was now working at the American GISP-2 camp to assess oxygen isotopes in the ice, recalls an occasional bit of revelry and a genial sense of competition. "We invited the Europeans over to a Fourth of July party, and we played golf in the snow," he recalls. White ball? "Orange ball," he says. "And they had a tradition of playing volleyball, so we also played volleyball in the snow with them." The Americans didn't have much in the way of wine and beer, but the Europeans did, which gave their camp a certain appeal as a weekend destination.

Separating GISP-2 and GRIP made sense for a number of political and financial reasons. But there was a scientific justification as well. There was a lingering question of whether ripples and folds in the layers of ice at the bottom of the previous Greenland ice cores, caused by the flowing movement of the ice sheet over uneven bedrock, were to blame for evidence that suggested strange alterations in ancient climates. If two separate drilling projects produced similar results, it would likely settle the question. And if folds and imperfections in the ice weren't the cause, it meant there would be unassailable proof that earth's climate could change abruptly. Some scientists had already developed a term for the phenomena of sudden changes. They were starting to call it *threshold behavior*. Force the climate in a certain way, perhaps even in a modest way, and it would trigger a vast reorganization of energy on earth, with effects that might go beyond anything previously imaginable.

If you were going to the center of Greenland's ice sheet as a scientist in the summer of 1990, you would do the following: Take a flight from Sondrestrom, on the southwest coast of Greenland, from an airport that was built by the Americans during World War II and was previously known as Bluie West Eight. You leave from the southern edge of the runway, in front of a hangar where the 109th Air National Guard is based. The 109th—a group of pilots and technicians who live and work in upstate New York during the colder months of the year, but who spend most of the spring and summer at Sondrestrom—have a talent for landing ski-equipped military planes on the ice sheets of Greenland and Antarctica. The seats of these planes, loud and powerful Lockheed LC-130s, are usually configured in two sets of rows that face each other and run lengthwise along the front half of the cabin. The back half of the plane is typically reserved for cargo, tied down on sturdy pallets.

In 1930, on his urgent trip by dogsled with Fritz Loewe, Alfred Wegener spent forty days—"the worst journey in Greenland he had ever heard of"—going from the west coast to the center of the ice. In 1990, on an LC-130, it takes about one hour and forty-five minutes.[19] In the cabin it is too loud to talk. So you might pull a wool cap down below your eyes to sleep, especially if you've been traveling for twenty-four continuous hours to get to Greenland from Europe or the western United States. If you're not tired, you might wander about the cabin, or gnaw on candy bars, or snap dull photographs of the ice sheet from the plane's small, scratched windows. But then a landing on skis makes for a crunching, rumbling, bumpy awakening. There's a sudden need to zip into cold-weather gear and fish out sunglasses, lest you be unprepared. When the sun is up in a cloudless sky, the doors of the plane open, the staircase drops, and the dazzling white panorama can seem an acid assault on the eyes.

It almost never rises above freezing at Greenland's summit. In midspring, it hovers around zero degrees Fahrenheit, or sometimes lower. When you deplane and walk several hundred yards in the snow—toward the row of flags and tents, toward the "big house" on stilts where the mess hall is housed, toward the geodesic dome farther in the distance where the 110-foot tower peaks through an aperture in the roof and the drilling takes place—you may get the impression that you're not atop Greenland's summit but in a shallow bowl, as though the horizon all around slopes up slightly on the edges. It is an optical illusion. GISP-2 is just about as high as Greenland's ice sheet gets.[20] And below you, in layer upon layer of snowfalls, hundreds of thousands of years' worth, the ice is as deep as ice can be.

Under the camp's geodesic dome, ice drilling goes on twenty-four hours a day, with the drillers working in shifts. But the scientific work is being done adjacent to the dome, where scientists have excavated by chainsaw an under-ice network of snow caves, covered with timber and then snowdrifts, that branch off from the drilling room. The main snow cave, known as the science trench, is one hundred feet long and thirteen feet wide, with small chambers and alcoves off to the side. Within this trench, a series of tables and stations make up what's known as a "core-processing line" to evaluate the miles of ice that will come up from the depths. On any given day, about fifteen people work the line. If you're toiling down there, you find that it is not always interesting; indeed, you quickly discover it is often unpleasant. To Mayewski, who's in charge of making sure GISP-2 runs smoothly, this is "industrial science, factory-style work," the kind of

labor that makes for a repetitive drudgery that after ten or twelve bitterly cold hours is numbing in mind and body.

You find that the first few days at GISP-2 are a haze of headaches,
fatigue, and breathlessness. You're at 10,450 feet. But once your heart
and lungs acclimate, on a typical day, it goes like this: You wake up in
a cold tent, get yourself to the big house for breakfast, and then roll
down to the science trench for a long, monotonous, frigid day. Sometimes you need to put on a poncho to keep the drilling fluid that
clings to the ice cores—butyl acetate—off your clothes and hands. It's
loud in the trench with the constant sawing. Music is playing through
a stereo at high volume to alleviate the boredom and fatigue. Sometimes it's K. D. Lang; more often it's U2 or Pink Floyd. Someone once
suggested playing classical music instead, but when it was tested out,
those in the trench discovered that their productivity dropped by
about 50 percent. That was the end of the classical music.

As the drillers bring the cores to the surface, they remove them
from the core barrel, place them in a tray, and send them down by
elevator to the science trench and the processing line. The cores rest
for a couple days so that the drilling fluid evaporates from their surface. Then the cylinders are cut into segments of one meter, and
sawed in half lengthwise on a horizontal band saw, so that the flat and
pristine surface at the center can be examined. Sometimes, the cores
arrive fast, and in profusion. They start piling up, which means it
looks to be a long day ahead.

Before anything else, the cores have to be dated.[21] The date is the
fundamental thing, for without a date, the crucial juncture of climate
and time is lost.

There are a variety of ways to count years in ice. At GISP-2, as the
cores moved along the line, not long after cutting they arrived at a
table for measuring electrical conductivity. This was the first stop. A
scientist named Kendrick Taylor moved two electrodes along the flat
center of the core to record electrical readings on a computer—a line
like a cardiogram snaked across his screen. The electrical readout
characterized the difference between summers and winters.

Another way was to cut samples in the upper layers for isotopes and chemical analysis, bag them, and have them measured later (in the United States or Europe) with a laboratory instrument, so that accurate readings for summer and winter could be checked.

There was a third way to count, though: A researcher in the trench could scrutinize the ice over a light table. A core would be moved there, and one or more scientists would pore over its defects and stratigraphy. Because summer and winter produce different kinds of snow, they also produce two stripes of evidence. "Layers that were cloudy looking and obscured were right next to layers that were very clear," Joan Fitzpatrick, a geologist in the trench for three summers, recalls. "And a pair of those represented one year's net annual accumulation."[22] So, pair by pair, the years would go backward. The shallower parts of the core—the parts nearer to the surface—usually had thicker ice years that measured several inches; the deeper parts of the core, where the weight and flow of the ice sheet stretched the ice, had thinner ice years. Thus the job of counting backward became more challenging as the depth increased. This was why the counters looked for large segments of volcanic dust—from Iceland's Laki in 1783, for instance—so as to peg their work to historical touchstones. At the light table, a scientist would use a paper logbook that was one meter long, just like a core. And in the book they transposed what they saw—signs and clues and gradations that were inscrutable except to those who knew how to spot telltale hints of summer and winter, or dust storms and ancient wildfires, or old volcanoes. The cores were counted once, twice, even a third time, with layers sometimes debated in the trench amongst the team until agreement was reached. Eventually, the years were marked off.

The investigations that came next were endless. Pieces of core were doled out for isotopic studies for oxygen and deuterium and other elements. The "gas people," as they were known, were given ice with bubbles to look at remnants of oxygen and CO_2 and methane, another potent warming gas, which could attest to the conditions of ancient wetlands. Vestiges of dust and salt in the cores were extracted to evaluate wind and atmospheric conditions when storms blew across Asia and the North Atlantic fifty thousand years ago. Meanwhile, slices were cut from each core, slices thin as a sheet of cardboard, so researchers could look closely at the microstructure of the ice, the inner fabric of the glacier, which was represented by microscopic grains that revealed the orientation of the ice.

Another crucial measurement came much later, after the coring work was done. Several scientists would gauge the temperature at thousands of points within the borehole made by the drill to the bottom of the ice sheet. Years before, Paul-Émile Victor's team had noticed that the ice sheet seemed to retain its cold from ages past. This insight meant that the ice sheet has a "memory" of temperatures from long ago. If you happened to be working in the center of Greenland in 1990, you also happened to be working during a year that fell in a period on earth known as an *interglacial*. Summers are warm; winters are mild. Just about everywhere but Greenland and Antarctica has left the ice ages behind. Yet the temperatures deep in the hole, with hints of what happened ten thousand and twenty thousand years ago, tell a different story.

In the science trench at GISP-2, one of the glaciologists counting back through the layers was a young professor from Penn State named Richard Alley. For as long as he could remember, Alley had been interested in rocks; on childhood vacations to the Badlands in South Dakota, and to Yellowstone in Wyoming, he found wandering that landscape to be a profound experience. He declared himself a geology major as soon as he arrived as an undergraduate at Ohio State, and found himself drifting toward glaciology. A few years later, while working on his doctorate, Alley decided to look deeper at the process by which buried firn turns to ice, and he traveled to Antarctica to help with a coring project. He arranged for some of the cores to be sent to his school, the University of Wisconsin, and returned to the United States. He got a call at six A.M. on a spring morning in 1985 to tell him his ice had arrived, and he came down to open the box. "I was supposed to have cylinders but what I had was little refrozen puddles," he recalls. The freezer on the ship from Antarctica had failed.

Once the disappointment receded, he says, "I knew I had to find more ice."[23]

Alley had the good fortune to go with a group traveling to Greenland the following summer. It was his first trip to the island, and GISP-2 was still in the planning stages. He joined a team of drillers who were looking for good sites to drill. For three weeks he worked at a remote location near the south central region of the ice sheet called Site A. There was nothing there—just a few engineers and scientists drilling core samples about three hundred feet deep to see if it would be a good spot for a deeper core. "Sometimes the radio worked and sometimes it didn't," Alley recalls. Mostly he remembers that the drillers had told him Greenland was warm, unlike Antarctica, so he had brought insulated Sorels—the wrong kind of boots—and eventually needed to strap large pieces of foam around the soles to ward off the −30 degree Fahrenheit chill when he was working in the snowpits at the camp.[24] The upshot was that Alley got a good ice core to study. Also, he got enough experience processing cores to be included in the team when GISP-2 started up a few years later.

In his first few summers at GISP-2, as the Americans got their drilling machinery working properly, Alley educated himself on ice layering and counting. That summer of 1989 he spent time reading Carl Benson's old research papers from decades before on measuring snow pits in Greenland, and he talked at length with Tony Gow, another scientist at GISP-2, whose experience with ice cores went back to working with Henri Bader in the late 1950s, at the Army Corps of Engineers. When the ice cores finally started coming up with the drill, Alley turned out to be exceedingly good at recognizing the telltale signs of winter and summer ice and counting back the years. "When we finally hit a good time marker—Laki in 1783—I was spot on," he says. "And this had come from just digging pits and reading geology and reading Tony Gow and reading Carl Benson and some of these other folks."

The summer of 1992 was Alley's third summer at GISP-2. By that point, the drill was about halfway through the ice sheet, and it was bringing up ice that was between ten thousand and thirteen thousand years old. There was a sense that they were getting close to the abrupt changes that had come up in previous cores. When core 1678 arrived—meaning it was drilled from 1,678 meters down, or about 11,700 years ago—it was sliced in half and went first to Kendrick Taylor, who measured the electrical conductivity. When Taylor ran his electrodes along the ice, it elicited a bizarre response and a conductivity (because of an excess of trapped dust) that dropped nearly to zero. "It was flabbergasting," Alley recalls. "We're sort of looking at each other. And then it was my turn to look at it."

In the trench Alley counted the core and saw extreme variations in several yearly layers that suggested a drastic increase in snow accumulations. Eventually, he and his colleagues would put the total warming—not a temporary swing, but a wholesale, sudden switch in the climate system—at about 18 degrees Fahrenheit. Later, too, Taylor would reflect how this piece of ice demonstrated that an environmental threshold was somehow crossed, eliciting a sudden temperature change. "I used to believe that changes in climate happened slowly and would never affect me," he wrote. But this ice proved otherwise. If such a drastic switch happened again, Taylor posited, the change would occur "so rapidly that it would be impossible to rearrange agricultural practices quickly enough to avoid stressing world food supplies. So rapidly that many species would not be able to adapt, because their habitat, already greatly reduced by human activities, would be eradicated."[25]

The European team at GRIP finished their core first, in July 1992, by drilling 3,028 meters, or nearly 9,934 feet, to the bottom of the ice sheet. They invited the American team for a celebration soon after, and the Americans arrived with a sauna they pulled over the ice sheet by snowmobile. That evening the Europeans started a raging bonfire by burning a huge stack of six-foot wooden trays that had been used to hold ice cores in the GRIP science trench. They knew they wouldn't need the trays anymore. "At the end of the season it's ridiculous to haul all that wood back, it's too expensive," Todd Sowers, an American climate scientist who was working at GISP-2, says. "So they made a pyre that was just enormous. It was an insanely fun party."[26] The American team went back to their camp the next morning with headaches and an awareness that they still had a ways to go. They reached bottom at a depth of 3,054 meters, or just over 10,000 feet, a full year later, in July 1993.

In the final account, the race between the two camps didn't matter. Herman Zimmerman, a National Science Foundation official, later called the American effort "pound for pound, dollar for dollar, the best project that NSF did during the 1990s."[27] All told, GISP-2 cost about $25 million. And whatever its true value, those involved with the project understood immediately that it delivered something new to the world, even if the recognition of its significance was so far confined to a small guild of glaciologists and climatologists. To Tony Gow, the meticulous, year-by-year reading of the layers in the ice made it "the quintessential core that was ever taken."[28] Mayewski considered the data record provided by the cores "the Holy Grail of climate research."[29] It turned out that the dramatic climate shifts that Richard Alley had perceived in the science trench were also obvious in the ice eighteen and a half miles away, at GRIP. "The almost perfect match back to 110,000 years ago between records from the two cores 30 km apart," a summary of the results later explained, "should dispel any lingering doubt about the climatic origin of the events."[30]

The Danish team had a freezer to store their GRIP cores. They shipped the ice back to Copenhagen. The Americans, meanwhile, in keeping with the spirit of the GISP-2 project—new drill, new camp, and a new generation of ice-core researchers—decided to build a big new freezer. To store ice cores, colder is almost always better, since lower temperatures can better preserve the ancient record within the ice. What the National Science Foundation eventually funded was a freezer in a big government warehouse in Denver divided into two main spaces. In one, cores would be permanently stored on racks in a big room chilled to –33 degrees Fahrenheit. Next door, a large work space—for sawing and cutting, for light tables and microscopes— would be kept at a more comfortable –11 degrees Fahrenheit. The task for designing this lab fell mainly to Joan Fitzpatrick, a U.S. Geological Survey scientist who had been at GISP-2. In the new facility, she decided, she would replicate the Greenland science trench as closely as possible.

One of Fitzpatrick's first jobs was consolidating most of the existing American ice cores in Denver. Few people in the world knew these cores still existed, but a small number of glaciologists and paleoclimate researchers viewed them as they would a precious mineral deposit. Since the late 1950s, the cores had been under the care of Chet Langway, who kept them first in a cold storage facility in New Hampshire and then moved them to Buffalo, New York, after he took a job as a geology professor at the state university there. The collection comprised thousands of feet of ice: from Camp Century and Dye-3 in Greenland, from various locations in Antarctica, and from mountain glaciers on several continents. By the time Fitzpatrick got to them, the cores were situated in the middle of a walk-in freezer in Buffalo that was also used by the university's cafeteria services. In 1992, Fitzpatrick and a number of scientists traveled to Buffalo to catalog and pack up the cores. Then they sent them to Denver by train, on two refrigerated boxcars.

Once it was up and running in Denver, the National Ice Core Laboratory—NICL, or "Nickel," as it came to be called—bore a startling resemblance to the vision Henri Bader had conjured decades before. It was a library of ancient ice, a place where visiting academics could come look at cores that interested them and ask for a "cut" of a relevant piece so they could take it back to their own labs for analysis. When GISP-2 finished in 1993, Fitzpatrick fielded the task of bringing its ice cores back to Denver. She flew to the site in Greenland, packed the cores in thick foam containers, and flew them from the drilling location to the airport at Sondrestrom, on Greenland's west coast. "We had a chartered DC-8 waiting there," she recalls. "I transferred all the ice onto that plane and flew all the way back with it directly to Denver. That was a nice way to travel, because I could be there to make sure it stayed cold the whole time." In fact, Fitzpatrick had asked the pilots to keep the heat off, and had to ignore the complaints from the freezing crewmembers, since nothing mattered more than the health of the ice. When she got to Denver, she and her colleagues stacked two miles of GISP-2 cores on the shelves in one-meter segments.

On a warm spring morning in 2014, two decades after she retrieved it from central Greenland, Fitzpatrick brings out core 1678 from the cold room. She removes it from its storage tube, takes off the clear plastic bag that protects the ice inside the tube, and places it on a light table. "This is the transition ice," she says, meaning the transition in global climate. The date is "11.7," or 11,700 years ago. Core 1678—a half-cylinder—is broken in places where chunks have been cut for research. It is a strangely unattractive piece of ice: discolored, clouded with dust, uneven. "See how wide the bands are spaced here?" asks Fitzpatrick, pointing to one end of the core. "And then: Boom. All of a sudden they get tighter here." She points again from one side of the ice core to the other to emphasize how it shows temperatures and snowfalls began leaping to a new state that was ultimately about 18 degrees Fahrenheit warmer.

"So that's it," she says. "Ice age here. Not ice age there. We think this was in the space of a few years. And the whole point is, we all once thought that would take *thousands* of years."

After GISP-2, James White and Richard Alley worked on voluminous reports for the National Academy of Sciences about what came to be known as "abrupt" climate change.[31] They also began to lecture around the country to explain the implications of the cores from central Greenland. The reasons for abrupt climate changes were then (and remain) in the realm of debate. Many scientists agreed that such massive planetary disruptions must be a result of profound changes in the ocean, atmosphere, or ice sheets—or perhaps all of those systems changing in concert. The most prevalent and widely accepted theory was that established patterns in ocean circulation that bring warm water from the tropics up to the North Atlantic were disturbed during certain moments in the past, perhaps as a result of being "blocked" by melting ice that dumped cold fresh water into the northern oceans. Another possibility was that significant changes in atmospheric circulation brought on big temperature jumps.[32] What seemed likely was that whatever the change in the climate system—what scientists called a "perturbation"—it was followed by a series of positive feedbacks that greatly magnified the rate and intensity of the original shift.

Alley and White didn't assert that rapid climate changes would undoubtedly happen again in the near future. But they urged the scientific establishment to absorb the idea that such changes *could* happen again, and that "the paleo record," as it was called, showed that earth's climate system had a tendency to switch drastically from one mode to another. Even without settled explanations for the change, there was now strong evidence: Earth's climate had no problem changing quickly and furiously, regardless of human expectations.

It was likely this message made an impression on the scientific establishment. It was also likely it had only a glancing effect on public consciousness, much like those old climate studies on CO_2 from the 1970s, the ones that were ignored. There may be no single reason why some scientific ideas catch on quickly in the public imagination while others do not. But the notion of abrupt climate change might have been too strange, or too unpleasant, to contemplate. In a popular book he wrote a few years after returning from Greenland, Alley, who increasingly became an elder statesman amongst American glaciologists as his career progressed, gravely noted that the cores showed "that the ice-age came and went in a drunken stagger, punctuated by dozens of abrupt warmings and coolings." Such changes happened so fast—*ice age here, not ice age there*—that if they occurred again human and natural systems would likely be unable to respond. To Alley, the changes "could pose the most momentous physical challenge we have ever faced, with widespread crop failures and social disruption."[33]

Moreover, abrupt climate change wasn't just a matter of higher temperatures. The Greenland ice cores opened up the possibility of a variety of unwelcome "surprises" in the future—surprises that could become manifest in all kinds of natural systems, especially as the world's climate was transformed by carbon dioxide. For example, it was possible that temperatures wouldn't jump in the future as drastically as they had in the past. Average temperatures might instead increase slowly, incrementally, year by year, decade by decade, based on the inevitable greenhouse effects of increasing CO_2 concentrations. And yet, the *impacts* of the warming might be abrupt anyway. A massive glacier in Greenland or Antarctica could pass a point whereby its melting or total collapse would become unstoppable, triggering sea level rises that permanently flooded the great cities of the world. Sea ice floating on the Arctic Ocean could suddenly reduce its cover drastically, triggering all sorts of tempestuous effects on earth's oceans and atmosphere. As a result, some urban economies—even some national economies—could go from humming prosperity to panic in a few fleeting years. "We may have centuries of safety," Alley would later explain, emphasizing the degree of uncertainty. "We may be close to the edge already."[34]

Abrupt climate change was therefore not a prediction. With the Greenland cores as proof, it was a different, darker way of thinking about the possible future.

Building GRACE, Germany, 2002 (NASA)

14
Sensing

Forty years after he crossed the Greenland ice sheet on skis, Fridtjof Nansen sailed on an ocean liner from Gothenburg, Sweden, to New York City. "The dean of living explorers," as one newspaper described him, had come seeking support from American officials for a new project that was being hailed as "the greatest exploration of his long career."[1] Nansen was old now and fatigued from a life of travel; his thick moustache and the fringe of hair ringing his bald head were silken and white. This very year, 1929, would prove to be the last of his life. His mood upon arrival in New York was nonetheless upbeat, and his enthusiasm and stature—he was now a Nobel laureate as well as a global celebrity—impressed those who crossed paths with him.

His intensity still burned. Wearing a wide-brimmed black hat and speaking in nearly flawless English, Nansen answered reporters' questions about his new organization, the Aeroarctic association. His goal for the following year, he said, was to launch a scientific expedition from the air to study the geography, oceanography, and meteorology of the Arctic. The work would be conducted from an aerial perch aboard the huge German dirigible known as the *Graf Zeppelin*.

These kinds of airships—kept aloft by hydrogen-filled tanks and limited to a top speed of about eighty miles per hour—had already been used for polar travel. Several had sailed over the North Pole a few years before.[2] But Nansen wasn't interested in that kind of journey. Instead he made the case to journalists that the weather and ice of the Arctic had profound effects on Europe and the lower latitudes. "We're thinking of study up there for the population of the rest of the world," he explained.[3] Using an aerial approach for purely scientific purposes, Nansen also seemed to think, would mark a serious technological leap. "Should the *Graf Zeppelin* flight prove successful," the American journal *Science* noted, "it would doubtless be the forerunner of future Arctic flights, on still larger airships, thus exploring thoroughly all the Arctic regions in a far more complete manner than is possible on the ground."[4]

As usual, Nansen had seen the future before almost anyone else. And while his death several months later shook colleagues who were helping to plan the *Graf Zeppelin* journey, the project moved forward anyway. In July 1931, the airship left from Leningrad, flew over the Siberian Arctic, and then reached a point 560 miles from the North Pole before turning east and returning south to Leningrad and Berlin. A team of technicians, using panoramic cameras for surveying and custom-built instruments to measure earth's magnetism, carried out a hectic five-day scientific program. The journey, as it turned out, did not set off a revolution in Arctic science—at least not right away. In the decades following the Graf Zeppelin's flight, mapmakers increasingly relied on aerial photography for charting northern islands and the coastline of Greenland, but these instruments were mainly used for military purposes. In the years of World War II, pilots intent on finding targets for bombing runs made detailed photographs of enemy

terrain and factories. In the Cold War era that followed, similar techniques were used by American U-2 planes, developed in the mid-
1950s, which could cruise at seventy thousand feet, snap detailed
photographs of Soviet installations below, and usually evade detec-
tion.

There was no precise moment when scientists began borrowing the
techniques of military intelligence for their own work. But by the
start of the satellite era—beginning with Sputnik in 1957—a number
of American and European researchers tried to envision the stratosphere as a vast new playing field. In 1960, NASA started launching
observation missions, first with the TIROS satellites, and later with a
series of more sophisticated "birds" built under the Nimbus program,
which began in 1964. For the most part, the goal was gathering
weather data; the early Nimbus satellites—about twelve feet tall and
weighing about eight hundred pounds—took photographs of the
earth while traveling at an orbit that ranged from two hundred and
fifty to six hundred miles up.[5] Within a decade, these kinds of satellites were outfitted with exquisitely sensitive cameras and instruments
that used microwave and infrared sensors to "read" the surface of the
planet beyond the spectrum of visible light. The new equipment
could relay information to engineers on the ground about earth's temperatures, clouds, winds, and sea ice. As a matter of course, those
working in the emerging field of "earth observation," as researchers
began to call it in the 1960s, described the purpose of these satellites
as "remote sensing." It was an interesting phrase. To find out about
the farthest reaches of the planet, the term suggested, you might not
even have to leave your office chair.

Remote sensing held tremendous promise for polar research—
these were "powerful new tools" for measuring snow and ice, in the
words of one scientific report of the era.[6] Sensors could conceivably
get carried into an orbit over the poles by satellite, or they could be
mounted under airplanes flying over desolate regions of Greenland
and Antarctica. They could help survey regions where it was dangerous to launch fieldwork studies or regularly assess areas where it had
proven almost impossible to gather a broader understanding of
changes in ice, snow, or oceans. In 1968, Henri Bader, the father of

ice-core drilling, chaired a symposium in Easton, Maryland, about the future possibilities of remote sensing in the polar regions.[7] A year later, at a NASA-sponsored meeting in Williamstown, Massachusetts, a group of academics put forward recommendations for new sensing instruments that would be able to measure the sea levels of oceans and the movements of ice masses on Greenland and Antarctica from the sky.[8]

It would become clearer, in retrospect, that those imagining the potential for remote sensing in the late 1960s were dreamers. They knew some of their hopes exceeded the ambit of existing technologies. Still, by the late 1970s some of the new equipment—ice-penetrating radar, for instance, carried aloft by aircraft—was proving useful in studying the Greenland ice sheet. Before drilling the Dye-3 and GISP-2 holes, engineers conducted flyover missions with this type of gear, which could essentially see through the island's white blanket and map the topography of Greenland beneath the ice.[9] The idea was to improve the chances of finding an ideal place—level bedrock, which should produce smooth and even layers in the ice sheet—to drill. In this regard, remote sensing became a friendly partner to deep core drilling: To find the best place to dig up the past, one could first look down from high above.

In theory the new sensing technologies could do far more than that, though. Going back to the age of Nansen's first crossing on skis, the mysteries of Greenland that originally interested explorers were the ice sheet's origins, extent, and weather. Later came the effort to decode the secrets locked in its deepest layers. By the late twentieth century, researchers had largely realized these goals, but now glaciologists had arrived at the brink of a third mystery: Was the ice sheet growing or shrinking? What sounded simple was in fact the hardest of problems, one that had resisted a sure answer for decades. Due to the size and remote nature of the island, going out into the field to measure its total ice was akin to standing in the midst of a dense forest on an autumn morning and trying to count every fallen leaf. The question had nevertheless tantalized Alfred Wegener and Fritz Loewe, who wondered if Greenland's ice sheet would eventually face the same fate as the enormous glaciers that had once covered much of

Europe and North America, before they'd shattered into the oceans. The question had compelled Carl Benson, who in digging snow pits all over Greenland in the mid-1950s had tried to estimate whether annual snow accumulations on the ice sheet outweighed its annual losses from meltwater and icebergs. The problem had fascinated Paul-Émile Victor, who after his early expeditions in the mid-1950s helped plan another expedition around Greenland to calculate the "mass balance" of the island's ice.[10] Henri Bader, too, made an estimate for the ice in Greenland, in 1961. At the time, he thought the ice sheet might be growing rather than shrinking.[11]

In truth, there had never been a satisfactory answer. In the late 1960s, Børge Fristrup, a Danish glaciologist who wrote a history of the ice cap, considered the previous attempts to answer whether or not the ice mass of Greenland was in balance and concluded that "it has been assumed that the ice is in a state of equilibrium . . . in fact little is known about whether this is indeed so."[12] Not much had changed by the mid-1980s, when a NASA report urging the development of remote sensing satellites put it more bluntly. "Despite 25 years of intensive field work in Greenland and Antarctica, and the expenditure of billions of dollars, we are still unable to answer the most fundamental glaciological question: Are the polar ice sheets growing or shrinking?"[13] The question was not only scientifically compelling; it was important for practical reasons. At a minimum, a diminishing Greenland ice sheet would mean rising sea levels and imperiled coastal cities.

Scientists working in Greenland around this time had little reason to believe that the ice was in immediate jeopardy; some of the glaciers on the west coast actually seemed to be advancing, meaning the huge ice mass could be expanding. What's more, the notion of abrupt climate change had not yet become a concern. During the drilling of Dye-3, the Danish scientist Dorthe Dahl-Jensen remembers, "At that time, there was little understanding at all of the melting. We were still talking about: When will the next ice age come?"[14] While working on GISP-2 a few years later, Joan Fitzpatrick recalls, "Certainly the people who had proposed the whole drilling project used this as a justification: What if? What if global warming continues? What will be the impacts on the Greenland ice sheet and how will it respond? But it was all very much in the abstract."[15] Without a bigger picture, in other words, and without a sense of urgency, there was a pastiche of observations, speculations, and computer projections. These were only educated guesses as to how ice might behave in a hotter world.

One of the scientists who authored the NASA report asking why billions of dollars had not yielded an answer about the state of the ice sheets was a glaciologist named Bob Thomas. A native of Liverpool, England, Thomas had earned an undergraduate degree in physics but decided he was far more interested in geology. In 1960 he signed up for a scientific tour and traveled by ship from England to the Falkland Islands, and then to Antarctica. The passage there took two months. "We were dumped off for a couple of years at a base on the west side of the Antarctic Peninsula, a place called Galindez Island, around about the Antarctic Circle, just north of the Antarctic Circle," he would later recall.[16] There were thirteen people there. Mostly he took meteorological readings, but from his colleagues and a few visitors he began to learn how to measure glaciers. It happened to be a transitional era in polar studies, a time when premodern customs lingered on. "We still used dogs for transport, so it was good old explorer stuff," Thomas recalled—although by the time he returned to Antarctica, in 1966, for another two-year stint (this time with the British Antarctic Survey), there were also a few snowmobiles at hand. He was focused by that point on the movement of ice sheets and the behavior of ice *shelves*, the tongues of large glaciers that extend past the edge of land and float on the ocean, sometimes for many tens of miles. A lot of the time Thomas's work had involved putting large aluminum or bamboo stakes into the ice and tracking the movements of these stakes over time, which gave insights into how glaciers crept forward and back. Those who knew Thomas considered his energy and stamina almost superhuman. In another era, he might have been the type to sign up with Nansen or Peary to cross Greenland's ice. As it was, he returned to England after his tours of Antarctica, got a PhD at Cambridge, and entered a life of academic research, mostly in the United States.

In the early 1980s, Thomas landed as a program manager at NASA headquarters in Washington, D.C. His job was to fund promising research ideas that advanced the state of knowledge about the earth's *cryosphere*, or its ice-covered areas. One day at NASA headquarters in 1988, a team of NASA technologists, led by a man named William Krabill, gave a presentation to a few administrators. They were trying to see if there might be some interest among upper management—and, possibly, some funding—to expand their research work. Bob Thomas was there that day and listened closely.

Krabill was based at the Wallops Flight Facility, a NASA airfield on Virginia's sandy eastern shore, which lies several hours east of Washington, D.C. He and a few colleagues had been experimenting over the past year or two with a device called a laser altimeter. Essentially, a laser altimeter is an instrument that's used to measure the height of something far away; in Krabill's work, he was measuring Virginia's Chincoteague Bay from a plane flying above and sending laser pulses down. By calculating the time it took for the light pulses to leave his instrument, reflect from the water surface below, and return to the aircraft, the altimeter could get a fairly exact calculation of the plane's distance from the water surface. Listening to Krabill talk about this airborne sensing equipment that day in 1988, Bob Thomas became interested. Actually, he became *very* interested. He suspected that climate might already be having an effect on the polar ice; he was also thinking of his years doing fieldwork on ice sheets that were "so bloody inaccessible," as he liked to put it, that their titanic size made it impossible to see how they were faring. That day, Thomas told Krabill that if he could make his laser altimeter work to within a resolution of twenty centimeters—meaning that from a plane flying, say, twelve hundred feet above, it could make measurements to help him calculate the elevation of ice below to within eight inches—he would make a major contribution to science. Krabill recalls that he vowed to get the resolution even better, to ten centimeters. "Thomas gave us some seed money," Krabill says, "and we did some local test flights. Then we took off for Greenland in 1991."[17]

Breakthroughs don't necessarily happen because of one big technological jump. More often it's because a cluster of new technologies and ideas suddenly coalesce around a difficult problem, along with a headstrong person. That seemed to be the case with remote sensing. Without question, a reliable aircraft—a Navy P-3 surveillance aircraft—that could fly long distances was essential to making Thomas's plan work. Also, Krabill's laser altimeter, a sophisticated tool bolted to the floor of the plane, could now help researchers make fairly precise ice elevation measurements over tens of thousands of square miles. But the device was only useful for measuring the changes in the ice sheet because a host of new navigation tools had also become available. "GPS came along," Thomas later explained, "and for the first time we knew where an airplane *was*."[18] In other words, GPS could tell Krabill exactly where the plane was in the sky, which was crucial to measuring the ice. What's more, it ensured that a pilot followed a precise route over the ice sheet—not just during a single year of research, but over several years. This was vital as well. To understand how the ice sheet was changing, you couldn't measure its elevation once. You had to make surveys of exactly the same routes over Greenland a few years apart and then compare them. If the height of the ice dropped in a place that was measured repeatedly, it would suggest the ice cap was thinning. And that in turn could mean Greenland's ice was shrinking rather than growing.

The mission became known as PARCA, which stood for Program for Arctic Regional Climate Assessment. And there was no certainty that Thomas and Krabill's approach would succeed. "Bob basically bet his career on this," John Sonntag, a NASA engineer who was part of the team flying routes over Greenland, recalls.[19] Hundreds of hours were logged over Greenland in 1993 and 1994, and again in 1998 and 1999—long and sometimes monotonous flights over the most desolate stretches of icescape and rock. Krabill and his laser altimeter surveyed the surface below and collected data. The flight tracks of the PARCA study, superimposed over the ice sheet by the flight planners, began to make Greenland look like a map of crisscrossing interstate highways.

In 1999, the PARCA group, led by Krabill, published an article in

Science that noted "Rapid Thinning of Parts of the Southern Greenland Ice Sheet." A year later, in the same journal, they reported that their measurements of the ice sheet over a six-year period showed it was losing about 51 cubic kilometers of ice per year.[20] This was akin to an ice cube, 2.3 miles long on each side, falling into the ocean every year. "That was the first time a mass balance assessment of a large ice sheet had ever been done," Sonntag would later explain. For those who cared about the earth's cryosphere, it was a significant scientific breakthrough. "What that taught us was the ice sheet was not in balance, even then, in the 1990s," Sonntag says. "But it also showed us that the ice sheet was growing a little bit in the center but it was thinning a lot around the edges. And the thickening in the center did not offset the thinning on the edges. In other words it was, in the net, delivering sea level rise."[21]

As Thomas and Krabill were working on measuring the mass balance of Greenland by plane, a number of scientists were contemplating a bigger step. By the 1990s, research projects in the Arctic were classified by American science agencies into three different categories. The first involved traveling to a site to do fieldwork. This might entail measuring snow accumulations by digging snow pits, for instance, or using stakes to track the movement of individual glaciers. Regardless of what tools technologists could invent, that kind of work remained crucial, since it could lead to deep scientific insights about the ice sheet. That was precisely what deep drilling had accomplished and— even further back—what the researchers at Eismitte and Station Centrale had achieved. The biggest shortcoming was that field experiments couldn't cover large areas. Also, they usually could not be sustained over long periods of time.

But there was now a second area of research—using an airplane to measure the ice on Greenland. This approach could not collect the same depth of detail as a team in the field. On the other hand, there were no supplies to drag over the ice, no snowmobiles, no tents, no stoves, no drills, blizzards, or crevasses. Most crucially, an aircraft could range effortlessly over a tremendous area and take one flight after another, season after season. While the expenses could be significant, so were the potential breakthroughs.

Satellites promised to take Arctic research to a third level. Circling the earth from three hundred or five hundred miles up, a remote sensor would likely provide scientists with even less detail than aircraft surveys. But satellites would cover a far greater area and have an observational frequency—orbiting the earth fifteen times a day for years—that meant a constant stream of data. If a glacier at the end of the earth melted, or if some big piece of ice broke off the wildest and most desolate nub of north Greenland, someone—*something*—would notice.

Bob Thomas's boss at NASA was a glaciologist named Jay Zwally. From the mid-1980s on, Zwally had lobbied within the agency for building a satellite that could measure the polar regions through laser sensing. After making the case repeatedly for the importance of a constant stream of data, he recalls that "the response would be, 'Ice sheets are a long-term climate problem. You only need to measure them every ten years or so.'"[22]

But in the mid-1990s, concerns about the possible impacts of global warming intensified. As the data from Bob Thomas's Greenland flights filtered in, news about abrupt climate changes in the past, gleaned from the deep cores drilled in Greenland, was spreading around academic circles. Zwally began making progress with his idea. In 1999, NASA committed about $280 million to a satellite ice project. Led by Zwally, the program was christened ICESat, which stood for Ice, Cloud, and land Elevation Satellite. The main goal was to use a laser altimeter to measure elevation changes on the ice sheets of Greenland and Antarctica. The satellite would keep constant watch on the ice caps for five years—and it would never blink. At the time, Zwally noted that the technical challenge of the mission was to launch a one-ton satellite 365 miles above the earth, which performed a laser measurement "40 times a second [while] moving 16,000 miles per hour."[23] But NASA now viewed that kind of technology as perfectly doable. Following several years of planning and assembly—

along with a few delays—the mission was sent into orbit with a Delta-II Rocket launched from Vandenberg Air Force Base in California in January 2003.

It made sense that Thomas and Zwally were pushing ahead with remote sensing. They were glaciologists. Both had spent time in the field (Zwally had been involved in the Greenland drilling during the 1980s), and they knew from cold, hard experience the difficulties of working on ice sheets and the advantages of surveying them from high above. They understood a warming world would put the ice in a precarious position. What was more difficult to understand, though, was the project that a young NASA scientist on the West Coast was mapping out at around the same time. Mike Watkins hadn't ever been to Greenland or Antarctica. He had just earned a PhD in engineering from the University of Texas and come to work at the Jet Propulsion Lab in Pasadena, California. Watkins wasn't thinking much about ice, or about lasers, in the early 1990s when he first started working through an idea for a pair of satellites that would circle the earth together, one following the other, going around and around. Mainly, he was thinking about gravity.

Ever since the late 1960s, scientists at NASA had aspired to create a remote-sensing satellite that could take a good measurement of the planet's gravity field.[24] But other projects had taken priority—sensors for the atmosphere, ocean, and soil—and decade after decade, a gravity satellite project kept getting pushed aside. Still, the idea of measuring earth's gravity field with extreme precision remained a point of discussion, in part because such a measurement could have practical as well as scientific value. The planet's gravity is not consistent everywhere. The pull on an object moving overhead can be greater wherever the mass is denser—for instance, above mountain ranges like the Rockies or Alps, or over vast ice sheets like Greenland's. These gravitational variations can have subtle but important effects. They can influence the paths of satellites and ballistic missiles, for instance, which is something the U.S. military cares deeply about. They can also affect the oceans, since sea levels can be distorted in some locations by the gravitational effects of what lies far beneath the water's surface. Down below, there are deep trenches, submerged mountain ranges, and the remnants of lost continents that slid under the sea-floor hundreds of millions of years ago. If you could gather an improved measurement of the planet's mean gravity field, the data could prove useful in fields ranging from aeronautics to oceanography.

Watkins's proposal for the gravity project was called GRACE, for Gravity Recovery and Climate Experiment; it was greenlighted by NASA administrators in 1997, with Watkins appointed as the project scientist, and his former mentor, Byron Tapley, as the research leader. To those who weren't familiar with the project, it was not immediately obvious what a gravity satellite could discover about climate, let alone about ice. There were no cameras or radars directed toward earth on these machines. But Watkins was optimistic about what GRACE might do. He understood that earth's gravity fields vary not only from place to place but from season to season—and almost always because of the movement of water from one region to another. He believed that "time variability," as he thought of it, might yield some intriguing results. "I started to do simulations," he recalls, "and I'd ask, 'How much can we measure changes in the Greenland and Antarctic ice sheet? How well can we measure aquifer changes in groundwater?' And we started to realize that this was the thing that was going to break the mission wide open. And so we wrote the proposal, we said, we'll get the mean gravity field of earth, that's a slam dunk. But here's this *other* cool thing we can do."[25]

The trick was in how the satellites would measure the planet's changes. GRACE's two modules would orbit earth about every ninety minutes, chasing each other around the sky in measured pursuit, about 135 miles apart, one following the other like cat and mouse. They were designed to be sleek: Clad in solar panels, they had a resemblance to oversized gold bricks, each about the size of a small sports car. "We wanted a very compact design," Watkins recalls, "so that nothing moves, nothing changes, the center of gravity doesn't move. We never turn anything on or off on the spacecraft. They're just sailing along, sailing just as quietly as they can, so they're mostly only affected by gravity." But the gravity from below could vary. That was the key. Where there's more water mass in one place on earth, there is more gravity; where there's less water mass, there is less gravity. That meant that when the first GRACE spacecraft approached the space above the Greenland ice sheet, with its mass of about three quadrillion tons, it would respond to the subtle gravitational pull of the ice. It would be pulled slightly forward and away from its trailing partner, and the distance between them—those 135 miles—might increase by less than a human hair. But because the two modules were in constant contact with each other through a microwave communication link, the minuscule change in distance could be precisely measured. This could be measured over and over again, month after month, year after year. If the ice on Greenland kept pouring into the ocean and its ice sheet grew smaller, the gravitational pull would therefore change, too. Scientists on the ground could conceivably convert that distance measurement into a calculation of ice loss.

As Watkins conceived of it, GRACE would constantly *weigh* the ice sheet, which is why one of his colleagues began calling it "the scale in the sky."[26] And in this respect, it would be different than almost any other satellite. GRACE's movement—or more exactly, its change in movement from one month to the next, and from one year to the next—was the measurement itself. Or at least that was the plan. Watkins was sure the satellite would work in terms of its engineering and its capability to collect scientific data. "But we weren't quite certain what we were going to see," he says. "Is Greenland changing? Are there big changes in aquifers underground? What if those signals are really small?" He worried that if the ice sheets were close to equilibrium, meaning they were gaining as much mass from snow every winter as they were losing in summer, GRACE would not prove particularly interesting to glaciologists or climate scientists. *It might turn out to be a very boring set of observations,* Watkins thought.[27]

GRACE was authorized during a late-1990s era at NASA when some science missions were approved on the condition that they satisfy an agency directive to be "faster, better, cheaper." The joke around NASA at the time was that the first two parts of that motto were fine; it was the last part that made life difficult. What ultimately carried GRACE to completion and a launch date was a partnership that the

American scientists struck with several German science agencies to share costs. The German team agreed to pay for a rocket to launch the satellite and conduct GRACE's mission operations while it was in orbit. NASA's Jet Propulsion Lab meanwhile did the bulk of the satellite design and paid for the instrumentation. The two spacecraft were assembled by Airbus, just outside of Munich. Ultimately, the cost amounted to $97 million for NASA and about $30 million for Germany.

Using a Russian rocket, GRACE was sent into space without a hitch from the Plesetsk Cosmodrome, a launchpad about five hundred miles north of Moscow, on March 17, 2002. As planned, the two modules of the satellite moved into orbit and, using onboard tanks of nitrogen for acceleration and positioning, eventually separated by the requisite 135 miles. GRACE was meant to sail in a circumpolar orbit, meaning that rather than tracing the equator it would cruise over the South and North Poles. The two spacecraft soon began to transmit their measurements several times per day, usually to a ground station located in Svalbard, which lies in the Arctic Ocean north of Norway. From there, the data would be routed to the German science team, near Munich, and to Americans at the Jet Propulsion Lab in California. In those early days, as the raw data began coming back—data that no one had ever really seen before—scientists didn't immediately gape in wonder. Mostly, they scratched their heads and tried to figure out how to make sense of it.

On a clear night, if the alignment of the earth was just right and you knew exactly where to look—and then, if you walked into a field on the dark outskirts of some town, somewhere on earth—you could gaze up at a certain moment and watch GRACE fly by, two tiny fleeting dots. "It was *whoosh, whoosh*, like streaks in the sky," one engineer working on the project would later say. The satellites looked like streaks because they were moving at about 17,200 miles per hour.[28]

For the first few years after the mission started, scientists pored over the data transmissions and tried to estimate changes in the polar ice caps. It was not a simple matter. Isabella Velicogna, a physicist with the Jet Propulsion Lab working at the University of Colorado Boulder, built complex mathematical models to try to analyze all the factors—mountains under the ice, for instance, and the weight of the atmosphere—that might affect the gravitational forces that GRACE was sensing in its orbit. "If you want to look at the total mass of the ice sheet," she reasoned, "you have to separate it from any other signal."[29] Velicogna focused on trying to "peel away" all the noise except for the tugging effect of ice on the spacecraft. Other teams of scientists were doing similar calculations, and while some came up with different results, all of the estimates pointed in the same direction: The ice caps were in serious decline. When Velicogna published her first results for Greenland in 2006, she calculated that the ice sheet was losing well over one hundred billion tons of ice per year, and that Antarctica seemed to be losing substantial amounts as well.[30] The article floored a number of prominent polar researchers. "I remember reading their first paper, and I literally couldn't believe it," says Berrien Moore, a scientist who managed NASA missions on and off for several decades.[31] A few years later, Velicogna updated her findings and noted that according to GRACE readings between 2002 and 2009, "the mass loss of the ice sheets [was] not a constant but accelerating with time." Greenland had gone from losing about 137 billion tons of ice per year to losing about 286 billion tons a year.[32] It wasn't easy to get your head around that kind of number, but it was about equivalent to two Mount Everests falling into the sea every year. "It's not a 'Run for the hills' kind of story," Waleed Abdalati, a high-ranking scientist at NASA, said at the time, "but it's 'Wow, this is serious.' "[33]

To Abdalati, it meant that Greenland's ice, and to some extent Antarctica's, was pushing up sea levels—not yet by leaps and bounds, but by small and steady increments. And this seemed revelatory. For years, global sea levels had been measured two ways: by tidal gauge markers in harbors around the world, and by satellites that NASA had put up in the sky. Both types of measurements had shown that the ocean had been rising over the past few decades. Both types were in close agreement. Still, the evidence of this slow and steady rise—a few millimeters per year—couldn't explain *why* the tide was creeping up.[34] Oceanographers understood that in a warming world, sea levels could rise for two distinct reasons: The first was that the oceans would expand in volume as they gained heat; the second was that ice melting from Greenland and Antarctica, along with ice melting from mountain glaciers around the world, would trickle into the sea. What GRACE could do was calculate the mass of ice pouring off Greenland and Antarctica and mountain glaciers. In effect, it gave scientists a new way to solve an old problem. For the first time, they could factor in how much of the rise in sea levels was resulting from additional mass, from the melting glaciers, and how much from additional volume caused by warmer temperatures.

It was obviously bad news. And in the years following, it didn't get any better. GRACE kept circling and sending down its results, and Isabella Velicogna's charts on Greenland's annual ice decrease began to resemble a steep staircase going down, with losses comprising more than a trillion tons. As observations started coming in from ICESat, the satellite measuring changes in ice sheet elevation, the picture of Greenland looked similar to what GRACE was showing: Ice was going down, down, down. But it wasn't as though an army of researchers was yet being dispatched to examine the situation; it was still a small-scale effort at understanding a large-scale problem. Writing in the British scientific journal *Nature* in 2008, and quoting Ohio State University glaciologist Ian Howat, Alexandra Witze noted:

> Very few researchers have Greenland as the main focus of their scientific work. Decades from now, this could turn out to be one of the most short-sighted allocations of resources that began the twenty-first century. Climate change elsewhere in the Arctic has been swifter than anticipated. The remarkable shrinkage of the sea ice is "the largest change in Earth's surface that humans have probably ever observed," Howat points out. Trying to get any and every handle on how that affects the poised mass of [Greenland's] ice next door must surely be a priority, he says.[35]

The average summer temperatures in Greenland were now sometimes 7 to 9 degrees Fahrenheit warmer than normal. And all kinds of distressing evidence about the Arctic was filtering in from the ground and from the air.

There wasn't a single dramatic change—a disruptive moment—between, say, 2005 and 2010. But there were trend lines now, showing the Arctic was under a steady siege from warmer weather. And then one day in July 2012, Tom Wagner, a polar scientist and administrator at NASA's Washington, D.C., headquarters, came home from work and greeted his wife, Renee Crain, who was also a polar scientist.[36] In addition to overseeing a number of the agency's earth science programs, Wagner spent a good amount of time looking at data coming in from NASA's satellites.

Crain said to Wagner: "Hey, what are your satellites seeing in Greenland?"

Wagner asked why.

"There's melt going on all over the place," she said.

Together they logged on to a computer and watched YouTube videos about rivers of meltwater coming down off the Greenland ice sheet. Floods were washing away bridges and heavy equipment in towns like Kangerlussuaq, the Inuit village formerly known as Sondrestrom, where the country's main airport was located.

Wagner immediately got on the phone with a colleague at the Jet Propulsion Lab in California who monitored the Arctic with an orbiting satellite that uses a sensor known as a *scatterometer*. In glaciology, a scatterometer is useful because it can measure a thin film of meltwater atop an ice sheet. Wagner recalls, "I talked to our guy who looks at the surface of Greenland, and he said, 'Oh, when I looked at the recent scatterometry image over Greenland, I saw all red. I just assumed the satellite was broken.'" But it wasn't broken. The scatterometer showed the extent of water atop Greenland's ice sheet because water and snow have different ways of "scattering" radio waves back to the spacecraft. Red meant water.

"Anyway," Wagner says, "that's when we realized the entire surface of the Greenland ice sheet had melted."[37]

*   *   *

Glaciologists who entered their field in the 1970s or 1980s sometimes tell a similar story: They had begun their work under the assumption that changes to the ice sheets, if they happened at all, would be slow and modest. These were natural features of the earth that changed in long spans—tens of thousands of years—and not in the time it takes for a life to begin and end. At the start of their careers, some of these scientists had been warned by wizened university advisors that they were pursuing an area of study that could prove painfully dull and possibly a dead end. As GRACE and ICESat began sending back observations, however, it was already clear this was not at all true. Bob Thomas, writing in the journal *Science* with a younger NASA glaciologist named Eric Rignot, noted: "Perhaps the most important finding of the past 20 years has been the rapidity with which substantial changes can occur on polar ice sheets. As measurements become more precise and more widespread, it is becoming increasingly apparent that change on relatively short time scales is commonplace." So now they knew. Glaciers could suddenly accelerate and thin. And all the while ice sheets could appreciably deteriorate. The idea—the old idea—that polar ice caps existed in some kind of "steady-state" needed to be discarded.[38]

That made the work more interesting and more urgent. But it left ice scientists in a somewhat bewildering place, too. They were struggling to understand a situation they had never considered before—as if they had sailed into a peripheral region, like those marked on the edges of the maps of old cartographers, who in drawing features of an unexplored place (the north of Greenland, for instance) called it *terra incognita*, an unknown land. In the summer of 2012, one thing glaciologists could be sure about was that colleagues who studied climates of the past had concluded that immense ice sheets once covered Europe and North America, that those ice sheets had shattered and melted, and that at various points in earth's history the climate was warmer and sea levels were much higher.[39] Yet whether Greenland and Antarctica, in their diminishment, would follow some kind of historical pattern or timeworn design was a question without a definite answer. Ice cores couldn't explain that. There was no recorded history of how an ice sheet disintegrates. "We've never seen it," Eric Rignot would later remark. "No *human* has ever seen that."[40]

It therefore stood to reason that if remote sensing had solved the question of whether Greenland's ice was growing or shrinking, it had also led into another problem, one that seemed even more difficult. It was the mystery of how nature might move much of that ice into the sea. Perhaps it would melt from the heat, or shatter as it softened, or be eroded by intrusions of warming salt water. Or perhaps all three forces in combination would undo it. Meanwhile, there was the worry: Would the polar glaciers reach a condition of degradation that would make it difficult to stop the changes that had been set in motion? The ice sheets in Greenland and in Antarctica were so big, and so complicated, that it was difficult to say where a point of no return might be.

It was around this moment that some glaciologists began to talk more freely about the notion of something they were calling "ice sheet collapse." It was also at this point that some began to ask: How fast could it happen?

David Holland (left) at Helheim Glacier, August 2018 (Lucas Jackson/Reuters)

15

A Key

Almost from the moment they put ICESat up in the sky, NASA's scientists watched with concern as the satellite suffered technical problems. The machine had been equipped with three lasers that could send pulses of light down from space to measure changes in Greenland's and Antarctica's ice. Each laser was intended to last a year or two. The first laser failed after thirty-six days. The second one burned out after a hundred days. The space agency devised an emergency plan to prolong the life of the third, but when that laser failed, too, there was no point keeping ICESat aloft. In 2010, the spacecraft received instructions to leave orbit, and not long after, most of it burned up as it reentered the atmosphere. The parts that didn't vaporize—a few bits of space debris—dropped into the Barents Sea north of Siberia.

GRACE was still circling the earth in good health; every month it sent down information about the demise of Greenland's and Antarctica's ice. What GRACE couldn't always provide, though, was a fine-grained picture of where the losses were coming from. That had mostly been ICESat's job. Thus the engineers and scientists now found it harder to understand which glaciers in Greenland were thinning the most, or precisely how (and how fast) the process seemed to be moving. A few glaciologists considered the situation urgent. "We are slowly going blind in space," Robert Bindschadler, a NASA veteran, told a journalist at the time.[1] Yet by that point, space agency administrators were discussing remedies. One solution was nicknamed "Gap Filler," which soon became known as IceBridge. The moniker arose from the idea that NASA needed to "bridge" the "data gap" between the now-defunct ICESat and a replacement satellite that was probably many years away from launch.[2] "Headquarters approached us and said: 'How about you take on this gap-filler concept?'" John Sonntag, who eventually became the IceBridge mission scientist, recalls. "So we kind of invented it as we went along."

The plan for the new program was straightforward: Fly an airplane over the world's two great ice sheets for several hundred hours each year. Cover Greenland for several months in the spring; survey Antarctica for a few months in the fall. And on those flights, use an arsenal of tools to measure changes in the ice sheets below, along with the sea ice floating nearby. In many respects, IceBridge resembled the NASA flights over Greenland a decade before, which were the first to use lasers to take a measure of the ice sheet's decline. Right away the mission planners considered it necessary to map out a number of routes that could represent a "sampling" to determine ice losses. These were routes that could be repeated with precision every year, so that any changes could be easily tracked.

Within a few years, the IceBridge air campaigns had become routine. Every April, IceBridge was stationed at the main airport on Greenland's southwest coast. Originally known as Bluie West Eight during World War II, the airport had for decades been known by its

Danish name, Sondrestrom. But a small village of about five hundred people had grown up around the site, and in the early 1990s the town assumed its Inuit name: Kangerlussuaq. In turn, the airport had been renamed Kangerlussuaq international airport. Not far from the runway's terminus, an old military barracks—a rambling, two-story, no-frills concrete structure—had been converted into a dormitory for scientists coming in and out of Greenland. It was known as KISS, which stood for Kangerlussuaq International Science Support. On a typical day during the IceBridge campaign, John Sonntag would get up early at KISS, put on an olive green flight suit, and begin his morning with a visit to the town's weather office, a suite of rooms near the small grocery store in town. A Danish meteorologist, gazing at a number of computer screens, would give Sonntag Greenland's regional forecast and a packet of maps. Then Sonntag would weigh his options. "On any day you have a quiver of flight plans you can follow," he would explain, "and you pick the highest priority mission you have on a given day if the weather supports it."

The local weather in Greenland can vary from glorious to deadly. Sometimes, too, a forecast for low clouds or fog can scuttle a mission, since they can interfere with sensing equipment. To Sonntag, the highest priority missions were ones routed above locations where ice losses had been most extreme. These were the glaciers—Petermann in the northwest, Zachariae Isstrom in the northeast, and Jakobshavn in the west—that observations showed had been thinning or could be most at risk. So in the early morning, Sonntag would spend a few minutes mulling over possibilities until he picked his route. Then he would inform the pilots. Usually, with temperatures at Kangerlussuaq hovering around zero, the aircraft engines would need to be warmed up. Preflight checks had to start. The rest of the science team, shivering, would gather on the runway near the plane by eight A.M.

A typical IceBridge flight—eight hours in duration—is like flying at very low altitude from New York to Houston and then back again without landing. Except of course it isn't. For one thing, you're flying a big military plane—a P-3 or C-130—over glaciers that flow for miles from the central ice sheet to the sea on winding courses that run between dark, desolate mountain peaks. "These are places that only a few hundred people in the world have ever seen," IceBridge pilot Jim Lawson observed one day, almost to himself, mid-flight. You're wearing flight headphones most of the time so you can talk over the din in the cabin, and also so you can hear Sonntag, sitting in front on the flight deck, call out the sights below. One day you pass over Nansen fjord on the east coast, near where the Norwegian landed in 1888 to begin his crossing. Another day you swing down the west coast, by the calving glaciers of Eqip, where Paul-Émile Victor's team landed in 1949. Amidst the wide white expanse of the ice cap, you might pass over Dye-3, the drilling site from the early 1980s, or above Summit Camp, the place where the GISP-2 core was drilled in the early 1990s. Somewhere nearby, you buzz over the blank stretches of ice where Eismitte and Station Centrale once existed.

It seems as though history becomes scenery, and scenery becomes data. In the airplane cabin, the science team consults its screens as information rolls in from their sensing instruments. An altimeter bolted to the aircraft floor sends laser pulses down to map the ice elevation below. An ice-penetrating radar peers through layers of ice to bedrock. A digital camera records high-resolution images of the icescape below. Even with a good weather forecast, it may get turbulent up in the air. Sonntag recalls how he once encountered turbulence over the peaks of east Greenland unlike anything he'd ever experienced. "The wind flows down these mountains, and it starts to eddy and swirl," he says. "I remember looking back into the cabin after one bump—the coffeepot had shattered, laptops are on the floor, someone's lunch is on the ceiling. A couple people are on the floor. Even parts of the floor panels had come off." He cut the mission short that day, turned back, and returned to Greenland's west coast for dinner and an evening jog. On another IceBridge flight over Antarctica's Weddell Sea he didn't have that option. At twenty-five thousand feet, in pitch dark and hundreds of miles from any airfield, the plane lost two of its four engines. "They just *stopped*," Sonntag says, "and I looked out the window and saw that the props weren't spinning." The plane began losing altitude. "And then the pilots rang their alarm bell, which sounds like a school bell. And the pilot does not ring that bell unless things go seriously wrong." Instructions from the flight

deck came down: *Get your exposure suits on.* "So we get on these rubber suits, what we call the 'Gumby suits,' thinking we are going to
ditch." Before they ended up in the Weddell Sea, though, the pilots
discovered that a lower altitude could thaw a frozen fuel line, which
happened to be the source of the problem. They restarted the engines. In twenty years of polar flights, Sonntag says, "that was the
closest I've ever gotten to . . . you know."[3]

Usually, the challenges are more mundane. Buffered from the cold,
windy misery of the ice sheet, measuring Greenland by IceBridge
means living fifteen hundred feet above the ice at a speed of about
three hundred miles per hour—all day, six days a week, ten weeks in
a row. Monotony and fatigue can become the biggest burdens as the
survey overflies thousands of miles of ice and rock. Inside the plane,
the larger purpose of the long days in the air, to better understand the
vanishing glaciers, can start to recede. Approaching Helheim Glacier
in east Greenland one afternoon—a glacier named for the word in
Norse mythology that describes the world of the dead—Sonntag an-
nounces through the headsets, "Okay, science crew, get ready." It is
the signal that the technicians in the cabin need to focus on their
instrument readings. Down below, the glacier is a snow-covered river
of ice, four miles wide and twenty-five miles long, which flows out
from the central ice sheet like a chute, curving between snow-flecked
mountains toward the black ocean, where it breaks apart.

One of the technicians in charge of data storage, Aaron Wells, sits
up from a nap and rubs his eyes. The members of the science crew
have heard Sonntag's instructions, and someone has quipped into
their headset: "I can't tell what the Helheim looking at." Wells responds with a sigh and logs on to his laptop. He says into the headset,
"I don't know what the Helheim doing with my life."

For the past fifteen years, since the GRACE satellite measurements
began, Greenland has been shedding about 286 billion tons of ice, on
average, every year.[4] A large amount of that ice is pushed into the
ocean through "outlet" glaciers like Helheim. While the losses can be
obvious to sensing instruments, they can often be imperceptible to

the eye. In the last few years, however, Sonntag has noticed clear incontrovertible changes. Meltwater lakes pool on the ice earlier in the spring season, and at higher elevations, than ever before. The lakes cover the white surface like giant beads of aquamarine, attesting to the fact that the warming Arctic air is now creeping onto Greenland's ice sooner—and deeper into the middle of the ice sheet. "But the biggest change is the Jakobshavn Glacier," Sonntag remarks. "It looks different every year. And especially in comparison from the early 1990s to now, it doesn't even look like the same place."[5]

Flying over the glacier one day, Sonntag points to Jakobshavn below: a fracturing river of ice, flowing from a channel on the western edge of the ice sheet. The glacier ends at a wall, its calving front, which from above resembles the shape of a seagull. The cliff face measures about three hundred feet high and is about six miles across; it is the place where Jakobshavn's icebergs break off and begin their journey down an ice-choked fjord and into a larger body of water known as Disko Bay, where tourists can take iceberg cruises to ogle at their stupendous size and variations in color—alabaster white, riven with deep blue hollows. In time, these icebergs are carried out to the North Atlantic, where they melt. Like Helheim, Jakobshavn is an outlet glacier, and in its workings you can understand a fundamental aspect of how Greenland's ice sheet loses mass, as well as how the world's oceans rise in height. Between 2000 and 2011, Jakobshavn alone was responsible for a millimeter in global sea level rise.[6] In August 2015, it calved several icebergs over two days that cut nearly five square miles of ice from the glacier's front.[7]

Mid-flight, in the IceBridge cockpit, Sonntag pulls open his laptop to display some color-coded maps. They compile the past few years of remote-sensing data. On his screen, Jakobshavn's glacier appears as a massive drain, increasingly sucking the western quadrant of the ice sheet toward it. "This is ominous," he explains. "Because there is nowhere for that ice to go but into the ocean. And when there's nowhere for the ice to go but the ocean, it means sea level rise."[8] As Jakobshavn loses ice, Sonntag says, it *retreats*. One way to visualize the retreat is to imagine that as the end of this river of ice breaks off into icebergs, the river itself pulls back from the coast. Or you might think of someone's outstretched arm, a fist on the end to resemble an ice cliff, being pulled back toward their body.

Jakobshavn is perhaps the world's most studied glacier. Its scientific history reaches back several centuries, at least to 1719, when a whale fisherman coming through the area named Feykes Haan noted a fjord "that is always full of ice, with frightfully tall icebergs, but from where they come is unknown."[9] Hinrich Rink, the first scientist to exhaustively study Greenland's ice sheet, arrived here in 1850. He made his way to the calving front—at that point in time, many miles farther into the bay—to take notes about the position and behavior of the glacier. A few decades later, Amund Helland was the first scientist to measure its speed. By 1900 or so, many European scientists understood Jakobshavn to be special—unusually fast, unusually productive, perhaps faster and more productive, in fact, than any glacier in the world. Its icebergs were credited for sinking the *Titanic*. But they were also picturesque, and a familiar presence to Knud Rasmussen, who spent his childhood in a house on the north bank of Jakobshavn's fjord, watching icebergs from the glacier move slowly past his front window every day.

In 1913, Alfred Wegener came here for a seven-day canoe trip. It was just a few weeks after his expedition party had finished crossing the ice sheet and was rescued by a passing ship. "The entire fjord in front of the glacier," Wegener wrote in his notebook, "all the way through the banks of the fjord's opening, [is] densely covered with icebergs and calving remnants."[10] Even then, he saw that Jakobshavn's calving front was moving steadily back toward the ice sheet. And when Wegener visited again in 1929, just a year before he made his fateful trip to the center of the ice sheet, he noticed that Jakobshavn had retreated even farther. In subsequent years—through the era of Paul-Émile Victor's explorations, for instance, and Henri Bader—the glacier continued sliding backward. And then, as the remote-sensing era began in Greenland in the late 1990s, interest in Jakobshavn sharpened further. The glacier's retreat, measured by satellite and aerial radar, continued—thousands of feet per year. But its speed and iceberg production multiplied. By the turn of the millennium, it was considered the fastest and most active big glacier in the world.

Around that time, it caught the attention of Bob Thomas, the NASA scientist who had spent his early years in Antarctica driving dogsleds and who had bet his career in the 1990s on the idea that a remote-sensing airplane mission could measure the Greenland ice sheet. Thomas was interested in why Jakobshavn's speed had increased. In its retreat, the glacier had lost an ice shelf that had extended out into the fjord—that was at least part of the explanation. Many years before, Thomas had been one of the first glaciologists to theorize that an ice shelf, a thick floating ledge of ice that extends past the edge of a glacier, is enormously important to how a glacier behaves. As he saw it, the shelf acts like a buttress that holds up the wall of a cathedral, which is to say the shelf prevents the glacier from moving a lot of ice out into the ocean. And yet once that shelf—that buttress—breaks off, ice from the glacier pours out. Some glaciologists preferred to compare the importance of ice shelves to something else: The effect of a cork popping out from a wine bottle tipped on its side. In this metaphor, the cork is the ice shelf, and the glacier's backlog of ice is the wine.

Thomas had started wondering what role the water temperatures in the fjord played in breaking up the ice shelf. Maybe warmer water had accelerated its disintegration? He started talking about it with a young researcher named David Holland at Columbia University's Lamont-Doherty laboratory. Holland liked Thomas's approach to research, especially his belief that theory couldn't get you too far in science. It was necessary to observe the natural world exhaustively, by remote sensing and by field visits, year after year, so as to answer the difficult questions about its behavior. As Holland remembers it, in 2003 Thomas called to say: "This glacier at Jakobshavn has just fallen apart. It's going crazy."

"That's *fascinating*," Holland replied. But he was being ironic. Holland was an oceanographer and mathematician, and had no idea what

Thomas was talking about. Most of the time Holland worked on computer models that tried to make sense of the ocean's effects on the
atmosphere, and vice versa. He had grown up in Newfoundland,
Canada, and when he wasn't thinking about math or oceans he was,
by his own admission, thinking about hockey. He knew quite a bit
about the polar explorers from the nineteenth century who had come
to Canada searching for the Northwest Passage and met tragic ends;
in some respects, his studies of their journeys had led him toward his
own career in polar research. At the time, though, Holland knew very
little about ice sheets or Greenland.

"Jakobshavn is a glacier in *Greenland*," Thomas explained. "I'm
going to get a dogsled team. I'm going there in the winter. You should
come." Thomas explained to Holland that he was intent on researching the fjord that the glacier emptied into and that he planned on
bringing a CTD—a standard device that oceanographers drop into
the water to measure its conductivity, temperature, and depth. "It was
a very romantic idea Bob had," Holland explains. "The last measurement in the Jakobshavn Fjord was done in the first international polar
year, in 1882 or so. And that was done by this guy Hammer from
Denmark." Holland's point was that even as Jakobshavn's glacier
front had been measured repeatedly, the waters around it had not.

It turned out that Thomas never went to Greenland that winter
with a dogsled team. But Holland visited a few years later, in spring
2007, with his wife and fellow researcher, Denise. At that point, the
prospect of collecting observations about the water, as well as the
calving front, had caught his interest. He and Denise arranged for a
pilot to fly them by helicopter from a small airport on the west coast
to a location about thirty miles inland. The pilot dropped them off on
a rocky scarp of land near the glacier's calving front. Together, they
began measuring the temperature of the fjord and watching the ice.
They stayed for about a week. And when they were ready to go back,
Holland called the helicopter pilot with a satellite phone, and the
pilot said, "I'm not coming today. Good night." Holland considered
this mystifying at the time, but he later found it amusing—a good
example of the surprises and frustrations of polar fieldwork. Eventually, when he made it back to the United States, he and Denise agreed

to start a long record of data collection. They intended to see if an observational record would help better explain the glacier's behavior.

By the time of Holland's first visit it was known that the to and fro movement of glaciers could be affected by many things apart from ice shelves. Gravity and friction—for instance, whether a glacier is moving through a narrow mountain pass—play a part. So do bumps and hills in the bedrock deep underneath the ice. Some recent studies had showed that glaciers could be sped along by meltwater when it flows *underneath* the glacier, between bedrock and ice. Holland had even come to think that chaos could play a role, too, as ripples from obscure long ago events set in motion unpredictable circumstances. "The weather could have been different 150 years ago on one day," Holland would say, "and this glacier could be very different because of it."[11]

In sum, glaciers were about as complex as a system could get. But on that first trip to Jakobshavn, Holland didn't think we understood glaciers very well at all. And he reasoned that we might only get so far with GRACE or the IceBridge missions. If we wanted to make accurate predictions about the future of the world's ice sheets, we would also need to get out in the field and watch Jakobshavn as it collapsed and retreated, even if it took twenty or thirty years. To measure Jakobshavn from the water's edge, then, would be to exercise persistence in pursuit of a higher ideal. It would test the notion that in understanding Greenland's fastest and most famous glacier, you might unravel some of the mysterious laws governing the behavior of all the dying glaciers in the world.

Every year since, Holland has returned to Jakobshavn to build up an observational record of the ocean and the ice. He has approached the task from a variety of angles. On the banks of the icy fjord near the calving front, he and Denise set up devices at several locations to collect seismic information. The anticipation is that a rumbling seismic "signature" might lend more clarity to what happens before, during, and after the glacier breaks off into icebergs. One of these camps is known informally as Camp Fox, since Holland once saw a fox there;

another is known as Camp Huge Rabbit, because he once spotted a large Arctic hare hopping about there. His main camp—the one near where the helicopter drops him off every June—is situated on a spectacular overlook just south of Jakobshavn's calving front. He has set up two insulated, prefabricated "igloo" huts there that can each house two people. He has also installed a small radar system, along with solar panels and small wind turbines, so he can focus it on the calving front twenty-four hours a day, seven days a week.

The flight to the campsite is thirty minutes by helicopter. The Hollands bring tables, chairs, tents, and sleeping gear. They haul solar panels and batteries for the new radar, which is packed tight in a blue steamer trunk. A drone system for aerial surveying is brought along. There is food in coolers, food in crates, and large stores of coffee, tea, bread, and Nutella. There is a Coleman stove, as well as generators, a portable toilet, stocks of water in big blue jugs, gasoline for generators in big red jugs, and butane for the stove in big yellow jugs. The chance a polar bear will come through is slight—"Why would a bear come here?" the helicopter pilots asks. "There's no food, no plants, no life"—but they bring bear alarm fences and rifles as a precaution.

At the camp, Holland walks around with a sense of purpose and a low center of gravity. As a high school student in Newfoundland, he played hockey seriously, even after he lost his two front teeth to a puck. "I've never been hit by a crowbar," he says, "but a slap shot in your face? Being punched is a joke compared to that." Any ferocity he might have demonstrated in the rink is hard to discern out here. In the field, Holland is jokey and genial. Usually he wears a black fleece zip-up sweatshirt and sports a baseball cap with a fabric flap to protect his neck from sunburn; even in the glaring sunlight of the Greenland spring, he rarely wears sunglasses. Instead he prefers spectacles—horn-rimmed, which give him the appearance of an IBM computer scientist from the early 1960s. Attending to small tasks, Holland walks constantly from one end of camp to another as he consults a small, bound booklet that he keeps in the front pocket of his cargo pants. He removes the booklet constantly to scribble out small notes—scientific insights, sometimes, but usually bits of information such as phone numbers, logistical reminders, or dates. There may be a tool he should have brought but didn't, and needs to have for the following spring.

By the first evening—the end of the first week of June—the station is mostly set up. Camping by Jakobshavn's calving front is akin to camping on a spot overlooking a river, except in this case the river is many miles wide and clogged with both large icebergs and shattered ice, the latter of which is so profuse that it forms a frozen stew, known as "mélange," which covers the fjord waters here. The sun never sets. The visitors watch the glacier's calving front with a constant sense of expectation. It rumbles ominously sometimes during the night, with a sound like distant thunder. In daytime it tends to emit an occasional and sharp rifle crack, the recoil of a split somewhere deep within itself or the evidence that a small chunk of ice has been shed. Though the front seems high at three hundred feet, the glacier actually descends below the mélange and fjord another three thousand feet or so. Jakobshavn is among the deepest fjords and glaciers in the world. Therefore a true calving event, one that cleaves the glacier from top to bottom, is bigger and more explosive—not a rumble, not a crack—and unleashes the energy of several atomic bombs. One morning a few years back, the Hollands awoke at three A.M. and had the good fortune to capture on video the breaking of a massive berg that measured twice the length of the Empire State Building. "It was very noisy, very spectacular," David recalls. "We were impressed."[12]

The glacier's riverbanks, instead of sand or grass, comprise a hilly moonscape of dust and rocks, ranging in size from skipping stones to boulders as large as refrigerators. Decades before, the big stones must have been moved about effortlessly, and then left behind, by the retreating movement of the glacier. In fact, this area happens to be called "new land"—that is, land that was covered in ice as recently as twenty or thirty years ago but has now been unveiled thanks to the recession of the ice sheet and glacier. Gray, colorless, and devoid of vegetation, new land is grim. Tens of thousands of years ago, when glaciers retreated from northern Europe and North America, many parts of the world looked like this—a place that exists before life arrives.

As it is, new land makes simple tasks unpleasant. When a cold wind blows down from the ice sheet, microparticles of till, a fine rock flour ground down by the glacier, kick up into the air. A film of grit descends on the lenses of eyeglasses, on the keys of a computer keyboard, on the skin of a boiled egg at breakfast. It settles on the cusps of molars. Meanwhile, the dry lunar wastes of new land seem even stranger in contrast to what's around: The sound of water, flowing and rushing, circles the campsite. It comes from rivulets that trickle down the edges of the ice sheet a few hundred yards behind the campsite, as well as from fast-moving rivers in front of the campsite, hidden from sight, which run a few hundred yards down the hill and pulse under the edges of Jakobshavn's glacier itself. What seems like wind, in fact, is just water running, far off and nearby, hard to place but difficult to ignore, an enormous, unceasing white noise, which is merely the sound of ice melting, all day, all night, everywhere.

A few years back, Holland decided to informally call his efforts in Greenland HELISHE. "The whole idea was that this was going to be difficult, and we actually liked the word *hellish*," he explains. HELISHE was not a true acronym, but Holland didn't care, since it amused him anyway. H is for hydrography; I is ice; S is seismic; E is for experiment. "It was important to have the acronym first," Holland says, "and the words fit pretty good."

HELISHE is conducted every June at Jakobshavn and every August on the other side of Greenland, on the east coast at Helheim Glacier. The experiment has two main areas of inquiry. The first involves the study of what Holland calls "the mechanical part"—the calving process that produces icebergs. "It's a fracture process not really understood," he says, "but people are making progress, so we are trying to contribute to that." The other aspect of the study, meanwhile, involves the arrival of those warm waters that not only melt the face of the glacier but can also erode the glacier's perch on hard land—what's known as its "grounding line"—many thousands of feet below the surface. "The waters have a reactionary force, and the glacier responds," Holland says.

These two parts of the Jakobshavn research subdivide into a variety of different endeavors. In addition to Holland's seismic sensors at Camp Fox and Camp Huge Rabbit, his radar at the main camp focuses on the calving front to measure the moment-by-moment movement of the glacier, which in June 2016 seemed to be moving about sixty feet per day. Every summer, moreover, Holland spends time dangling from helicopters in a dry suit so he can drop probes into the icy mélange of the fjord in front of the glacier; his intent is to measure the temperature and salinity of the water at various depths. More recently, he has also begun dropping probes into gaps in the glacier itself, behind the calving front and closer to the ice sheet, so as to measure "stretching properties" of the ice and thereby gain insight into what may be happening as the ice flows and strains before breaking off. All of this work is finally complemented by a week in a boat, which follows the week of camping by Jakobshavn. Holland cruises around Disko Bay, far from the fjord and the calving front, to deploy temperature probes and other robotic tools. The point is to gather more data on the salinity and temperature profile of the surrounding waters at various depths and see how they infiltrate the fjord and affect the glacier.

"The thing about the helicopter and the boat is that it's human intensive," Holland says. "You do get data, but it takes a lot of work." It can also be dangerous. Even if a captain could pilot a small boat through the icy mélange to get to the calving front—a place where a big iceberg may break off—it could be deadly. Over the last few years, Holland has settled on a more efficient way to get information all year round. With the help of a Greenlandic marine biologist named Aqqalu Rosing-Asvid, he augments his field observations by outfitting temperature sensors on local seals and halibut that swim in the fjord. The seals tend to go down about twelve hundred feet or so. "So they're constantly covering the top half of the fjord," Holland says. "And by some miracle of nature, the halibut cover the bottom half of the ice fjord." That means they swim in the icy depths down to three thousand feet or so.

Using a satellite network, the seals phone in their data several times a day—information that is routed to a rooftop antenna near Holland's office at NYU, in Manhattan's Greenwich Village. "The cost per measurement is really small," Holland says. "And the data—well, I don't know if any technology can compete with seals for that purpose."

One afternoon, a pilot arrives in a small helicopter, and Holland takes short flights across the fjord, to Camp Fox and Camp Huge Rabbit, to upgrade the seismic instruments. Meanwhile, his students keep the radar focused on the calving front and set up the solar panels and batteries. In the near distance, Jakobshavn rumbles and cracks. As the sun dips toward the horizon late at night, the temperature drops into the thirties, but when the sun is high during the afternoons the approach of summer seems obvious. It's getting warmer. The mosquitoes are coming out. The ice sheet can sense that, too. Day by day, the sound of rushing water grows louder.

The past decade has already yielded evidence to Holland that warming waters act as a kind of control knob to accelerate the speed of the glacier.[13] But the "mechanical" aspects of iceberg formation—how and why an iceberg breaks off—remain largely mysterious. Several recent studies suggest that once a glacier front reaches a height of about three hundred feet, which is about the height of Jakobshavn's front, the cliff can no longer sustain itself, and the ice inevitably breaks apart, or calves, from a kind of material failure. Sometimes this failure might be accelerated by meltwater on top of the glacier that eats its way into the glacier from above and *hydrofractures* the end of the iceberg.[14] But Holland believes there are likely other reasons glaciers calve and that we don't yet understand the conditions that might determine the process of breakage. Amongst polar researchers, the pursuit of these "calving laws," as they're sometimes called, comprises a kind of holy grail of glaciological research. To find them would be akin to being able to predict the conditions for, say, an earthquake in San Francisco or Tokyo. At the moment, that prediction happens to be impossible.

"As I see it," Holland says one afternoon, "the nature of the problem is one of sea level, and sea level is affected on fast time scales when warm water comes near glaciers in Greenland or Antarctica."[15]

There is a crucial aspect to what he is saying that can be easily missed: He is looking at changes on *fast time scales*. To a glaciologist, *fast* might mean decades or even centuries. In either case, however, it happens to be the reason why glaciers known as "outlet" or "marine terminating" carry such importance. They flow from the edges of ice sheets and end at the ocean. And in a hotter world, they have the potential to quickly move massive amounts of ice from ice sheets into the ocean.

But no one seems to know precisely how quickly. Holland's belief is that the work at Jakobshavn—whether it takes another ten or twenty years—will provide information to create better computer models. Without these observations, the task of projecting Green-land's future, or estimating how high seas will rise in the next one hundred to two hundred years, may not be feasible. One evening, sitting in the dining tent at the calving front, Holland says, "In ocean-ography, the models are basically complete, and in meteorology they're essentially complete. But in glaciology they're incomplete. So, you write down an equation, and someone could come over with a flashlight and say, 'Where did you come up with *that?*'"

They'd be right to do so, he believes. Even though climate models have become better at predicting future temperatures and precipita-tion patterns, he says, ice sheet models have not reached nearly the same level. The reasons why glaciers move and break (buffeted invis-ibly by oceans, air, gravity, bedrock hills, and the properties of ice) have proven extraordinarily complex. And Holland doesn't believe that we can build good models from theories and ideas—we need to look first to nature, so that it can teach us how it works. Pointing outside the tent to the calving front, Holland says, "This is the way it is, and your model has all these ways it could be different. If you're going to build a computer model that has all these different choices you've made, which are a thousand million different choices, why would I believe the one you just came up with?" The best way to make a good model is to see what happens, he insists. Then we can build it into the code.

Perhaps thirty or fifty years from now, he says, the glaciology mod-els will be robust.

*   *   *

He knows that is a long time—maybe too long—to wait. For one thing, the stakes for the glacier at Jakobshavn appear to be significant. It drains about 7 percent of the Greenland ice sheet, and its continuing retreat and collapse could someday boost sea levels by more than a foot. But in looking at Jakobshavn, Holland is also thinking about a glacier in Antarctica, located in the western part of the continent's massive ice sheet, called Thwaites. It was named many years ago for Fredrik Thwaites, a geologist from the University of Wisconsin.

Thwaites is sometimes described as the Doomsday Glacier.[16] Like Jakobshavn, it has recently come in contact with warming ocean water. And like Jakobshavn, it is precariously poised on a deep bed that gets even deeper as it moves inland, making a sudden retreat (and a massive dump of icebergs into the ocean) much easier. The collapse of Thwaites, however, would likely have a far greater impact on sea levels than Jakobshavn, since it is about the size of Great Britain. The result would be at least two feet of sea level rise, and a possible and subsequent collapse of the entire West Antarctic ice sheet, which holds enough ice for about twelve feet of sea level rise. In 2014, two scientific studies based largely on remote-sensing data, one led by Eric Rignot at the University of California, Irvine, another led by Ian Joughin at the University of Washington, declared that Thwaites appeared to be in or close to the early stages of collapse. There was no settled conclusion on how fast it would happen—many decades or even many centuries seemed possible. Nevertheless, the glacier was starting to move massive amounts of ice into the ocean. "The collapse of this sector of West Antarctica," Rignot declared, "appears to be unstoppable."[17]

Holland has been to Antarctica seven times. For a long while, he was interested in a retreating glacier next to Thwaites called Pine Island. He spent ten years planning an expedition there, only to later realize that Thwaites was probably less stable, although both glaciers may in fact be in perilous condition. "That's an important glacier," Holland says of Pine Island, "but I think Thwaites is really the focus.

We picked the guy next door." The problem is that you can't just re-turn to these places on a whim for a closer look. The difficulty of get-ting to Thwaites—possibly the most remote glacier on earth, and a place only a few dozen human beings have ever walked upon—makes anything beyond remote sensing exceedingly difficult. In part, this is why predictions about its speed of decline contain large margins of error. "You cannot make a laboratory model for Thwaites," explains Holland. "And if you build a computer model, you have to make so many choices that the permutations are on a scale of a billion."

Holland thinks that we may not be able to predict the effects of Thwaites in time. The difficulty in gathering facts about it leads, in turn, to deficiencies in our computer models for ice sheets. "It's hard to imagine we can gain knowledge quick enough," he says. "If Thwaites was to go bad, and it collapses, I think we'll be observing it." Perhaps only then will we understand it. Meanwhile, in Jakobshavn, Holland has found a real-world model for collapse. He can get here in a few days and visit several times a year. He can watch every aspect of the ice and measure every aspect of the surrounding waters.

"I think this is a faith article—that Jakobshavn is an analog for Thwaites," he says one evening, walking out to the high rocky bluff where his radar is trained on the calving front. Still, he doesn't seem to be someone who acts on faith. He starts taking out his notebook from the front pocket in his cargo pants to jot down a reminder. The dust is kicking up a bit again. The calving front is quiet for now. "I don't have any other cards to play," he says.

Ice sheet river, summer 2006 (Sarah Das/Woods Hole)

16

Meltwater Season

On some days, Holland liked to say that he was working on "one small piece of the Greenland puzzle." And no doubt it was true. Walking through Kangerlussuaq airport on a spring or summer morning, there were now so many researchers coming and going that it seemed the Greenland puzzle was far larger than anyone had previously imagined. Long ago, Robert Peary had called the island's frozen regions "the eternal ice." But any reference to the island, in the press or on television, seemed to carry the warning that *Greenland is melting*.[1] Perhaps this surprised the general public; to those watching the big melt up close, any sense of astonishment about the ice sheet's demise had long since passed.

Many of the scientists coming to Greenland were there to gather observations about the ice and make computer models more accurate, so that the future could be projected with greater confidence. Some, like those in David Holland's group, were passing through airports on their way toward any one of the hundreds of glaciers along the island's coasts. Others, such as J. P. Steffensen and Dorthe Dahl-Jensen, were leading international teams drilling deep cores out of the center of the ice, both to better understand the ice sheet's flowing movements and to fine-tune their historical knowledge of Greenland's climate. On a sunny morning at Kangerlussuaq airport, you could sometimes see Steffensen, smoking a pipe and driving a front-end loader, bringing coolers of ancient ice cores to a plane heading back to Copenhagen.

But there were so many others, too, arriving for meltwater season— scientists mustering supplies to camp on the ice sheet and measure water pooling along its surface, for instance, or to gauge streams rushing into *moulins*, the deep holes in the ice sheet that bring water into secret plumbing channels thousands of feet below. In the local cafés, you could meet glaciologists, hydrologists, anthropologists, geomorphologists, and sedimentologists. Greenland was like Los Alamos in the 1940s, the small city in New Mexico that served briefly during World War II as the nexus for all subatomic knowledge, except here it was all about the science and impacts of lost ice. You might chat with marine biologists studying the survival habits of narwhals and polar bears in a changing habitat; or you might discuss ancient history with archaeologists who happened to be digging into the remains of Norse ruins in the southwest. The latter were racing to finish their work before the permafrost that preserved those remains melted and the artifacts deteriorated. If and when that happened, the historical clues to the Norse disappearance in the 1400s might vanish forever.

All of these research projects, conducted between 2014 and 2018, happened to coincide with the hottest years ever recorded on earth. But in Greenland, the rise in temperatures was even more acute than rising global averages implied. Much as climate models had predicted, the Arctic regions appeared to be warming at twice the rate of the rest of the world. In June 2016, a visitor could relax outside in a T-shirt and shorts in Nuuk, Greenland's capital city, previously known as God-thab, where it was warmer—75 degrees Fahrenheit—than in New York City. The winters in Greenland were becoming more temperate, too. At the National Oceanic and Atmospheric Administration—NOAA, the U.S. agency that monitors the weather and oceans—scientists summed up their annual observations for 2017 by noting that "the Arctic environmental system has reached a 'new normal.'" The region now appeared to be in a permanent state of decline, with thinner snow cover, steady reductions in the mass of ice in the Greenland ice sheet, and dramatic losses in sea ice in the Arctic Ocean. In fact, over the past three decades, the sea ice cover had been reduced in area by about half—and less ice, in turn, meant that darker and more open ocean waters could absorb more sunlight. That change would melt the ice even more, and thus bring even more heat into the region.[2]

It was a painful feedback loop. And it seemed that no place in the north would be immune to the seeping warmth. In February 2018, at Cape Morris Jessup, a weather station at the northern tip of Greenland, instruments registered daily temperatures 45 degrees Fahrenheit above normal. Seeing as the station was only four hundred miles from the North Pole and was shrouded in the constant dark of Arctic winter, the news left many of the world's meteorologists dumbstruck. "The Arctic is the world's cooling system," a Finnish official, Stefan Lindstrom, remarked not long after. "If we lose the Arctic, we lose the world."[3]

To experts on the region, the "new normal" had the air of an encroaching emergency.[4] The loss in Arctic sea ice and the jumps in Arctic temperatures were so drastic that it seemed reasonable to begin asking whether abrupt climate changes, observed in ice cores extracted from the center of Greenland's ice sheet in the early 1990s, were showing up again today. "If you look at the modern warming of the Arctic, in a five-year period from 2007 to 2012, we see a doubling of the length of the summer in the eastern Arctic, and that is equivalent to a 5° centigrade rise in temperature in less than five years," says Paul Mayewski, who had been in charge of the GISP-2 drilling project twenty-five years before. "There is no doubt that that is an abrupt climate change event." Mayewski pegged these shifts to significant alterations in atmospheric circulation. But he saw the trend as "the first trigger in a series" of climatic changes that would in time migrate to other areas of the Arctic.[5] In early 2018, a series of academic papers called attention to evidence that ocean circulation patterns in the North Atlantic Ocean around Greenland were slowing, too, with potential devastating effects. These included higher sea levels on the East Coast of the United States, alterations in seasonal weather patterns, disruptions in fisheries, and more intense storms. Climate models had long predicted the slowdown in circulation as a potential risk for a hotter planet. "I think it is happening," Stefan Rahmstorf, of Germany's Potsdam Institute for Climate Impact Research, told *The Washington Post.* "And I think it's bad news."[6]

In Greenland, many Inuit seemed to think they would be able to withstand the switch in weather and sea levels.[7] They explained to reporters that they had lived in the world's harshest environment for nearly a thousand years, and that they would endure far longer than the rest of us. The melting ice and shifting climate had been obvious to them for decades already. Harbors that had been iced in for six months a year—harbors once so impenetrable they had delayed Alfred Wegener for weeks and led indirectly to his death—were suddenly navigable through the year, which had the positive effect of boosting the local fishing industry. The retreating ice sheet had exposed areas of new land in the south that for a decade had been attracting the interests of mining companies, intent on digging for gemstones and rare earth metals. A few intrepid souls, also in the south, had been trying their hands at agriculture—growing strawberries, in one instance—in a country where the climate had historically destroyed the lives of those who could not hunt.

Still, the local traditions were in jeopardy. On an April afternoon near Thule, just a few hundred feet from Rasmussen and Freuchen's old trading station, a hunter named Sofus Alataq, dressed in a hooded sweatshirt and Adidas warm-up pants, came through the snow with his dog team. He lived in Qaanaaq, the northernmost town in Greenland, which had been set up by the U.S. Army during the construction of Thule Air Base in the 1950s. Alataq's sled was laden with clothing, rifles, and polar-bear furs, which he used for warmth in extreme cold. But in English, he said with disappointment that most of his hunting season had been for seal, not bears, and that recent changes to the climate had made sledding on the frozen bays and fjords more difficult. "Less ice, warmer weather, spring is sooner," he explained.[8]

In Ilulissat, the coastal town near Jakobshavn Glacier, another hunter from Qaanaaq, Toku Oskima, put it slightly differently. She explained that the changing weather had made it more difficult to preserve meat in caves; and every winter, the thickening of ice along the coasts happened later and slower, therefore limiting the days she could take her sled out. She seemed confident she would adapt. But Oskima said, "The warmth of the sun is getting heavier."[9]

Every year, about half of Greenland's ice losses happen on the edges, through glaciers like Jakobshavn. But another half is lost through melting on its surface.[10] This melting comprises turquoise lakes and rushing aquamarine rivers and thin lapis creeks. Unlike the island's biggest glaciers, we already know a fair amount about how much the ice sheet's surface is warming. For nearly thirty years, a Swiss scientist named Konrad "Koni" Steffen has been taking readings on temperature, wind, solar radiation, and melting at a station known as Swiss Camp, on the central ice sheet. Located about fifty miles east of Jakobshavn's calving front, Steffen's camp has weather towers that collect data on the surface environment several times a minute; the information is then transmitted to him in Europe and the United States (he has offices in both Zurich and Boulder) every hour. But the observations aren't limited to one location. Over the past few decades, he has set up a system of eighteen installations around Greenland that measure weather on the ice sheet. Every spring he checks on these towers by setting out from Swiss Camp, moving from one site to another either by snowmobile or by turboprop. "It is really by chance that I ended up studying the ice sheet," he says. In 1990, he was planning to conduct a study on glaciers in Tibet, but the Chinese government wanted to levy high charges on his science team. Steffen thought the costs were unreasonable, so he pulled out, and that led him to try Greenland. He built a camp that year on the ice sheet at an elevation of about thirty-eight hundred feet—several permanent, thickly insulated tents set on a large plywood platform. And then he began to monitor the environment.

Steffen is the modern era's equivalent of Carl Benson, the scientist who in the 1950s went driving around the ice in a weasel, digging snow pits to measure conditions on the ice sheet. "He is my model," Steffen remarks, noting that he met Benson in Alaska in the 1970s and sometimes consults his old papers from the Army Corps of Engineers, where Benson recorded accumulation rates and melting around the ice sheet. One difference in the work of the two men is that Steffen has reaped the benefits of GPS and electronic sensors. Another is that Steffen has experienced changes in the ice that Benson could not have imagined. In the early 1990s, Steffen recalls, "we almost lost the camp because we got too much precipitation." The tents were buried by snowfalls, so he had to enter through the roof. "It was only in the year 2000 that the surface started to go lower down," he says, but that year marked a steady and unrelenting downward trend. His plywood platform and tents collapsed several times as the ice sheet continued to melt and plummet in elevation. It was like going on vacation and returning home only to find that the floor of your bedroom had dropped by several feet, and that your possessions and bedding were now in shambles. In the wreckage of tents and equipment at Swiss Camp, Steffen was witnessing the same changes in ice sheet elevation that IceBridge and ICESat were measuring from above.

He and his students rebuilt the camp twice, driving foundational supports deep into the snow and ice to future-proof the platform. But in time, as the ice sheet continued to melt and thin underneath the station, the supports were exposed. If you happened to be flying low on NASA's IceBridge mission during the months before the summer heat turned the landscape to slush, Swiss Camp appeared like a tiny oil drilling platform, sitting atop stilts in a vast ocean of white. By 2017, Steffen's measurements indicated that the ice had declined by nearly forty feet at Swiss Camp since he arrived, and the trend lines for the future prompted him to put the camp up for sale for one dollar.[11] He didn't really expect a buyer. Even if someone wanted to, there was no easy way to get to Swiss Camp except to ski there, as Nansen might have done—or to fly there, in a helicopter that costs five thousand dollars per hour, from the nearby village of Ilulissat. So the sale price became a joke that Steffen liked to tell at lectures. He meanwhile began making plans to move the camp to higher elevations on the ice sheet—perhaps thirty or forty miles farther in, where it would be situated in a colder and more stable region.

At this point, it was hard to imagine Greenland without Swiss Camp. The station over time had become a destination spot for visiting dignitaries seeking to find a kind of ground zero for global warming; Al Gore had been there, along with so many journalists, politicians, and European princesses and princes that Steffen could barely list them all. The true value of Swiss Camp was in the growing record of observations, however. Steffen calculated that between 1990 and 2018 average temperatures on the ice sheet had increased by about 2.8 degrees Celsius, or 5 degrees Fahrenheit. Over the same time period, the total area of the Greenland ice sheet that was vulnerable to surface melting had increased by around 65 percent. Carl Benson would barely have recognized the place.

In talking about the melting in Greenland, glaciologists sometimes like to describe the relationship between rising temperatures and diminishing ice with mock simplicity: "When it gets hotter," they say, "ice melts." And yet the surface of the ice sheet has turned out to be far more complex than it once appeared. In recent years, studies have observed that meltwater from the ice sheet doesn't necessarily run off Greenland into the oceans; occasionally it stays in the firn, trapped like water in a huge sponge. In 2013, a team of scientists even discovered an aquifer under the snow in east Greenland, containing extraordinary reserves of water that covered an area the size of West Virginia. Conceivably, the water could be released quickly, in a flooding gush, in the near future.[12]

In sum, the ice surface is beset by a variety of forces brought on by changing climates. The surface is getting darker, for instance, thanks to a combination of industrial soot, dust, microorganisms, and algae, which have settled upon a "dark zone" on the ice sheet's western region. This dust, black carbon, and biological life now form an ecosystem that flourishes during the warmer months.[13] But the danger is that the darker Greenland's ice gets, the more solar energy it absorbs. And the more solar energy it absorbs, the more it melts and the darker it gets. This feedback loop is known by glaciologists as the *albedo-melt loop* (albedo is a measure of surface reflectivity).[14] And unfortunately, it is reinforced by another self-perpetuating process. As Greenland's ice sheet melts and drops in elevation, just as it has at Koni Steffen's Swiss Camp, its ice becomes more vulnerable. Lower altitudes enjoy warmer temperatures; they allow lakes to creep farther up the ice sheet with each passing decade. Therefore it now appears that the more Greenland melts, the more Greenland melts.

At its current pace of erosion, Greenland's ice sheet adds about one millimeter of water every year to the world's oceans; at this rate, the ice sheet could last seven thousand years. Yet no glaciologist in the world seems to think this will be the case. An astonishing study published in December 2018 concluded that the ice sheet was now melting faster than at any time for at least the past 350 years, and that the "nonlinear" response of the ice to warmer temperatures would lead to "rapid increases" in Greenland's sea-level contributions.[15] With its mellifluous, singsong Swiss accent, Koni Steffen's voice tends to soften the bleakness of his outlook. Yet in his reflective moments, he predicts the melting ice over time will lead to a painful migration of 300 to 500 million people, globally, away from the coast.[16] "Greenland isn't pausing at 2100," he says. "It continues like this, the warming. And it gets worse and worse. Most people think we can model until 2100, and that 2 degrees centigrade is not so bad. But it won't stop there. And the melting won't stop there. The curve gets steeper, and steeper, and steeper." One evening in Greenland, he says darkly: "There will be a change coming, and obviously a change that we have not seen in thousands of years."[17]

In 2012, a glaciologist at the University of Colorado named Tad Pfeffer wrote a magazine article entitled, "Glaciology Needs to Come Out of the Ivory Tower." His main point was that his profession had found itself in an unlikely situation: "What happens when a field of scientific inquiry that starts out as a subject motivated purely by curiosity and driven by the simple desire to understand how the world works, is turned rather suddenly into a field with applications of the most urgent nature?" Pfeffer observed that what he and his colleagues were now doing—trying to project what would happen to ice sheets, as well as how much (and how fast) sea levels could rise—was crucial to untold millions of people. With their research redounding far beyond obscure academic journals, glaciologists had now become "stewards of the land."[18]

To be sure, some glaciologists already understood the stakes. Their most difficult challenge was trying to come up with an accurate projection for the near term—that is, a number by which sea levels might rise in the next few decades—and to demonstrate confidence about how the ice sheet melt might also affect, say, ocean circulation patterns. Contrary to many assumptions, climate scientists don't hold the conviction that computer models of the future suggest concrete reality. Often, they like to quote a British statistician, George Box, who in 1978 remarked that "all models are wrong, but some are useful." In truth, many computer models—such as ones done in the late 1960s and early 1970s by Syukuro Manabe—have proven nearly correct about warming trends, even if they haven't shown the same precision about ice sheets. But the larger point is that good models, especially those that build into their codes the latest discoveries and observations, can provide a general sense of what can happen. And in any event, coastal planners, economists, politicians, and ordinary homeowners aren't usually curious about the mechanisms of calving glaciers or moulins; they don't tend to worry about dark algae blooms spreading like an illness over the westernmost regions of the ice sheet. But they do desire knowledge of what the world will be like in the future. For those reasons, the endpoint of the century—the year 2100—has increasingly become a benchmark for the climate community. It seems to offer a practical guideline for, say, engineers, and how they might design infrastructure and buildings. In a more conceptual sense, the date signifies the end of the lives of our youngest children and grandchildren. A computer projection of that year is therefore a sketch of a world that they will—most likely—live long enough to glimpse.

As far back as the early 2000s, Eric Rignot was thinking about the difficulty of predicting this kind of future scenario for the ice. Rignot had grown up in France, near the town of Chambon-sur-Lignon, in a rural, mountainous region of cold winters and crystalline air. Early on, he was captivated by stories of the Arctic, and in particular he loved Jules Verne's novel *The Adventures of Captain Hatteras*, the story of an intrepid polar explorer who travels to the North Pole only to find an ice-free region and a volcano. Published in 1866, the story predated not only Nansen and Peary, but other explorers who actually made it to the pole.

The contrast with Rignot's career is curious. His work involves the risk-taking of an explorer, and his accomplishments in research—with NASA's Jet Propulsion Lab and the University of California, Irvine—demonstrate an intention to make observations in some of the world's most treacherous places. He has a luxurious French accent but is not a romantic. Rignot's goals are to challenge assumptions about the world's ice sheets and to speak volubly of the dangers of collapse. "My long-term objective is to try to come up with ways we can put *upper bounds* on these changes—upper bounds on how fast the ice sheets could change, and how fast sea level could rise," he explains. "If we can get there and tell people, 'Look, it's going to be *very* hard for nature to go faster than this,' I think I'll be able to say we got our job done. Nobody will come back in fifty years and say, 'You guys were just so conservative and you did a big disservice to society by giving this conservative view on everything.' No, you actually pushed it, and reality maybe is going to fall a little bit below that."[19]

To find an upper bound for ice sheet collapse—that is, the worst scenario that current data and the laws of physics could allow—requires a combination of things. You need years of field observations, reams of remote sensing and geologic information, and a deep knowledge of feedback mechanisms. As Rignot points out, every glacier in Greenland and Antarctica is unique, and to predict their future behavior requires observing the individual characteristics—width, depth, bedrock characteristics, seawater exposure—of each. You need ample funding, the best computers and coders for modeling, and a willingness to stick to the objective for years. You need imagination, too, so as to contemplate things that we know happened in an era before the advent of recorded human history, when ice sheets collapsed and sea levels rose catastrophically.

For an upper bound, it might not help to think in a straight and logical progression; you have to be willing to think in terms of climate surprises, rapid jumps or unexpected slowdowns, *nonlinear* changes. You need to make a guess as to whether or not human society is capable of altering how much carbon dioxide it puts into the atmosphere over the next few decades. You also need to ponder whether melting permafrost, the frozen soil of the Arctic, will hasten the escape of methane, a gas far more intense in its heat-trapping characteristics than CO_2, and whether that will play a large role in influencing future weather. Ultimately, you are boiling down thousands of systems of immense complexity—systems of ice, water, rock, and air; systems that are natural as well as human—into a single data point of surpassing density, simplicity, and urgency.

For the calculations, the first thing to consider is how hot it will be in the future, since warmer temperatures will expand the oceans by a significant amount. This is not especially difficult, since a number of climate models already exist that have proven fairly accurate for different carbon emission scenarios. The second thing is to evaluate the fate of the world's melting inland glaciers—in the Alps, North America, South America, and the Himalayas. Those add to sea levels, too. A third thing involves a consideration of how much water the ice sheets will lose, on net, by melting and evaporation on their surface, a process that will likely be closely linked to future temperatures.

After that the calculations get much harder. Like David Holland, Rignot believes that ice sheets can lose large amounts of mass very quickly by way of big glaciers that end at the ocean. Also like Holland, Rignot doesn't think our models yet explain the future behavior of those glaciers. He, too, has long been worried about West Antarctica's Thwaites and Pine Island glaciers. In 1998 he published some of the earliest studies on their retreat—"I think at that point I realized, we are onto some big changes," he says—and fifteen years later he was among the first to declare the region might have reached a point of "unstoppable" collapse. In Rignot's view, "We've started a process here which is not easy to reverse." His assessment is that West Antarctica could—*could*—add several feet of sea level to projections for the year 2100. "I think what we're seeing in the ice sheets is a collapse," he says. "It's just that we don't realize the time scales. It's like looking at a glacier. Does it look like it's moving? Then you speed up things by twenty times, and it looks like a river. If you look at the changes in the ice sheet in the last twenty years you say, this is going on pretty fast, and it's only the beginning. So, what's next?"

The upper bound encompasses not only Antarctica but Greenland. In 2008, Rignot began renting a boat and going up and down the Greenland coast, threading in and out of the network of fjords where glaciers that flow from the ice sheet terminate at the water's edge. He was trying to assess what is known as the *bathymetry* of these regions— the depth of waters in the fjords and how ocean temperatures and salinity might accelerate the melting. The work was tedious, difficult, and not always safe. "It was unpleasant," Rignot acknowledges, adding that he also saw it as necessary, since it had never been done before. Much of the time he and his colleagues were pushing huge chunks of floating ice out of the way with poles so he could get good instrument readings. "At first we used a small fishing boat—not a very sturdy boat," he recalls. "We had a lot of problems with that boat. It was too old. Every year something bad was happening to it. Eventually I had a bit more money, and we got a bigger boat, and then we started mapping the sea floor." In time, the data led to the conclusion that the fjords often went much deeper than had been previously assumed, and that the glacier fronts were more vulnerable to melting and retreat. In conjunction with IceBridge data, a more detailed picture of Greenland's future came together for him. His vision for the island is often at odds with the consensus that Greenland will melt slowly into the future. As he sees it, the great danger is that warming ocean waters in the Arctic can melt glacier fronts far faster than warming air. This poses a threat to some of the largest glaciers in Greenland—"the big guns," as Rignot calls them, which include Jakobshavn on the west, and several others in the north. Like tunneling animals, these rivers of ice have over time gouged deep canyons, and these canyons go all the way back into the central basin of the ice sheet. That means that even as such glaciers retreat, warming waters will follow them, and as they retreat into the canyons, they may continue to melt.

"So, these are the weak points of Greenland," Rignot says. "And it's not like we're looking at the big taps opening up in the next century. They're going to be opening up in the next decade or so. And after that it's going to be really hard to slow it down."

There happens to be a consensus estimate for global sea level rise in 2100—between a half meter (1.6 feet) and a full meter (3.2 feet). The numbers were put forward by the IPCC, the Intergovernmental Panel on Climate Change, which is made up of leading scientists around the world. Rignot considers the IPCC numbers too conservative. At the moment, he thinks one and a half meters (or about five feet) of sea level rise is more likely, but also that the upper bound is not yet quantified and may be higher. This may not become clear until observations and models are improved over the next decade or two. "But with sea level rise, we're not talking about something that is going to happen down the line," Rignot remarks. "The changes are taking place right now, and some of these changes are *very* significant. And we did not expect them so soon. So, wake up. Sea level rise is real. It's three millimeters per year right now. And people say, 'Oh, that's only three centimeters per decade.' But you're teasing a giant here. And there is no red button to stop this. What is going to happen by 2100 with the ice sheets is, in my opinion, already locked in. There's not so much we're going to be able to do to change that."

Still, for someone who focuses on the most dire possibilities for the ice sheets, Rignot is not hopeless. He likes to point out that the world doesn't end in 2100. And he thinks it's possible we will slow the process of collapse in West Antarctica or Greenland, as long as we move to reduce carbon emissions in time. "I think it's probably in human nature that we're going to react once we have our back to the wall," he remarks. But even the "one-meter-plus" by 2100, which is not necessarily the upper bound and may be too low, is problematic. The United States alone has more than 88,000 miles of shoreline; roughly five million people and 2.6 million homes are situated less than four feet above high tide. "With one meter of sea level rise," Rignot says,

"San Francisco airport is under water, and Oakland airport is under water. So, San Francisco doesn't have an airport? What is the impact of not having an airport in San Francisco?"

It's unthinkable to him. "But the last time I checked," Rignot adds, "nobody was building an airport higher up."[20]

It may be the case that the collapse of some glaciers in West Antarctica and Greenland is unstoppable. But for the moment there is great uncertainty when it comes to the ice sheets. It may take one hundred years or three hundred years or five hundred years for some glaciers to fall into the ocean, due largely to the way they will break, sliding backward in the process, sometimes pausing for years or decades on a bedrock bump, as sea waters around and underneath them warm. In any event, amongst glaciologists there seems a consensus that the situation is urgent now, even if it isn't yet at the point of being catastrophic. "We're not positive if we've already triggered it or if we're really close," Richard Alley says of a West Antarctica collapse. "I think almost everyone would agree that it's either one or the other." But Alley also thinks that if we have committed to losing West Antarctica, "Greenland just became way more valuable."

This can require a bit of explanation. It's largely due to the fact that Greenland's ice sheet sits within a bowl and is ringed by mountains, just as Wegener's expedition discovered almost a century ago. Even with its deep glaciers, the losses of ice are constrained to a certain degree by the island's geography, and by the limits in how glaciers can only push ice into the sea by threading through narrow mountain passes. That isn't to say that Greenland couldn't contribute many feet of sea level rise over the next few centuries. But with Greenland, says Alley, "we have a little more leeway."[21]

The summer temperatures of the Arctic are already too warm for Greenland's ice to endure. "At some point we will stick our nose beyond the mean annual temperature that is survivable," Alley explains. And at that point it would be difficult to stop the melting and the feedback loops, as the ice sheet got increasingly thinner and lower in elevation—a criterion for collapse. "We may be pretty close to that;

or we may have a few degrees yet," he says. "But if we stuck the temperature up, and we pulled it back down in a very few decades, Greenland wouldn't have time to thin enough to destroy its ice." In terms of sea level rise, that might mean—in the United States at least—that we lose Miami but save New York. Alley says, "Rather than being the end of the world, West Antarctica would just be the huge wake-up call." Then humanity's attention would have to turn to saving Greenland's ice.

Of course, there remains a more difficult question: How is it possible to get society to "pull" temperatures down in a few decades, before it's too late? Nature might buy us a little time. Even in a steadily warming world, climate can vary over the course of a few years. Greenland sometimes is subject to the weather that recalls its distant past, including big snowfalls (in 2018) and cold summer snaps. These events can slightly replenish the ice sheet and can even lead to temporary advances in some glaciers. What's more, if ocean circulations change, and the warm water flowing from the tropics to Greenland decreases, colder water may slow the disintegration process of some outlet glaciers.

There may also be solutions that fall under the description of geo-engineering. We may develop technological means to remove atmospheric carbon in the future—sucking it in through machines, and then burying it, for instance. Applying such technologies could conceivably delay the worst effects of climate change, though so far doing so on a scale with any significant impact remains untested and unproven. Another idea, now in a testing phase, involves shading the earth, and thereby cooling it, by dispersing an umbrella of atmospheric sulfate particulates, so as to emulate a volcanic eruption.[22] Meanwhile, a few scientists have suggested a radical engineering project: damming the Jakobshavn ice stream—effectively building a wall in front of the glacier and blocking warm ocean water from reaching its deeper levels—so as to slow sea level rise.[23]

At the moment, however, the only permanent solution to impeding rising sea levels seems to be reducing carbon emissions by burning less and less fossil fuel. And even with the unceasing efforts of an expansive network of climate activists, so far the progress is slight.

The signing of an international climate agreement to limit atmospheric emissions, conducted in Paris in 2015, seemed a significant step in the right direction, but in 2017, CO_2 levels actually rose faster and higher than any time in recorded history. That same year, the United States—the world's largest emitter of greenhouse gases, along with China—declared under President Trump that it was "getting out" of the Paris Agreement. Somehow, the idea that the continued existence of the world's coastal cities will depend on collective and political actions over the next few years, and not the next century, has failed to spread wide enough, deep enough, or fast enough.

To meet the Paris Agreement goals, which attempt to limit average global warming to 2 degrees Celsius, "the wealthier parts of the world would need to be zero carbon energy by about 2035," Kevin Anderson, a climate scientist at the University of Manchester, explains. "And the poorer parts, including China, would have to deliver zero carbon energy by about 2050. And by that I mean *everything*—cars, planes, ships, industry, all of the energy would be zero carbon by 2050, globally, for us to have a reasonable chance of the two-degree framing of climate change." It's not that we lack the technological tools, Anderson says. It's that the enormity of the task, and the sacrifices involved, haven't yet sunk in. "I think it will be hugely challenging," he adds.[24]

And yet the alternative—not to try or achieve the goal—would be dire.

In reflecting on the work he did in Greenland in the early 1990s, Richard Alley, who toiled in the GISP-2 trench in the center of the ice sheet, sees several deeper meanings in the ice-coring project. "It demonstrated the absolute reality of abrupt climate change," he notes, which was immensely important to our scientific understanding of climate history. "By knowing about abrupt climate change, and by knowing about the craziest stuff that's out there, in some bizarre sense we have gained a much stronger understanding of how big an experiment we humans are conducting on the climate." Alley often lectures on how CO_2 is the immensely powerful "control knob" on our climate. These cores, he adds, and the climate changes registered therein, present an even stronger sense that in the steady climb of atmospheric CO_2, "what we are doing is really a big deal." The ultimate results of our actions, in other words, might not only be severe. They might be sudden, too.

Still, one conclusion Alley has drawn from the ice cores is not really about any scientific discovery. It's a note of historical optimism that we might keep in mind when we think about reducing carbon dioxide emissions. Amongst all the trace gases and chemicals analyzed in the Greenland ice cores, a clear residue of lead showed up in certain segments. A few thousand years ago, for instance, Alley notes that "you can see the little blip of the Romans." This would mean the residue of ancient smelters, in Spain and elsewhere, which the Romans used to burn ore to render silver. The process released lead into the air as a by-product, which eventually was deposited in snow that fell on Greenland. In more recent cores, Alley says, we can see lead traces from the fumes of the industrial revolution, which began in the late 1700s. And then eventually, in cores from the twentieth century, the unmistakable fingerprint from leaded gasoline comes through.

And yet, something interesting happens in the 1980s. Lead traces in the ice mostly disappear. "We turned it off. We cleaned it up," Alley says, pointing to the switch in automobiles to unleaded gasoline after lead was banned by environmental regulations. "And the world didn't end, and the economy didn't end. And you can't look back at economic data and find a horrible disaster that happened when we decided we didn't want to poison ourselves with lead."

He pauses and then adds, "It's so beautiful, so *clean* in the ice core records. And you can just see: This lead is human caused. And then you see: This is when humans decided that we didn't want to do that anymore."

Ice Watch, *Paris*, 2015 (Martin Argyroglo/Courtesy Olafur Eliasson)

EPILOGUE

The Ice Clock

Over the course of four consecutive years, I made a habit of returning to a spot in Greenland at the edge of the ice sheet, on the west coast not far from Kangerlussuaq airport. Back in the year 2001, the Volkswagen company decided to build a road that led east from the airport to the inland ice; the idea, apparently, was that engineers would use the road to drive automobiles to a track on the ice sheet for testing in extreme conditions. The experiment lasted only a few years. By 2005 or so, Volkswagen had stopped bringing cars to Greenland, but their road endured. In the time since, it became a means by which tourists could access the ice sheet. It likewise became an easy way for scientists doing research on the ice to drive over from

Kangerlussuaq. On some summer days, you could see a research team in town, stocking up at the grocery store and heading over to the ice sheet in a rented Toyota Hilux pickup.

The Kangerlussuaq road, at about twenty-two miles, is the longest in all of Greenland. It may well be among the most jarring, too, with a cinder surface now runnelled and potholed by the punishing winter weather. The scenery is nevertheless dazzling along the way—small mountains, crystal lakes, and carpets of purple flowers. If you're lucky, you'll also encounter skittery caribou and grazing musk ox. In the distance, the ice sheet looms. Seeing it for the first time, it is not difficult to understand why it became an obsession for so many explorers. Even from many miles away, it seems so massive and strange—"the mysterious desert," Nansen called it when he first encountered it from a ship on the east coast—as to effortlessly swallow a man or woman. To somehow master its difficult expanse must have seemed akin to achieving something superhuman.

Coming toward it from the west coast, the ice sheet gives an impression not of a desert but of an ocean—not only because it seems to capture the entire horizon, but because it is sculpted into hillocks and hollows, like a roiling sea on a day of serious weather. Sometimes, the ice sheet has also struck me as the photographic negative of an ocean. Rather than darkness streaked with white foam, it is lightness streaked with silt and dust.

Even over the course of a few years, I could see it thin and recede. With each succeeding visit, the road would end where it always ended, and we would park, jump out of a truck or van, and walk over enormous piles of gray stones and dirt—*moraines*, as they're called—that marked where the ice had once terminated. Every year, we had to cover a longer and longer distance to the edge of the ice sheet. Rene Nielsen, a local guide who drove me there several times, told me in 2017 that the ice had retreated by about a thousand feet in the time since he'd come to Greenland in 2001. More important, the ice had dropped in elevation by about a hundred feet, which meant that rather than climbing *up* onto the sheet, we now climbed a hill of dirt and stones and walked *down* onto it. The ice losses of such a large drop in elevation, spread over hundreds of square miles, must have been stupendous in terms of the water that had been shed. "Sometimes I can see it move back from one day to the next," Nielsen said.

His words made me think of the ice as a receding wave. And at the time, I was trying to work out why traveling to the ice sheet always gave me the uncanny sense of traveling to the sea. On my last visit, I recalled a conversation I'd had with James White, the University of Colorado scientist who had worked on Greenland's ice cores in the 1980s and 1990s. "One of the most underreported, misunderstood facts about the earth's history," he had told me, "is that the planet trades ocean water for ice—for ice on land—and it does it in magnitudes that would just shock the crap out of most people. It's meters, it's dozens of meters." For millions of years, in other words, the relationship between ice sheets and oceans has been mutual, intervolved, repetitive. Ice sheets went down; sea levels went up, sometimes drastically. And vice versa. As this glacier near Kangerlussuaq pulled back, it exposed stretches of dusty new land in Greenland. At the same time, I knew, somewhere else in the world, on some faraway coast, it was drowning old land with its runoff.

What struck me on each trip was not only the diminishment of the ice sheet, though. It was what the glacier had left behind. Certainly the new land was curious, lunar in its contours and colors. But the retreat of the ice meant I could sift through an extraordinary array of rocks and dirt, unseen for thousands of years, which the glacier had spit out in piles as it pulled back. All over Europe and Alaska, alpine glaciers were doing the same thing. But besides large and small stones, they were offering artifacts back to the world that had been trapped in ice—or, you might say, trapped in time—for hundreds or thousands of years. Glaciers in Switzerland and Norway were coughing up ancient wooden shovels, knives, arrows, shoes, and in some cases the leathery carcasses of men and women. Looking at the future trajectory of Greenland, its ice will eventually reveal its stores, too. Already, crashed planes from World War II had been recovered from the ice, and in a fairly well preserved state. But the residue of science—the lost diary of Wegener, as well as his body, and that of his colleague Rasmus Villumsen—would perhaps be revealed in time, too. One could imagine the rest. There would be harnesses from Rasmussen's sled dogs, and the rations that Peary mislaid during one of his miserable crossings. There would be the bones of dead ponies, Grauni and Red, from Wegener's 1913 crossing, as well as the theodolite—the prized navigation instrument—that Wegener's colleague J. P. Koch dropped when he fell into a deep crevasse and broke his leg. If the ice sheet continued to thin, the cache of goods left behind on the ice road to Eismitte, supplies that might have saved Wegener's life, would be revealed. And the retreat would kick back into the modern world the empty jerry cans of wine discarded by the French team at Station Centrale, perhaps the old Edith Piaf records, maybe even a rusted weasel or two.

Almost certainly, Camp Century would become unburied. In 2016, a team of scientists led by William Colgan, of the Geological Survey of Denmark, conducted a study on the future surface melt rates of Greenland. They concluded it was possible that the wastes of the camp would be revealed before the end of the century, as the melt areas crept farther up the ice sheet. What's more, even if the camp didn't melt out completely by then, the toxic remnants left behind might be spread about the region by meltwater percolating through the thinning ice. These wastes "were previously regarded as properly sequestered, or preserved for eternity," Colgan noted.[1] But his study showed that Camp Century would remind us again of what happened in Greenland during the Cold War. And when it did, its chemical, biological, and radioactive legacy would not only create a significant environmental problem, but a political one as well.

By that point—perhaps around the 2100 mark—it could already be the case that the world's coastal regions, especially those located in its poorest and low-lying regions, will be struggling with catastrophic floods and mass migrations. This will likely be true even if sea level rises do not reach an upper bound reality.

It's important not to minimize the impact of such a calamity. Residents of the world's richer coastal countries—and, in the United States and Europe, its richer cities—will have far more options for mitigation and relocation, and will likely not face the same kinds of crises. But one implication of ice sheet collapse reaches beyond any of the global impacts that are typically discussed in policy and environmental forums. In 2013 a team of scientists led by Anders Levermann of the Potsdam Institute in Germany conducted a group study on what they called "the multimillennial sea-level commitment of global warming." The paper raised the issue of something known in the climatology community as "commitments." Because the atmosphere and ice sheet system can be slow to respond, the authors made the case that our carbon emissions had already ensured a long-term sea level rise—over two thousand years—of 2.3 meters, or about 7.5 feet. What's more, for every further one-degree rise in Celsius (or 1.8 degree rise in Farenheit) within this current century, there would eventually be an additional jump in sea levels of about the same amount.[2] Based on projections for temperatures later this century, it therefore seemed plausible that society could soon commit itself to twenty feet of sea level rise or more, almost all of it coming from the "threshold behavior"—meaning the collapse—of Greenland and West Antarctica.[3]

The study raised a curious question: Should anyone care deeply about life on earth two thousand years from now? Perhaps the notion comprises an intellectual argument, rather than an emotional or practical one. Still, for a society that considers itself morally and cognitively evolved, it seems reasonable to think we might be willing to consider both the short-term and long-term results of our actions. In the Levermann study, it was clear that even if the year 2100 turned out to be manageable, the more distant future looked exceedingly grim, with many of the coastal cities in the world wiped out. This scenario seemed even more bleak in a study done a few years later, in February 2016. In this instance, a team of scientists led by Oregon State University climatologist Peter Clark collaborated on a paper that looked ten thousand years into the future. Clark and his colleagues argued that focusing only on what would happen in the next eight decades—up to the year 2100, in other words—was misguided. We needed to think in both human *and* geologic time scales, and a continuing focus on 2100 had created "the impression that human-caused climate change is a twenty-first-century problem, and that post-2100 changes are of secondary importance, or may be reversed with emissions reductions at that time."[4] Clark and his co-authors argued that this was hardly true; without some form of intervention some of our emissions would stay in the atmosphere for thousands of years and continue to push atmospheric warming toward devastating impacts. (The urgent and effective solution, they argued, was a rapid switch to clean energy systems.) "The next few decades," Clark and his colleagues wrote, "offer a brief window of opportunity to minimize large-scale and potentially catastrophic climate change that will extend longer than the entire history of human civilization thus far."

So we have very little time—a few years, maybe a few decades. And as we go about our daily business, the window is closing. By their nature, ice sheets have an asymmetric quality. They took hundreds of thousands or millions of years to form. But at this rate they could collapse—they *would* collapse—in a mere few hundred or few thousand years. In the meantime, large glaciers within those ice sheets, such as Jakobshavn in Greenland and Thwaites in Antarctica, could break down much faster.

One question, at least as I saw it, was whether confronting these long-term challenges required a scientific response or a political one. I imagined that they ultimately required both, but that any action first had to be preceded by some kind of factual and ethical awakening that had not yet occurred. James White told me he thinks that "in the long run, the challenge of sustainability is the challenge of us growing morally as a species, growing up as a species, becoming better human beings than we were a hundred years ago or two hundred years ago." The development of nuclear weapons holds a useful lesson—the realization that dangerous technological trends, even if they carry a degree of uncertainty about the outcomes, should lead us to safeguards that help us save ourselves. But it seemed hazy, at best, whether that historical achievement would somehow translate to actions toward averting the distant impacts from melting ice sheets. In general we seem disinclined to think too far—or too selflessly—as a species. "I ask historians this question," Kerry Emanuel, a professor of atmospheric science at MIT, remarked. "Can we find examples in human history of a whole generation consciously doing something for the benefit of more than one generation downstream that doesn't benefit that generation itself?"[5]

Emanuel concluded that it was difficult to find examples of such forethought. Indeed, after looking far and wide, he admitted that he had not yet discovered any.

If the exploration and investigation of Greenland offers an encouraging counterpoint, it may be in how it shows that in our quest to develop new technology, especially for the purpose of gathering new knowledge, we have rarely come up short. The discoveries of climate change have in many respects been the result of using science and technology (sensors, satellites, and ice core analysis, for instance) to understand the impacts that science and technology have wrought upon our environment. And it may be the case that our evolving tools, the ones now used for the collection of data and computer models, offer us insights that will push us beyond inaction and toward a more rapid engagement with the oncoming crisis. In this regard, I sometimes think of the moment when Wegener confided in Loewe on the journey to Eismitte, a few weeks before his death, that he held the sincere belief "that mankind will ultimately be liberated by the growth of knowledge." The question is: Was he prescient, or merely optimistic?

The spring of 2018 was gearing up much like any other in Greenland. The IceBridge flights were sweeping over the ice sheet, collecting reams of data. David Holland was getting ready to visit Jakobshavn again in June and Helheim Glacier in August. His seals and halibut were transmitting data from the glacier front to his rooftop in New York City, and his radar was still recording the calving front every minute. Holland told me his seals had already helped him gain several new insights into the ice sheets. He had seen a direct relationship between the water temperatures in the fjord and movement of the glacier—cold waters prompted an advance, warm a retreat. In spring of 2018, Holland was chosen as one of several dozen scientists in a combined, $50 million U.S.-British effort to investigate Thwaites Glacier in Antarctica over the next five years. It was a hugely ambitious scientific project. But to me it had the ring of an old-time expedition as well, involving profound risks, complex logistics, and a courageous pursuit of the unknown. Holland's plan was to drill a 2,600-foot hole in the ice shelf of Thwaites and send down robotic tools to investigate the properties deep below the ice; if he could understand the ocean characteristics and melting rates on the underside of the glacier, it was Holland's hope to gain a better sense of how quickly and drastically Thwaites could collapse.

In the meantime, Eric Rignot was gathering observations around the world to improve his models for an upper bound. He was still spending several weeks every summer going up and down the coasts of Greenland to map the depth of the island's fjords and gauge the vulnerability of coastal glaciers. Rignot's investigation into the ice was now being complemented by a large-scale NASA mission known as Oceans Melting Greenland, or OMG. The goal was to investigate the coastline with sensors and sonar so as to gain a deeper knowledge of the interactions between ice and ocean. "We are encircling the entire island with ocean measurement," Josh Willis, NASA's main investigator on OMG, told me as the mission was kicking off, adding that it was conceivable that any new insights would shed light on similar conditions affecting the massive glaciers—including Thwaites, the Doomsday Glacier—in West Antarctica. "This is a bit of a testing ground, a proving ground," Willis remarked. "But we want to get better at predicting: What are these ice sheets going to do in a warming world?"

Similar efforts were being mounted to solve the questions around surface melting. Koni Steffen's weather stations around the island were still gathering data every fifteen seconds, and so was a newer network of Danish and Greenlandic weather stations. The dark zone of snow and ice was being investigated many different ways, including by drones. There were countless other scientists at work, too. One day at Summit Camp, in the center of the ice sheet, I visited a group studying the formation of clouds and how they affected the melting on the ice sheet. On a different day, I visited an international team, led by Dorthe Dahl-Jensen and J. P. Steffensen, drilling the new ice core in the center of the ice sheet at a camp known as EGRIP. In the subglacial drilling trench, a core was coming up from snow that fell six thousand years ago. One of the goals here, Dahl-Jensen told me, was to understand the "ice stream" beneath the camp. The stream was, quite literally, a massive flowing river *within* the ice sheet that carried ice toward several big glaciers—several "big guns"—far away in the north.

So it seemed that the information was piling up, higher and higher; in sum it would inevitably make predictions better. And a continued push to examine the ice from high above was helping, too.[6] Thanks to remote-sensing efforts, I could wake up on a summer morning in Greenland or in the United States, and, after a cup of coffee, I could log on to Greenland Ice Sheet Today. On a Web page fed by remote-sensing data, wet areas on the Greenland ice sheet were shaded orange, and those crept farther in as the summer progressed. Meanwhile, the dry and cold region of the ice sheet, shrinking smaller and smaller in the warmest months, remained white.

There was also GRACE, the gravity satellite. In October 2017, after fifteen years in space, it ceased working and burned up in the atmosphere. But NASA and the German space agency had built two replacement units, which were meant to replicate the original mission almost exactly. These were successfully launched into space in late May 2018. The year before, I had gone to see the modules at a big industrial facility on the outskirts of Munich. I walked around and around the satellites—two lean rectangular blocks, raised on racks in a clean room so that engineers and scientists in white coats could finish working on their innards, an imbroglio of snaking wires and foil-wrapped cables. Frank Webb, the chief scientist for NASA on the project, took me through GRACE's interior part by part, and the complexity seemed difficult to grasp. There were accelerometers to measure the least bit of change in motion. There were a variety of communication links so the twins could communicate with each other and with earth. The list went on and on. "The original GRACE satellites were not nearly as dense in electronics as these are," Webb remarked, almost apologetically, adding that each of the twins had gained a few hundred pounds since the last version, largely because they now included a new, experimental laser system that was going to be tested on this flight.

For the next ten or fifteen years, GRACE would no doubt tell us how fast Greenland and Antarctica were melting, and I imagined the news would not be good. But here on earth they were mute. They could only speak from three hundred miles up, through their response to gravity, the pull of water, the changes we were forcing on the earth's ice.

In all my travels, there was often a lingering worry. If more knowledge, and more data, and more satellites, and more computer projections couldn't make us act quickly enough about the polar regions and the risks of the future, then what would? It was hard not to notice that around the world, during any given month, there were men and women trying other methods to translate the science of the Arctic, as well as the anxieties of climate change, into recognition. In 2014, for instance, an Icelandic-Danish artist named Olafur Eliasson brought twelve enormous chunks of ice from Greenland, weighing a total of 100 tons, and had them delivered to Copenhagen's city hall square. He arranged the chunks in the shape of a clock, and then allowed his ice clock to melt. He repeated the art installation the following year in Paris. Visitors were encouraged to walk around and touch it while it dripped away: Ice, but also time. "The 100 tons melting in Copenhagen is the amount that is melting from the Greenland Ice Cap in $\frac{1}{100}$ of a second," Eliasson explained.[7] Another perspective on the situation struck me a few years later. In Venice, the artist Lorenzo Quinn sculpted two giant white hands, weighing five thousand pounds each, reaching up from the city's Grand Canal. The hands were grasping desperately at a nearby building. Clearly, Quinn saw a flood was coming, that we were near to drowning. "Something has to be done," he explained.[8]

In June 2015, I attended a conference at the Hotel Arctic in Ilulissat—very close to the Jakobshavn ice fjord—which brought together dozens of leading scientists to discuss the changes in Greenland and the Arctic. There were scientists from all over the world in attendance. Koni Steffen gave a talk, and so did David Holland. Other scientists offered presentations on remote-sensing innovations, ice core drilling, the current state of knowledge about ice sheet losses, and various plans to improve projections in the future. In the context of the history of Greenland, the progress seemed miraculous. To be inclusive, the conference organizers had also invited two Icelandic artists, Anna Líndal and Bjarki Bragason, to attend, and they sometimes listened in on the science sessions. In the evenings, they worked down the hill from the hotel, in Ilulissat's small downtown, living and sleeping in a small art museum for a week.

The Ilulissat museum is set in an old frame house with a worn cedar-shingled roof. Located close to the harbor, it is painted a deep, rich red. The museum walls are hung floor to ceiling with oil paintings by Emanuel Petersen, a Danish artist who visited Greenland in the 1920s and was smitten by the landscape and the Inuit customs. His paintings strive for the picturesque—sunsets over fjords, for instance, and square-rigged sailing ships arriving in the harbor. In his oils you can see scenes of Inuit fishermen kayaking in the ice fjord and sledders commanding dog teams during polar bear hunts. They seemed to me timepieces from the era of Peary and Rasmussen and Freuchen—reminders of a Greenlandic world now almost entirely gone.

Bjarki Bragason told me he liked the Petersen paintings; in particular, he liked how they contrasted with his own art. He was keenly interested in "disappearance," as he put it. Not long before, he had come across a hunk of ice from one of Iceland's glaciers when he was leaving the cinema near his home in Reykjavík. As he recalled, the chunk was being used by scientists to teach students how to chemically analyze old ice. He carted the chunk home on his bike—it weighed more than sixty pounds—and then took one half and cast it in wall plaster. The ice melted soon after, but in the empty plaster he was left with the idea of what the ice once was, the space it took, the fact that it existed at all. He took the other half of the ice, melted it with a heat gun, and saved the wrinkled black paper underneath. He took photos of both, which were on the wall in front of us in Greenland.

Nearby was Anna Líndal's work. Among other things, she had filmed herself swimming amidst melting chunks of ice in a glacial lake, because she wanted to show that the act and process of measurement was not merely for scientists and technologists. Her body could be a tool for measurement, too. Later, she showed me a video she had shot of several people exploring a glacier, tied closely together by a rope. It was dizzying. The explorers moved ahead, tethered together, but she had edited the video into a loop that went around and around, so the ice sheet journey never actually got anywhere. "That's what this conference is," she told me, "this endless look for evidence for how tomorrow is."

It was not my inclination to think we had enough information or evidence already. Nor was it my tendency to think scientists should somehow—could somehow—stop in their quest to know more about the ice. But I could understand her point. We had come so far, and to a meaningful extent, we already knew where we stood.

After visiting the exhibition, I walked over to Jakobshavn's wide ice fjord, near Rasmussen's old house, where his father's church still stands. Down a small hill from the church, a sign at the water's edge warns those standing by that they should beware of icebergs. The reason is that even though the bergs have calved already from the Jakobshavn Glacier, they can split again and crumble as they move through the fjord. Those breaks can suddenly raise the tide with a cresting wave to the shoreline, and thus injure or drown someone looking out.

I stood by the edge of the water anyway. I was alone. It was late. It was a pleasant night, cold and clear and bright, with the Arctic sun still burning and a knife-sharp breeze blowing off the water. There were several massive icebergs out in the water, lined up, shaped not like Matterhorns but like long rectangular blocks, some the size of a skyscraper laid on end. They were standing in a slow, sculptural procession. It was hard to detect the movement as they floated out to sea, but the journey, I knew, could take a year or two from calving front to ocean. I closed one eye and held my thumb out, steady as I could, for several minutes; I was measuring myself against one of the icebergs, to gauge its movement. It was subtle, almost imperceptible. But it was there. And there really was no doubt. They were on their way.

ACKNOWLEDGMENTS

This first reason this book exists is because one morning many years ago I floated an idea to my literary agent, Sarah Burnes, at a coffeehouse in downtown Manhattan. "I'm thinking of writing about Greenland next," I confessed.

"Greenland?" she asked.

"Greenland," I said.

I assured her I was talking about the big island in the Arctic. Then I explained why it was worth writing about. Half an hour passed, by which point I was walking her to the subway station. She had stopped talking; she was apparently thinking it through. "Greenland," she said again, before heading down the stairs to her train—but this time I noticed it didn't have a question mark at the end. "That sounds interesting." And with that vote of confidence, this book actually began. I'm indebted to Sarah for countless instances of help and good judgment in the time since, but above all her willingness to support ideas that require years of research and are well outside the mainstream— first Bell Labs, the subject of an earlier book, and now Greenland's ice—has been extraordinary.

The second reason this book exists is because Andy Ward, my editor at Random House, took a chance on this project, and on me. Andy's light touch and insights are everywhere here, and I am profoundly grateful. He encouraged me to look deeper into the early period of arctic exploration. He helped me weave this book together in numerous ways that cover up its seams and spackle. And he has the unerring talent of finding the weak spots in my prose and arguments, and gently demanding improvement. Other writers should be so fortunate.

* * *

I can't imagine how this book could have come together without the support I received from a number of organizations. During the course of my research, I spent a year on fellowship at the Dorothy and Lewis B. Cullman Center, at the New York Public Library's main branch. For writers, I think it's probably akin to winning the Powerball lottery. For five days a week, eight hours a day, I would rummage through obscure texts about the Arctic, with a break for lunch with the other fellows. The conversations at our dining table that year veered between books, food, travel, and the darkening political mood of the country. My only regret from that remarkable experience was that it couldn't have continued indefinitely.

I was fortunate to receive help from the Alfred P. Sloan Foundation, which has been a boon to science writers and storytellers over the years. The Sloan grant allowed me to finish this book—more specifically, to make an additional trip to Greenland and complete my manuscript on time. For that I am deeply thankful.

At NASA, a number of people helped make my field trips possible, including the flights to Thule and my time on Operation IceBridge. Special thanks to Tom Wagner, John Sonntag, John Woods, Jefferson Beck, Mike Watkins, and the late Piers Sellers. Also in Greenland, the National Science Foundation brought me along on several flights to the center of the ice sheet. My gratitude goes to their team at KISS in Kangerlussuaq and to the scientists at Summit Station who towed me around on a Nansen sled. On a different day at the center of the ice sheet, the Danish team at EGRIP, including Dorthe Dahl-Jensen, was similarly welcoming. Dorthe also spent a long morning with me in Copenhagen, several months later, answering questions and giving me a personal tour of her ice core labs.

Much of my research involved archival digging far from the ice fields of Greenland. At the Army Corps of Engineers' CRREL facility in New Hampshire I spent many hours conducting interviews; and all through my research I depended on CRREL's librarians and their storehouse of ice-related publications. Thanks also to Bent Nielsen and the Danish Arctic Institute, where I spent several days reading through their collection, and to Genevieve LeMoine, at Bowdoin's Peary-MacMillan Arctic Museum, who took me through the collections and made useful research suggestions. At the NSF ice core lab in Colorado, Joan Fitzpatrick and Geoff Hargreaves spent time explaining their work and carting out ancient ice from the freezer. The U.S. National Archives was helpful in providing photographs for this book, as was the Alfred Wegener Institute and Danish Arctic Institute. So too were Sarah Das, a glaciologist whose stunning photograph of a meltwater river graces this book's title page; Joe MacGregor of NASA; the artists Peggy Weil and Olafur Eliasson; and Daphné Victor, who permitted me to use a striking photograph of her late father, Paul-Émile.

I owe a special debt of gratitude to several publications whose support allowed me to report parts of this book as I progressed on my larger journey. At *The New York Times Magazine* I wrote several lengthy stories that explored aspects of Greenland, climate change, and glaciology. My thanks to my longtime friend and editor at the magazine, Dean Robinson, for shepherding those complicated pieces through to publication—and always with insight, grace, and patience. Thanks also to Jake Silverstein, Bill Wasik, and Jessica Lustig for supporting my reporting over the past few years. At *The New Yorker*, Vera Titunik encouraged me to go to Thule and write about the trip, which allowed me experiences I would have missed otherwise.

At Random House, I owe a debt to the art and design departments that helped bring this book together with such elegance. Special thanks as well to Evan Camfield, whose production team saved me from a number of embarrassing errors, and to Chayenne Skeete, whose superior organizational skills and optimism kept the manuscript moving along. It made the long haul seem so much easier. My thanks, too, to those at the Gernert Company who have helped my books reach new readers, both at home and abroad.

In ways that are impossible for readers to discern, many people behind the scenes made this book better, smarter, and more accurate. Claire Wilcox translated dozens of pages of French texts relating to Paul-Émile Victor's time in Greenland; her work was such a big improvement on my own French translations that I am eternally grateful. Janni Andreasson, Peter Freuchen's Danish biographer, offered answers to questions that I couldn't find elsewhere. My daughter, Emelia Gertner, helped with a variety of research tasks and assembled this book's bibliography from a patchwork of digital files and the mess that I call my home office. And my sister, Patricia Stern, courageously volunteered to help me transcribe a number of interviews, even after my warnings that she might be daunted by the jargon—GISP-2? Dye-3?—of Greenland glaciology.

So many scientists contributed to my understanding of the ice that it is difficult to list them all. Still, a few stand out. As I began this project I visited Richard Alley at Penn State, who offered encouraging words while also warning me that I'd chosen a hard path. He was right, but I don't regret the journey, and my frequent interviews with Richard in the years since have been always illuminating and often delightful. I'm also grateful to J. P. Steffensen, who spent several days with me in Greenland and who patiently answered more emails than he probably had time for. David and Denise Holland kindly allowed me to camp with them for a week by the Jakobshavn glacier, and David allowed me to ask a billion questions while providing all the Nutella I could eat. Kathryn Snell, at the University of Colorado, graciously helped me to be more precise in my explanation of stable isotopes. And Kendrick Taylor assisted enormously by reading a final draft of this book and correcting my scientific understanding in a number of crucial ways.

My appreciation goes out to friends and relatives whose curiosity about my work boosted my resolve to see this project through. Thanks in particular to Jim Kearns and Susan Panepento; Donna and Pete Boyd; Dave Henehan and Kyndaron Reinier; J. P. Olsen; Laurie Weber; William Green; Mike and Marcelle Doheny; the Columbus-Memorial Day Group (Adam and Danielle, April and John, Sean and Cathy, Tom and Kathy); Dave Allegra, Marc Aronson and Marina Budhos, and my friends in Copenhagen, Peter and Marianne Luke. Thanks also to my mother, Doreen; to my sister, Tish, and her husband, Dave Stern; to my brother, David Gertner; to my in-laws Roz and Henry Weinstein; to Laura Weinstein and Mark Lorenzen; and to Joe and Adrienne Weinstein.

This brings me, at last, to the largest debts of all. Even with my frequent trips away from home, my kids, Emmy and Ben, supported my writing, texted me on my travels, and did not—apparently—hold a grudge about my excess of work and stress. I owe them my eternal love and gratitude, as well as a trip to Greenland. Meanwhile, the last time I tried to thank my wife, Lizzie, I ran out of things to say. And that seems to be the case again. When I think of her I wonder: How did I get so lucky?

NOTES

Introduction

1. According to PROMICE—Programme for Monitoring of the Greenland Ice Sheet, operated mainly by the Geological Survey of Denmark and Greenland—the ice mass amounts to 2.7×10^{18} kilograms. I've converted it into English units.

2. This classifies Australia as a continent, rather than an island. Australia is larger than Greenland.

3. I'm indebted to Barry Lopez for this insight: "The land in some places is truly empty; in other places it is only apparently empty." Lopez, *Arctic Dreams* (New York: Charles Scribner's Sons, 1986), p. 383.

4. Ernst Sorge, *With Plane, Boat and Camera in Greenland: An Account of the Universal Dr. Fanck Greenland Expedition* (London: Hurst & Blackett, 1935), p. 88.

5. Lopez, *Arctic Dreams*, p. 12.

6. Jean Malaurie, *The Last Kings of Thule: With the Polar Eskimos, As They Face Their Destiny* (New York: E. P. Dutton, 1982), pp. 67–71.

7. Although the oldest ice recovered from Greenland's depths goes back about 150,000 years, there is growing agreement that the oldest parts of the ice sheet date back much further. The Danish glaciologist Dorthe Dahl-Jensen (author interview, Copenhagen, November 6, 2017) explains, "We say that ice has probably covered central Greenland for one million years." She also thinks it possible that glacial ice exists in Greenland's eastern mountains that may be far older—perhaps dating back seven million years.

8. Seth Borenstein, "Fast Melting Arctic Sign of Bad Global Warming" (Associated Press, August 14, 2017). As Borenstein also points out, the goal of keeping global warming to 2 degrees Celsius—a tenet of the 2015 international Paris Agreement—is already irrelevant in the Arc-

tic: "Last year [2016] the Arctic Circle was about 3.6°C (6.5°F) warmer than normal." The weather and circulation disruptions stem from the melting of Arctic sea ice as well as land ice on Greenland. For further reference, see Jon Gertner, "Does the Disappearance of Sea Ice Matter?" (*The New York Times Magazine*, July 29, 2016).

1. The Scheme of a Lunatic

1. For instance, the map that accompanied the publication of the second edition of Hans Egede's *Description of Greenland*, ca. 1818.
2. During most of Nansen's life, Oslo was known as Kristiania or Christiania. In 1925 it was renamed Oslo, its traditional Norwegian name.
3. The *Viking*, like many sealing ships of the era, was rigged for sails but also had a powerful engine. Sometimes, a crew would try more active methods to forge their way through the ice pack. They could try to cut channels into the ice by hand or with explosives, or ram the boat to and fro (as long as they had a reinforced hull). Sometimes, too, a crew could "roll" the ship out of the ice by running from one side of the deck to the other.
4. Fridtjof Nansen, *The First Crossing of Greenland*, volume 1 (Unabridged) (London: Longmans, Green, and Co., 1890), p. 289. Nansen's notes show that the *Viking* was caught in the ice between June 25 and July 17, 1882, between 66 and 67 degrees latitude. "We believed ourselves to be within some twenty-five miles of shore."
5. Ibid., p. xii.
6. Børge Fristrup, *The Greenland Ice Cap* (Seattle: University of Washington Press, 1967), p. 44. Fristrup cites the *Speculum Regale*, known in Norwegian as *Kongsspegelen*.
7. The famous—and possibly unreliable—passage from Íslendingabók, "the Book of the Icelanders," reads: "He gave the land a name, and called it Greenland, arguing that men would go there if the land had a good name." From Andrew J. Dugmore, Christian Keller, and Thomas H. McGovern, "Norse Greenland Settlement: Reflections on Climate Change, Trade, and the Contrasting Fates of Human Settlements in the North Atlantic Islands" (*Arctic Anthropology* 44, no. 1), 2007.
8. Umberto Albarella et al., *The Oxford Handbook of Zooarchaeology* (New York: Oxford University Press, 2017).

9. From Greenland, the Norse, following Leif Eriksson (the son of Erik the Red), likely launched the first European settlement of North America. The evidence is located in L'Anse Aux Meadows, in Newfoundland, Canada.

10. Gwyn Jones, *The Norse Atlantic Saga* (New York: Oxford University Press, 1986). Jones notes that Greenland's (and Iceland's) acquiescence to Norwegian rule around 1262 probably involved a guarantee of regular trade ships. "For the first few decades one ship, safeguarded by royal monopoly, made the Greenland run at frequent intervals, though apparently not every year. This was the *Groenlands knorr*, the Greenland carrier; but she seems not to have been replaced after her loss in 1367 or 1369. Thereafter communications were scanty."

11. In his memoir of his missionary work in Greenland, Hans Egede suggests several reasons for the change, but "chiefly by the great difficulty and innumerable dangers of such navigation." Egede, *A Description of Greenland* (London: T. and J. Allman, 1818), p. 10.

12. Greenland's cultural prehistory is now traced back through archaeological remains to about 2500 B.C. when two cultures, the Independence and Saqqaq, arrived from Arctic Canada. Both died out—as did the Dorset people, who came later in two waves (800 B.C. to A.D. 1; and A.D. 750 to 1300). It is possible, perhaps likely, that the Dorset people interacted with the Norse. As far as is known, the Thule (or Inuit) culture, which likely arrived around A.D. 1200, did not intermingle with the Dorset. Modern-day Greenlanders are descendants of the Thule. See, for instance, *Nomination of Aasivissuit-Nipisat, Inuit Hunting Ground between Ice and Sea*, a 2017 book-length nomination for the inclusion of a large swath of ice-free land in west Greenland on the UNESCO World Heritage List, p. 38.

13. Tim Folger, "Viking Weather," *National Geographic*, June 2010.

14. The demise of the Greenland Norse, which is tangential to the main theme of this book, remains a puzzle with thousands of clues. For the past few decades, the extinguishment of the Norse has been attributed to climate change—that is, to a *drop* in local temperatures during the early 1400s and to increased storms—as well as to their failure to adapt their European habits to resemble those of the more resilient Inuit. Jared Diamond made this case in detail in his 2005 book *Collapse*. But more

up-to-date archeological research, especially work being done under the supervision of Thomas McGovern (author interview, Lambertville, N.J., October 26, 2015) suggests the Greenland Norse were highly adaptable (switching their diets from meat to seafood, for example) and resilient. Indeed, a number of recent academic analyses argue that the Norse's disappearance may be more complex than previously thought, and perhaps an instance of "near-miss sustainability" that had more to do with their unsteady trade relationship with Europe and, possibly, the residual effects of the black plague on Norway's population, which interrupted trade with Greenland. There is also a question of whether the Norse suffered an extinction event (circa 1420) while in Greenland—or whether they instead slowly and systematically immigrated to Scandinavia when conditions became too difficult. Two recent articles in the mainstream press—see, "Why Did Greenland's Vikings Vanish?" (Tim Folger, *Smithsonian*, March 2017) and "Why Did Greenland's Vikings Disappear?" (Eli Kintisch, *Science*, November 10, 2016)—consider both possibilities. While some academics make the case that an exodus from Greenland was more likely, McGovern believes that the Norse suffered some kind of devastating event. My own sense is that the latter argument is for the moment more persuasive. Otherwise it seems difficult to see why Egede and generations of Danish and Norwegian officials would pursue the "rescue" of the Norse. How is it possible they would not have been aware of a large wave of immigration of the Norse *back* to Scandinavia and the effective shutdown of a colony that had been a valued trading partner for so long?

15. The chronicle of exploration of Greenland in the 1500s and 1600s is long and complex and somewhat tangential to the discovery and exploration of the ice sheet. Not all the visits were commissioned by the Scandinavians. For instance, in 1577, the English seaman and privateer Martin Frobisher visited the west coast of Greenland, and in the centuries afterward, a steady increase in fishing, whaling, and sealing traffic led to a heightened exploration of the Greenland coast and country.

16. Nansen, *The First Crossing of Greenland*, volume 1, p. 280. Nansen's later book, *In Arctic Mists*, also details the history of sea voyages in the Arctic, going back to the centuries preceding the Norse colonies in Greenland.

17. Louis Bobé, *Hans Egede: Colonizer and Missionary of Greenland* (Copenhagen: Rosenkilde and Bagger, 1952).

18. In the eighteenth and nineteenth centuries Greenland and Iceland were officially "dependencies" of Norway, which at the time was joined with Denmark in a union that observed the Danish king. This changed after the Treaty of Kiel (1821) when Norway and Denmark severed their link and Norway entered into a union that instead observed the Swedish king, Charles XIII. At that point, Greenland and Iceland were ceded to Denmark. Except for a brief period during World War II, Greenland remained a colony of Denmark until 1953, at which point it became a *province* of Denmark. The process toward autonomy and home rule has since followed a progression of steps—adopting Greenlandic as the official language, for instance, and renaming towns and cities in the original Inuit. In 1979 Greenland established its own parliament. And in 2009, it passed a home rule referendum that will in time reduce its financial subsidies from the Danish government and put it on a path toward full economic and political independence.

19. Egede, *A Description of Greenland*, p. xcvii. The "Sketch of the Life of Hans Egede" in the introductory pages is uncredited.

20. The ruins of the church at Hvalsey have been preserved. In 1723 Egede traveled with about a hundred natives to Kakortok Fjord to see it. Bobé, *Hans Egede: Colonizer and Missionary of Greenland.*

21. Egede, apparently, was led by the Inuit to the ruins of the Eastern Settlement, but he thought it was the Western Settlement because of its location on the west coast. Rather than trying to cross the ice sheet to look for the Eastern Settlement, he would have done better to go north by two hundred miles or so to search out the ruins of the actual Western Settlement. Meanwhile, he continued to believe the Eastern Settlement was alive. He writes in his memoir: ". . . from many tokens and remainder of probable evidence it may be inferred that the old colony of the Eastern district is not yet quite extinct." Egede, *A Description of Greenland*, p. 14.

22. Ibid., p. 36.

23. In 1829, this method of getting from the west coast to the east was attempted by Lieutenant V. A. Graah, who set out with "two kayaks and two umiaks, the latter manned by Eskimo women rowers." He reached

65° N before being stopped by coastal ice. He wintered on the east coast before returning home. See Jeannette Mirsky, *To the Arctic!: The Story of Northern Exploration from Earliest Times to the Present* (New York: Alfred A. Knopf, 1948).

24. Nansen, *The First Crossing of Greenland*, volume 1, p. 466.

25. In maps from the late 1700s—for instance, "Carte du Groenland, dressée et Gravée par Laurent, 1770s"—the Eastern Settlement and Erik the Red's farm were still marked on the east coast of Greenland. By the mid-1800s, however, it was clear to officials and mapmakers that the Eastern Settlement had been located on the west coast, not the east.

26. Henry (Hinrich) Rink, *Danish Greenland: Its People and Its Products* (London: Henry S. King & Co., 1877), pp. 42–43.

27. Nansen, *The First Crossing of Greenland*, volume 1, p. 468.

28. Egede, *Description of Greenland*, p. 223.

29. Nansen, *The First Crossing of Greenland*, volume 1, p. 144.

30. E. E. Reynolds, *Nansen* (London: Geoffrey Bles, 1932), p. 24.

31. Ibid., p. 20.

32. Ibid., p. 25.

33. L. P. Kirwan, *A History of Polar Exploration* (New York: W. W. Norton & Company, 1959).

34. Nansen actually only brought one wooden boat for his crew; the captain of the *Jason*, meanwhile, offered Nansen the use of one of his ship's small wooden boats that were used for hunting seals. Knowing that two boats would make the journey to shore easier, Nansen accepted the offer.

35. W. C. Brogger and Nordahl Rolfsen, *Nansen in the Frozen World* (Philadelphia: A. J. Holman & Co., 1897), p. 52.

36. Nansen, *The First Crossing of Greenland*, volume 1, p. 166.

37. Ibid., p. 222.

38. Ibid., p. 226.

39. Ibid., p. 230.

40. As Nansen recounted it: "I meant to use it to help us with our boats and baggage in the floes, and if we could get it so far, on the way up on to the Inland ice. I was not sanguine that it would be of much use to us, but when we were obliged to kill it it would give us many a meal of good fresh meat." Nansen, *The First Crossing of Greenland*, volume 1, pp. 139–40.

2. Hauling

1. Nansen, *The First Crossing of Greenland*, volume 1, p. 442.
2. Ibid., p. 251.
3. Ibid., p. 276.
4. Ibid., p. 342.
5. Significant aspects of Agassiz's ice-age theory were based upon the work of geologists who formulated parts of the glacial theory earlier: Ignaz Venetz, Jean de Charpentier, Karl Schimper.
6. Nansen writes of Agassiz's work in the 1840s: "The necessity of more extensive and elaborate investigations at the only accessible spot where the glacial forces were now actually at work on the largest scale was soon made plain, and there consequently followed a new series of attempts to penetrate to the interior of Greenland." Nansen, *The First Crossing of Greenland*, volume 1, p. 470.
7. Jamie Woodward, *The Ice Age: A Very Short Introduction* (Oxford, U.K.: Oxford University Press, 2014), p. 25.
8. Mirsky, *To the Arctic!: The Story of Northern Exploration from Earliest Times to the Present*, p. 251.
9. Ibid., p. 269.
10. Ibid., p. 291.
11. Ibid., p. 292.
12. The history of pemmican, which apparently derives from Native American origins, is detailed in a number of books and is also the subject of some historical explication by Josephine Peary, Robert Peary's wife, in *My Arctic Journal*, her account of the first Peary expedition. In *The Fat of the Land* (New York: The Macmillan Company, 1960), the writer and explorer Vilhjalmur Stefansson traces the first reference to the food to a European account of the 1540 Coronado expedition to the American Southwest. He also notes that "the extreme supporters of pemmican recommend it as the most concentrated food known to man, or possible within the modern concepts of physiological and chemical science." While this seems debatable, the appeal of pemmican to nineteenth- and early-twentieth-century polar explorers, who valued imperishable high-calorie foods, is historically obvious and undeniable. Though ingredients can vary, pemmican usually consists of thin strips of beef, bison, or venison, dried and then pulverized, which are mixed with suet and berries

or raisins. Nansen apparently underestimated the percentage of fat that should be mixed into the pemmican he ordered for his expedition; this explains his men's rampant desire for butter and their constant, gnawing hunger.

13. Nansen, *The First Crossing of Greenland*, volume 1, p. 300.

14. Ibid., p. 343.

15. Nansen, *The First Crossing of Greenland*, volume 2, p. 350.

3. Simple and Easy

1. Baumann also informed Nansen—much to Nansen's surprise—that he had been awarded his science doctorate while he was on his expedition.

2. The *Fox* was bringing workers home from a mine that harvested cryolite—once a vital mineral for smelting aluminum—in southern Greenland. In his note carried by the Inuit kayakers, Nansen also asked the captain of the *Fox* whether he could come to Godthab to retrieve him and his team, a request that was politely rejected, owing to the lateness of the year and the risk of delays.

3. The London *Times* published Nansen's translated note on November 14, 1888. But in his book *The First Crossing of Greenland*, Nansen puts the precise arrival of news of his crossing as November 9. Various Peary biographies—those by Hobbs and Weems, for instance—place Peary's awareness of Nansen's triumph in September. But September 1888 would be impossible, as would September 1889, seeing as Peary was well aware of the crossing (see note 8, below) by January 1889.

4. Marie Peary Stafford, *Discoverer of the North Pole: The Story of Robert E. Peary* (New York: William Morrow & Company, 1959), p. 78.

5. There was also the possibility of going from the northwest coast to the northeast coast, and then to the east coast, before heading back to the northwest coast—thus making a triangular journey across the ice sheet.

6. Robert Edwin Peary, *Northward Over the "Great Ice,"* volume 1 (London: Methuen & Co., 1898).

7. *New-York Tribune*, December 7, 1888, p. 6.

8. *New-York Tribune*, February 18, 1889, p. 3. Peary wrote the note on January 30, 1889.

9. Over the course of 1891, Peary secured enough funding from scientific societies and wealthy patrons to reserve a ship for the trip north. Meanwhile, Nansen's account of his ice sheet crossing, an engrossing book of more than a thousand pages in length, was published. "I bought Nansen's book while in New York recently," Peary wrote his mother, "and we have been reading it. It is a pretentious affair in two thick volumes, with numerous illustrations and maps." Peary asserted incorrectly that Nansen had "profited much by my experience"—meaning the hundred-mile excursion onto the ice sheet that Peary had taken in 1886. Peary's irritation might have been due to the fact that in the book, Nansen questioned Peary's measurement methods for the hundred-mile excursion and thought it possible that he had not traveled quite so far into Greenland's interior. "The distance of a hundred miles from the margin of the ice cannot, therefore, be considered as established beyond all doubt," Nansen wrote.

10. Frederick Cook, "Dr. Cook, 'The Prince of Liars,' Writes from His Prison Cell," New York American, July 19, 1925.

11. "Mrs. Peary and Her Furs," The New York Times, March 14, 1897.

12. "Going to Greenland: Civil Engineer Peary Ambitious to Find the North Pole," The New York Times, February 27, 1891.

13. Robert Peary, letter to Langdon Gibson, March 8, 1891. Bowdoin College archives.

14. Peary, Northward Over the "Great Ice," volume 1, p. 43.

15. Chauncey C. Loomis, Weird and Tragic Shores: The Story of Charles Francis Hall, Explorer (New York: Alfred A. Knopf, 1971), p. 3.

16. Peary, Northward Over the "Great Ice," volume 1, p. lxxviii.

17. John Edward Weems, Peary: The Explorer and the Man (London: Eyre & Spottiswoode, 1967), p. 84.

18. After Peary had lost the toes, with "the bones . . . protruding through the raw stumps on both feet," he lay immobilized and in agony on a bunk in a long-abandoned military fort he had located on Ellesmere Island, an uninhabited island so frigid, and so far north, that it would not be fully explored until 2011. In a story he later liked to tell audiences (a story that was probably apocryphal), he scrawled in pencil on the wall next to his bunk a quote in Latin from the Roman philosopher Seneca:

Inveniam viam aut faciam—I shall find a way or make one. In later years, newspaper reporters would dutifully report on Peary's robust manner and "erect" carriage, but due to the loss of the toes (a fact never shared with journalists, lest the handicap be viewed as disqualifying him from further adventures) he walked in almost constant discomfort, and sometimes searing pain—"hot, aching, and throbbing," he wrote in his journals—and with a strange, gliding gait.

19. William Herbert Hobbs, *Peary* (New York: The Macmillan Company, 1936), p. 11.

20. Ibid., p. 20.

21. An entry in his college diary is typical: "This past week since Monday I have worked from five in the morning until eight at night." Ibid., p. 21.

22. Ibid., p. 45.

23. David McCullough, *The Path Between the Seas* (New York: Simon & Schuster, 1977), p. 327.

24. Weems, *Peary: The Explorer and the Man*, p. 64.

25. Aniceto Garcia Menocal, *Report of the U.S. Nicaragua Surveying Party, 1885* (Government Printing Office, 1886), p. 14.

26. Weems, *Peary: The Explorer and the Man*, p. 66.

27. Menocal, *Report of the U.S. Nicaragua Surveying Party*. "On the 19th, Mr. Peary suggested to be allowed to return to the 'divide' with three men . . . to this I readily consented, provided he would return in time to take the steamer of the 27th."

28. Peary, *Northward Over the "Great Ice,"* volume 1, p. 65.

29. Josephine Diebitsch-Peary, *My Arctic Journal: A Year Among Ice-Fields and Eskimos* (New York: The Contemporary Publishing Company, 1894), p. 24.

30. The doctor was Frederick Cook, Peary's future archrival.

31. Robert N. Keely, Jr. and G. G. Davis, *In Arctic Seas: The Voyage of the "Kite" with the Peary Expedition* (Philadelphia: The Thompson Publishing Co., 1883), p. 92.

32. Peary, *Northward Over the "Great Ice,"* volume 1, p. 52.

33. J. Diebitsch-Peary, *My Arctic Journal*, p. 77.

34. Peary wrote an entire book, entitled *Secrets of Polar Travel*, on the "Peary System."

35. *New-York Tribune*, January 24, 1897. He was not the first Westerner to appropriate Inuit garb and customs. "Peary had taken the best ideas of those explorers who had gone before, and invariably improved upon them," polar traveler and Peary biographer Wally Herbert also noted.

36. They are now considered Inuit, but are more properly called the Inughiut, a title that gives them a regional distinction from the Inuit. In this book they will be referred to as Polar Eskimo and Polar Inuit.

37. J. Diebitsch-Peary, *My Arctic Journal*, p. 41.

38. Peary, *Northward Over the "Great Ice,"* volume 1, p. 483.

39. The obvious analogy to Peary's behavior might be the character of Kurtz, in Joseph Conrad's *Heart of Darkness*.

40. Peary, *Northward Over the "Great Ice,"* volume 1, p. 481.

41. Donald MacMillan, a protégé of Peary's, later wrote: "Eighty pounds to a dog is a good load for the average sledging surface encountered on a long spring trip . . . Given the hard surface of a fiord, and my ten dogs could easily pull two thousand pounds." But obstructions or rough ice could make the pulling more arduous. "Therefore, the question as to how much dogs can pull is a difficult one to answer, depending upon the qualities of the sledge, upon the distance to be traveled, upon the strength of the driver, upon the strength of the dogs, and again and always upon the sledging surface." *Four Years in the White North* (New York: The Medici Society of America, 1925), p. 45.

42. Eivind Astrup, *With Peary Near the Pole* (Philadelphia: J. B. Lippincott Company, 1899).

43. Peary, *Northward Over the "Great Ice,"* volume 1, p. 301.

44. Sheila Nickerson, *Harnessed to the Pole: Sledge Dogs in Service to American Explorers of the Arctic, 1853–1909* (Fairbanks: University of Alaska Press, 2014), p. 20. The fantail system is different from the two-by-two dogsledding methods more common in the present day. Nickerson writes: "The fan hitch system, attaching each dog to the sledge by a separate tugline, enables the dogs more room to maneuver around obstacles such as upheavals of rough ice and is used in open, treeless areas." The biggest problem to this approach is the frequent tangling of the dogs' traces, a problem that bedeviled Peary and his men. MacMillan would write: "The inevitable braiding of the traces into a rope as large as

one's arm [requires] the untangling of which at low temperature neces-
sitates hours and hours of extreme discomfort." MacMillan, *Four Years in
the White North*, p. 44.

45. A daily average is used because the dogs are often fed every other day or
every third day. Peter Freuchen would later write: "Humans and dogs eat
exactly the same things, but the amount consumed by the natives is
trifling compared with what the dogs demand." Peter Freuchen, *Arctic
Adventure: My Life in the Frozen North* (New York: Farrar & Rinehart,
1935), p. 41. In *Greenland by the Polar Sea* (London: William Heinemann, 1921) Knud Rasmussen pointed out that "walrus meat is excellent food for the dogs, but it has the great drawback of containing 65–70
percent of water. This makes it very heavy for transport, and whilst one
can reckon a pound of pemmican a day for each dog, one must reckon of
walrus meat or skin about three pounds a day, or from five to six pounds
every second day" (p. 41).

46. It wasn't only Peary who was compelled to act on this idea; Inuit travelers would often do so, too. On the Second Thule Expedition, for instance, Knud Rasmussen slaughtered many of his sled dogs and fed them
to the rest of the pack. Noting their good service to him and their loy-
alty, he did so reluctantly—as did Peary. Rasmussen wrote: "It is disgust-
ing work, fit only for an executioner's assistant, to flense these animals."
Rasmussen, *Greenland by the Polar Sea*, p. 115.

4. North by Northeast

1. Robert Peary, *Northward Over the "Great Ice,"* volume 2, p. 365.
2. The work typified Peary's unapologetic exploitation of the Inuit. "The
women had never heard of an eight-hour [work day] law," Peary wrote,
"and cheerfully acquiesced when our necessities required them to sew
from ten to twelve hours a day and even longer." Peary, *Northward Over
the "Great Ice,"* volume 1, p. 173.
3. J. Diebitsch-Peary, *My Arctic Journal*, p. 79.
4. Matthew Henson played a major role in almost all of Peary's explorations, from his early days in Nicaragua to his final, controversial attempt
to reach the North Pole. His achievements in the Arctic are the focus of
several books, including his own memoir, *A Negro Explorer at the North
Pole* (New York: Frederick A. Stokes Company, 1912), and a later biog-

raphy, *Dark Companion* (Robert M. McBride & Company, 1947). Henson ("my colored boy," as Peary sometimes called him) was held in high esteem by the Inuit, who respected his modest temperament and fluency in their language. They moreover admired his skills—far superior to Peary's—as a dogsled driver. Like Peary, Henson fathered a son, Anauqaq, with an Inuit woman; Anauqaq later became close friends with Kali, one of Peary's two Inuit children. The paternity and lives of Anauqaq and Kali are discussed in detail by Allen Counter in his book *North Pole Legacy: Black, White, and Eskimo* (Amherst: University of Massachusetts Press, 1991).

5. Elisha Kent Kane, *Arctic Explorations in the Years 1853, '54, '55* (Hartford, Conn.: American Publishing Company, 1881), p. 349.

6. Ibid., p. 356.

7. The exploratory work—the reason Kane had actually sailed to Greenland—also proved treacherous. When Kane's men could muster the strength, they made sledging journeys, a week or two in duration, which took them northward from the *Advance* to discover spectacular glaciers on Greenland's northwestern coast. Time and again, though, they encountered appalling low temperatures and storms of astonishing severity. On the return journeys to the trapped *Advance*, the small expedition parties sometimes tended to contemplate suicide or murder, or else fixate on whether the frostbite they were experiencing was merely severe or possibly fatal. Two of Kane's men died from complications after they had to undergo amputations to remove frostbitten limbs.

8. Pierre Berton, *The Arctic Grail: The Quest for the Northwest Passage and the North Pole, 1818–1909* (Guilford, Conn.: Lyons Press, 2000), p. 393.

9. Hall, who clung to the theory that the Franklin Expedition members might still be alive, was also traveling to the Arctic so he could find out from the native settlers, firsthand, what might have happened to this group. As the Arctic historian Michael F. Robinson recounts (*The Coldest Crucible: Arctic Exploration and American Culture*, Chicago: University of Chicago Press, 2006), "The testimony of local Eskimos and the bleached and scattered bones on the island finally convinced [Hall] that all of Franklin's men had perished in the arctic" (p. 75).

10. Loomis, *Weird and Tragic Shores: The Story of Charles Francis Hall, Explorer*. Nearly a century after Hall's death, Loomis's team traveled to

Greenland and dug into Hall's grave and opened the coffin without ex-
huming the body. Hall was frozen partly in ice and wrapped in a flag. "At
first we thought that the body, still well fleshed, was perfectly preserved,"
Loomis wrote. But Hall's internal organs "were almost entirely gone."
The team salvaged a fingernail and hair from the body; a neutron-
activation test later done by Toronto's Centre of Forensic Sciences "revealed 'an intake of considerable amounts of arsenic by C. F. Hall in the
last two weeks of his life.'" Loomis concluded Hall might have therefore
been the victim of murder, but that the high arsenic levels could also
have resulted from medical treatment or self-medication.

11. The *Polaris* might not have been the worst-case scenario for polar expeditions. Another disaster involved U.S. Army Lieutenant Adolphus W.
Greely, who sailed to the Arctic a few years after Kane and Hall. Peary
knew Greely's story to its most trifling details. In August 1881, Greely
landed with two dozen men on Ellesmere Island, six hundred miles
below the North Pole, where he established a military camp called Fort
Conger. He intended to spend several years making scientific observations and, by using Conger as a base, send small teams of men out to set
a new record for reaching a point in the farthest north. Greely's party
was dropped off with the understanding that a supply ship would return
the following summer. But a supply ship did not come the following
summer or the summer after. In August 1883, under Greely's direction,
the men left Fort Conger and moved south to a small island off Cape
Sabine, located about eighty miles from where Peary was stationed at
Red Cliff House. There, Greely hoped to rendezvous with a supply ship
or rescue vessel, or find that the army had left depots of fuel and food.
But he found no depots, and Camp Sabine became a subzero horror.
That winter, the Greely party, one by one, began to die. When a rescue
ship arrived in June 1884, only seven men, including Greely, were still
alive; they were rescued days away from certain death. While those who
returned to the United States were hailed as heroes, some were also accused of cannibalism, a charge that Greely denied until his death fifty
years later.

12. Peary, *Northward Over the "Great Ice,"* volume 2, p. 76; also, J. Diebitsch-
Peary, *My Arctic Journal*, p. 219. Knud Rasmussen visited these sites,
too—for instance at the start of the Second Thule Expedition, when he

came to Hall's grave and regarded the copper plate that marked the burial site. He noticed two other graves nearby. "Our minds were impressed by the atmosphere of this little Arctic cemetery," Rasmussen noted, "for the men whose earthly remains rest in this place lost their lives in an attempt to reach the places which are now our goal." (Rasmussen, *Greenland by the Polar Sea*, p. 76.)

13. Peary, *Northward Over the "Great Ice,"* volume 1, p. 169.

14. Ibid., p. 190.

15. Regions of the ice are also laced with deep curving ridges, forged by the wind and known as *sastrugi*, which can make sledging hellish.

16. Peary, *Northward Over the "Great Ice,"* volume 1, p. 297.

17. On the first journey across the ice, the team was not compatible. Two members in particular—Verhoeff and Gibson—were hostile to Peary and aggressive toward Matthew Henson, Peary's African American assistant. (Racial hostility was no doubt a factor.) In Peary's account of the ice sheet crossings, moreover, he mourned the fact that Verhoeff died while exploring a glacier alone just before the conclusion of the trip. It seems more likely that he worried the death reflected poorly on his leadership. The men disliked each other intensely, and Verhoeff was viewed by the entourage as unpleasant, unpredictable, and troublesome.

18. Astrup, *With Peary Near the Pole*, p. 241.

19. Peary, *Northward Over the "Great Ice,"* volume 2, p. 107.

20. Peary, *Northward Over the "Great Ice,"* volume 1, pp. 297, 310.

21. Peary, *Northward Over the "Great Ice,"* volume 1, pp. 318–19. Peary's contention was almost certainly incorrect: The Inuit had seen the northern shores centuries before.

22. The *kamiks* were sewn from sealskin; the insulated liners were sewn from deerskin.

23. Many decades later, a population of several dozen musk oxen were introduced in southwestern Greenland and flourished—to a population that presently numbers over twenty thousand. Indeed, the herd in the southwest is now much greater than the original herd in the northeast.

24. Peary, *Northward Over the "Great Ice,"* volume 2, p. 333.

25. In a historical study of cairns, David B. Williams (*Cairns: Messengers in Stone* [Seattle: Mountaineers Books, 2012], p. 96) wrote: "The British are well known for exploring the world, but what they really did was

travel around and stack stones." The Inuit in Greenland made cairns to mark sites and paths as well—long before Peary—which they called *in-ussuk*.

26. Peary, *Northward Over the "Great Ice,"* volume 1, p. 373.

27. Peary, *Northward Over the "Great Ice,"* volume 1, p. 380.

28. The orders given to the *Kite* were to "bring Mrs. Peary back under any circumstances." Whether Jo would have *agreed* to return to the United States if Peary did not come back from his ice sheet crossing in time is unclear. She wrote, "While I do not think there is the slightest doubt that my husband will be here before the latter part of August, and while I fully believe that if he is not here then he will never come, yet I could never leave while there was the faintest chance of his being alive." J. Diebitsch-Peary, *My Arctic Journal,* p. 177.

5. A Pure Primitive Realm

1. The explorer and ethnographer Jean Malaurie notes: "Before the 1893 expedition, Peary gave 168 lectures in ninety-six days, in order to raise $13,000." From *Ultima Thule: Explorers and Natives in the Polar North* (New York: W. W. Norton & Company, 2003).

2. Most crucially, Peary had perceived a large northern outcropping of land—now known as Peary Land—to be an island, rather than a peninsula. His misapprehension later contributed to the deaths of explorers Ludvig Mylius-Erichsen, Jørgen Brønlund, and Niels Peter Hoeg Hagen, who had used his map to guide their own travels a few years later. Twenty years later, during Knud Rasmussen's First Thule Expedition (discussed later in this chapter), Peary's erroneous observation was noted by Peter Freuchen. The native Inuit who were with Freuchen on that journey found it deeply amusing that Peary—"the Great Peary," as they called him—had been so mistaken. They could now see he was fallible, and perhaps less than great.

3. *The New York Times,* September 13, 1892, p. 4.

4. Weems, *Peary: The Explorer and the Man,* p. 128.

5. His belief in the movement and currents of the polar ice was influenced by his own observations, but primarily by the Norwegian meteorologist Henrik Mohn.

6. "Dr. Nansen's Arctic Expedition," *The Times* (London), November 15,

1892, p. 7. The newspaper had also talked to the Norwegian explorer about the *Fram* during an earlier lecture tour he made through England in early 1892. "Dr. Nansen is confident that it is indestructible," a reporter noted. See "Dr. Nansen's Projected Polar Expedition," *The Times* (London), February 2, 1892.

7. Fridtjof Nansen, *Farthest North* (New York and London: Harper & Brothers, 1897), p. 162.

8. Nansen, *Farthest North*, p. 435.

9. Jo returned with Peary to Greenland several times; in fact, their daughter Marie—"a little blue-eyed snowflake," as Peary called her—was born in Greenland in 1893. Jo was also aware of Peary's Inuit children and his Inuit "wife" Ally, whom she met on a ship several years after Marie's birth.

10. In 1895, the return trip Peary made over the ice sheet from the north coast of Greenland to the Smith Sound region was one of the most harrowing in the annals of Arctic exploration. Constantly on the brink of starvation, Peary weighed the calculus of survival. "Two methods were open to us: one was to eat our own meat rations and have to drag our load ourselves for the last half of the return journey; the other, to live on our biscuit ration, scant as it was, reserving the meat for the dogs, to prolong their effectiveness as far as possible, and thus be compelled to drag our sledges only the last quarter of the return journey." (*Northward Over the "Great Ice,"* volume 2). During this time, he considered what Nansen's trip had proved in terms of the capabilities of men, rather than dogs, in pulling the sledge over the ice sheet. Peary calculated that he was six miles less than the distance traveled by Nansen over the ice in the course of forty days. Except Peary's outlook was far more bleak: "We have," he surmised, "nineteen days' half-rations of biscuit, tea, and milk." Nevertheless, he survived.

11. His exploitation went well beyond furs and meteorites. Peary's assumption of the role of colonial overlord, dispensing goods and favors to the Polar Inuit (Inughuit) in exchange for their help and acquiescence, also involved arranging relationships, sexual and otherwise, between the Inuit and between the Inuit and white visitors. In addition to Peary's relationship with Allakasingwah—Ally—resulting in two children, there were other intrigues. In *Muskox Land* (Calgary: University of Calgary Press,

2001), an exhaustive history of Ellesmere Island, Lyle Dick notes: "Peary assumed the role of patriarch, dispensing Inughuit women to his employees as if they were his personal property." Dick also writes: "Peary's 'philanthropy' extended to offering women in pre-existing conjugal relationships to other Inughuit men" (p. 382).

12. The Polar Inuit, rather than burying their dead—a challenge in the frozen ground—cover them with heavy stones and leave them to decompose, often in a stone house or tent set at a distance from an encampment. The bodies are often surrounded by valued possessions and hunting tools. The story of how Peary brought back the six Inuit, acting in part on the request of Franz Boas, a curator of the American Museum of Natural History in New York, is beyond the scope of this book. A compelling narrative of the saga can be found in both Kenn Harper's *Give Me My Father's Body: The Life of Minik, the New York Eskimo* (South Royalton, Vt.: Steerforth Press, 2000) and in an *American Experience* documentary, *Minik: The Lost Eskimo*. Minik, one of the surviving Inuit, eventually returned to Greenland, and then came back to the United States again. His life in Greenland as a young adult intersected with Knud Rasmussen and Peter Freuchen, the subjects of this chapter. He even lived with the two Danes in Thule for a short while.

13. These meteors, 4.5 billion years old, were sacred to the Inuit in large part because they had been a useful source of iron for tools for hundreds of years. They are displayed at the American Museum of Natural History to this day; the largest, known as Ahnighito, weighs thirty-four tons, making it the third largest discovered meteorite in the world. A special support beam for the meteor goes down to the Manhattan bedrock beneath the museum, so that its spectacular weight does not collapse the floor. amnh.org/exhibitions/permanent-exhibitions/earth-and-planetary -sciences-halls/arthur-ross-hall-of-meteorites/meteorites/ahnighito.

14. Peary, *Northward Over the "Great Ice,"* volume 2, p. 524.

15. During his crossing Nansen had collected useful insights on the composition and hydrology of the ice sheet. Peary, while less interested in research, helped aid the work of dozens of reputable scientists, many of whom paid handsomely to accompany him on his ship for the Greenland journeys. These researchers explored the flora, fauna, glaciers, and geology of the coast as Peary went about his difficult ice sheet excur-

sions. In this regard, Peary was less a scientist than a scientific trailblazer. As it happens, the work of both men also opened up the field of polar archaeology—for instance, Eigil Knuth, a Dane who accompanied Paul-Émile Victor during his first crossing of the ice (see chapter 9), spent a long career, from the 1930s on, uncovering artifacts in Peary Land and west Greenland and theorizing about the origins of settlers in Greenland who preceded the Inuit. Knuth's hero was Nansen.

16. Peter Freuchen, *I Sailed with Rasmussen* (New York: Julian Messner, 1958), p. 23.

17. On some old maps, the fictional strait of Jakobshavn is marked thus: "It is said that this Strait was formerly passable but now choked up with ice." See map: Hans Egede, *A Description of Greenland*, frontispiece. The "wall" of the glacier—that is, the calving front—was actually much closer to Jakobshavn village in Rasmussen's era. The glacier lost its ice shelf and has receded dramatically over the course of the last century, a development discussed later in the book.

18. The journey from the calving front to the North Atlantic can take several years, by which time the iceberg may be significantly eroded, chipped, and shrunken. It is speculated, but not proven, that one of Jakobshavn's icebergs sunk the RMS *Titanic* in April 1912.

19. Knud Rasmussen, *Across Arctic America: Narrative of the Fifth Thule Expedition* (New York: Greenwood Press, 1969), p. vi.

20. Peter Freuchen, who recounts the story, notes that as a much larger man, he could lift more weight than Rasmussen, but there was no conceivable way he or anyone he knew could have done what Rasmussen did. He says, "I had found my master." From *I Sailed with Rasmussen*, p. 127.

21. To be considered among the best dogsled drivers in Greenland is a mark of immense distinction; it is an honor not unlike, say, being the best marksman amongst a crackerjack battalion or (in a more modern era) being the best coder in all of Silicon Valley. Rasmussen did sometimes need to exhort his dogs. During the Second Thule Expedition, for instance, he grew hoarse from shouting at the dogs to pull over soft melting snow and "swamps" of ice water. See Rasmussen, *Greenland by the Polar Sea*, p. 169.

22. Freuchen, *I Sailed with Rasmussen*, p. 30.

23. Knud Rasmussen, *The People of the Polar North: A Record* (London: Keegan Paul, Trench, Trübner & Co., 1908), p. 78.

24. Freuchen, *I Sailed with Rasmussen*, p. 37.

25. Stephen R. Bown, *White Eskimo: Knud Rasmussen's Fearless Journey into the Heart of the Arctic* (Boston: Da Capo Press, 2015), p. 282.

26. The relationship between Denmark and Greenland—a kind of protective, patriarchal colonization—led the Danish government to minimize outside tourism or involvement in Greenland for most of the nineteenth century and part of the twentieth century. The goal was to avoid outside exploitation and cultural interference, even as missionaries and businessmen (working for various Danish shipping companies) were allowed in. Denmark did not control the northern Cape York and Smith Sound region, only the area south of Melville Bay, which was sometimes called "Danish Greenland."

27. Bown, *White Eskimo*, p. 43.

28. Rasmussen, *The People of the Polar North: A Record*, Author's Preface.

29. Ibid., p. 11.

30. Ibid., p. 232. Peter Freuchen would recall that Rasmussen once pretended to be a vegetarian to ingratiate himself, during his student days as an aspiring thespian, with an acting troupe.

31. "It sounds bad but tastes good," Rasmussen would say of such delights. (Rasmussen, *Greenland by the Polar Sea*, p. 21.) Short, vivid tutorials on the preparation and consumption of kiviak (the wings of the fermented bird are torn off along with skin and feathers; then the red meat is eaten) are easily found on the Web, for instance via YouTube.

32. MacMillan, *Four Years in the White North*, p. 126.

33. The practice of infanticide is chronicled in multiple studies on the Inughiut. In a typical account, Rasmussen writes of a man named Samik, a former colleague of Peary's: "Once, it is true, he had a son, but his wife had died soon after it was born. The little one had lived a few days, but, as there was no one to be found who could give it milk, it had quickly faded away. Samik could not bear to see his son suffer, and one day, as he lay there whimpering, he had picked him up in his arms and strangled him, out of compassion." (*The People of the Polar North*, p. 66.) Peter Freuchen recounts similar stories, such as that of Itusarssuk, who killed four of her own children after her husband drowned in a kayak: "The

oldest girl, about twelve years of age, helped her mother hang the younger ones." Eventually, "the girl herself fastened the noose around her neck." (Freuchen, *Arctic Adventure: My Life in the Frozen North*, p. 66.) During the Fifth Thule Expedition some years later, Rasmussen also noted secret histories of cannibalism during famine years in the Canadian Arctic.

34. Rasmussen, *The People of the Polar North: A Record*, p. 117.

35. He also met Ally, Peary's Greenlandic consort and the mother of two of his children.

36. The two books on the Literary Expedition published in Denmark—*The New People* (1905) and *Under the Lash of the North Wind* (1906)—were edited into one English translation, *The People of the Polar North*, published in 1908. In addition to Rasmussen's work, Mylius-Erichsen wrote up his travels in *Grønland* (1906), a work not translated into English.

37. Rasmussen, *The People of the Polar North: A Record*, p. 113.

38. Inge Kleivan and Ernest S. Burch, Jr., "The Work of Knud Rasmussen Among the Inuit: An Introduction," *Inuit Studies* vol. 12 (1988). While Nansen and Peary and other explorers had done studies of the Polar Inuit (Inughuit), too, their work was not on par with Rasmussen's. In the same issue of *Inuit Studies*, ethnographer Rolf Gilberg ("Inughuit, Knud Rasmussen, and Thule: The Work of Knud Rasmussen Among the Polar-Eskimos in North Greenland") notes that the Polar Inuit's "first contact with Whites in historic times (1818) was of short duration and had little impact on the culture. Only few data on the Inughuit culture were published by Ross (1819). Through the 1800s many expeditions (Saunders, Inglefield, Kane, Hayes, Hall, Nares, Young, and others) were searching, without results, for [explorer John] Franklin in the Inughuit area. . . . With the Peary expeditions (1891–1909) things were different, as he brought goods with him to pay for the services he required."

39. In discussing Peary and Nansen, Rasmussen would say he was like "the little Polar fox, which everywhere on the Arctic coast follows the footsteps of the big ice-bear." Rasmussen, *Greenland by the Polar Sea*, p. xxii.

40. Ibid., p. xxii.

41. Kaj Birket-Smith, *Journal de la Société des Américanistes* vol. 25, no. 2 (1933), pp. 371–74.

42. Freuchen, *Arctic Adventure: My Life in the Frozen North*, p. 16.

43. Ibid., p. 21.

44. By a number of accounts, Rasmussen and Mylius-Erichsen did not get along during the Danish Literary Expedition of 1902–1904, and it seems likely their friendship did not continue in the time afterward. Years after the *Danmark* debacle, Freuchen conceded that the late Mylius-Erichsen, while no doubt an interesting and intelligent man, was a poor expedition leader because of inexperience and (one might conclude) a contentious personality.

45. The bodies of Mylius-Erichsen and Hagen were never recovered. There is some speculation that Brønlund cannibalized the other two, though no corroborating proof has ever been discovered. "The search party took a vow never to discuss what they found near Lambert Land: a vow they kept, every man, until they died." *Copenhagen Post*, January 23, 2017.

46. Freuchen also began supplementing his income by covering news stories for *Politiken*, a Copenhagen daily newspaper. Like Rasmussen, he was drawn into a controversy, beginning in 1909, around whether Frederick Cook had actually beaten Robert Peary to the North Pole. Freuchen interviewed Cook when he sailed to Copenhagen, soon after Cook publicized his claim to the Pole, and decided "his whole story is a damned lie."

47. Freuchen, *Arctic Adventure: My Life in the Frozen North*, p. 34.

6. Thule

1. John McCannon, *A History of the Arctic* (London: Reaktion Books, 2012), p. 67. There is a lack of clarity over whether Pytheas glimpsed Greenland or Iceland or the west coast of Norway. The Freuchen biographer Janni Andreassen notes that the Thule name actually dates back to 1903, when a traveling companion of Freuchen and Rasmussen— Harald Moltke—described a Polar Inuit settlement as "Thule."

2. "Mrs. Peary and Her Furs," *The New York Times*, March 14, 1897, p. 11. The Peary apartment had "furs on the walls, furs on the floors, furs on the couches and cushioned seats."

3. "What I have done in the past," Peary maintained, "and shall continue to do in the future, is to put [the Polar Inuit] in a little better position to carry on their struggle for existence, give them better weapons and implements, lumber to make their dwellings dryer, instructions in a few fundamental sanitary principles, and one or two items of civilized food, as coffee and biscuit—allies to rout the demons [of] starvation and cold." (Peary, *Northward Over the "Great Ice,"* volume 1, p. 508.) Before Peary arrived, the Polar Inuit had encountered guns and other modern tools through whaling ships and explorers such as Charles Hall and Elisha Kent Kane. But as Rasmussen would note (*Greenland by the Polar Sea*, p. 7), "Long before Peary appeared a lively bartering with the Scotch whalers certainly took place; but a thing like a gun was a great rarity."

4. Ibid., p. 6.

5. Freuchen, *Arctic Adventure*, p. 32.

6. The pelts were indeed valuable. On an early Greenland trip, for instance, Rasmussen had returned with two hundred fox pelts worth twelve thousand kroner, or about twenty-five hundred dollars. Several investors, notably a Danish industrialist named Marius Nyeboe, saw the potential of the trading station and backed Freuchen and Rasmussen in the venture. Rasmussen's wife, Dagmar, who hailed from a wealthy family, also contributed funds to the trading station effort, even as it meant her husband would be away frequently, and for years at a time.

7. Rolf Gilberg, "Inughuit, Knud Rasmussen, and Thule." As it would turn out, Rasmussen would also funnel some of the profits from Thule station to fund research work that traced the global origins and migrations of Inuit tribes.

8. Freuchen, *I Sailed with Rasmussen*, pp. 160–61.

9. This was Thorild Wulff, the Swedish botanist who died on the Second Thule Expedition.

10. MacMillan, *Four Years in the White North*, p. 127.

11. Freuchen, *Arctic Adventure*, p. 167. Navarana's original name was Mequpaluk. She changed it after "marrying" Freuchen. It is likely that she was as young as fourteen when Freuchen began courting her.

12. Ibid., p. 167.

13. Freuchen, *I Sailed with Rasmussen*, p. 185.

14. Freuchen, *Arctic Adventure*, p. 289. In essence, the aluminum kroner coins were like company scrip, which functions as a form of credit. The idea of the buyer determining the price of an item should not be interpreted as ignorance; it merely reflects a different value system. Sometimes the practice appears in modern Western culture, too. In 2007, the

band Radiohead released a new record, *In Rainbows*, which was sold to buyers based on what they wanted to pay—i.e., what they thought it was worth.

15. Rasmussen, *People of the Polar North*, p. 77.

16. Rasmussen's favorite book was Xeonophon's *Anabasis*, in the original Greek, a tale about an expedition of ten thousand soldiers that he sometimes liked to page through while on his dogsled. MacMillan, *Four Years in the White North*, p. 393.

17. Ibid., p. 58.

18. The personalities and chronologies here overlap and are admittedly complex. Mikkelsen's goal in 1909 had been to seek out the scientific papers, and possibly the bodies, of the three members of the *Danmark* expedition who had died two years before, in 1907. Freuchen had been a member of the *Danmark* expedition, too; the three men Mikkelsen was searching for were the ones who had perished while Freuchen had been living in the weather station hut, alone except for the howling wolves, for six months. The three dead men were also friends of Rasmussen's.

19. Freuchen, *Arctic Adventure*, p. 191.

20. "Peary Has Record Left in Greenland," *The New York Times*, September 21, 1913.

21. Rasmussen had a tendency to become more resourceful as a situation worsened, and also more charming. Freuchen liked to tell the story of how, during a winter of scarcity at Thule, Rasmussen had a very small amount of coffee, but every afternoon would reuse it, replacing old coffee beans with peas until finally there were no beans at all. He used "his familiar mesmerism to silence criticism." Everyone drank the coffee—or, rather, the "coffee." No one complained.

22. Mikkelsen survived his expedition and wrote a fascinating account, *Lost in the Arctic* (New York: George H. Doran Company, 1913), about his travails. He had come upon Mylius-Erichsen's cairn and had taken the note secreted inside; but he had not replaced it with his own note—a break with custom that would have helped Rasmussen and Freuchen understand when he had been there and which direction he had gone. When Rasmussen met Mikkelsen later in Copenhagen, Mikkelsen apologized for not replacing the note in the cairn.

23. Freuchen, *Arctic Adventure*, p. 218.

24. Freuchen—often an unreliable narrator—would later write that he left Peary's note inside the cairn, but news reports of the era suggest otherwise. *The New York Times* noted that Peary's record had been found and retrieved by Freuchen and Rasmussen. "The record was returned to Rear Admiral Peary by the Danish Government through C. Brun, the Danish Minister to the United States." September 21, 1913, p. C5.

25. "Peary Has Record Left in Greenland," *The New York Times*, September 21, 1913.

26. At the start of this period, Rasmussen traveled to Copenhagen to visit his family and brought back reports on the war and the situation in Europe.

27. Rasmussen and Freuchen were no longer the only outsiders living in the region: A number of American scientists, led by Donald MacMillan, had been stranded by pack ice and were staying at a nearby settlement called Etah. Freuchen and Rasmussen often visited their outpost, or had them come to Thule as their guests. The story of MacMillan's expedition and his stay in Etah is literally a book unto itself. (MacMillan, *Four Years in the White North*.) While Freuchen seems to have gotten along well with MacMillan, a disciple of Peary, he had a falling out with Elmer Ekblaw, an American scientist who spent an unhappy month with Freuchen in North Star Bay, when a lack of food and Freuchen's apparent flightiness irritated him. What's more, there was some concern on the part of Rasmussen and Freuchen that the American team would set up a competitive trading post for furs that would make their own business more difficult. To a minor extent this happened, but the competition disappeared after MacMillan returned to the United States.

28. Rasmussen, *Greenland by the Polar Sea*, p. 71. A few months later on the journey, Rasmussen came upon an old depot—from 1908, making it nine years old—from Peary in a place called Cape Salor. They found six tins of pemmican and ate it greedily. In addition to the meat and fat, the pemmican had "the wonderful addition of lots of raisins and sugar." Rasmussen concluded that "no marzipan cake could have tasted better" (p. 136).

29. They continued to discover relics from the past after finding the Peary note. A few days later, Rasmussen dug a note out of a watertight copper

case protected in a cairn that had been written on May 25, 1876, by
L. A. Beaumont of the Nares Expedition, who had been attempting to
reach a far northern destination. Beaumont, Rasmussen knew from history texts, had nearly died from his exertions. Several weeks after that,
Rasmussen's team came upon a small cairn and note from J. B. Lock-
wood, an explorer from the 1882 Greeley Expedition who at the time
had actually reached a more northern point than any explorer before
him.

30. Emergency rations meant small amounts of oats and pemmican. Knud
Rasmussen, "The Second Thule Expedition to Northern Greenland,
1916–1918," *Geographical Review* vol. 8, no. 2 (August 1919), p. 286.

31. Rasmussen, *Greenland by the Polar Sea*, p. 228.

32. Freuchen, *Arctic Adventure*, p. 329. Some recent research by Danish
biographers of Lauge Koch suggest that the deaths that occurred during
the Second Thule Expedition involved a more desperate and complicated scenario than Rasmussen's and Freuchen's stories suggest.

33. Freuchen, *Arctic Adventure*, p. 329.

34. In the intervening years, Rasmussen had conducted or authorized several other expeditions. Thus the chronology: The First Thule Expedition
crossed the Greenland ice sheet with the north coast destination of Independence Fjord; the Second Thule Expedition crossed the ice sheet
and surveyed the north coast of Greenland; the Third Thule Expedition
set food depots down for explorer Roald Amundsen; the Fourth Thule
Expedition involved ethnographic research in east Greenland; the Fifth
Thule Expedition covered the northern coast of North America.

35. "Eskimos All American, Rasmussen Trip Shows," *The New York Times*,
April 12, 1925.

36. Bown, *White Eskimo*, p. 263. After his return to Europe, Rasmussen
would begin working on a popular account of his travels, known in English as *Across Arctic America*, which compiled his insights about links
between the world's polar Eskimo tribes in language, traditions, and
myths.

7. TNT

1. The *Danmark* expedition was led by Mylius-Erichsen.

2. Mott Greene, *Alfred Wegener: Science, Exploration, and the Theory of*

Continental Drift (Baltimore: Johns Hopkins University Press, 2015), p. 20.

3. Martin Schwarzbach, *Alfred Wegener: The Father of Continental Drift* (Madison, Wisc.: Science Tech, Inc., 1986), p. 16.

4. Greene, *Alfred Wegener: Science, Exploration, and the Theory of Continental Drift*, p. 98.

5. Alfred Wegener, *Greenland Journal 1906–1908*, p. 60, December 25, 1906. Accessed digitally at environmentandsociety.org/exhibitions/wegener-diaries/original-document.

6. Ibid., p. 210, October 30, 1906.

7. Ibid., pp. 37–38, December 15, 1906.

8. He was also considering the possibility of future South Pole and North Pole excursions. He wrote: "I am living here completely absorbed in thoughts of a future German expedition." Ibid., p. 93, January 21, 1907.

9. Ibid., May 7, 1913.

10. Nansen, *The First Crossing of Greenland*, volume 1, p. 509.

11. Nansen did not yet argue that the disintegrating ice sheet held a similar influence on our oceans and shorelines, since the changes in Arctic climate from warming, and the reductions in the ice sheet, were not yet measurable.

12. Nansen, *The First Crossing of Greenland*, volume 1, p. 510.

13. Alfred Wegener, *Greenland Journal 1912–1913*, September 17, 1912.

14. "Captain Koch's Crossing of Greenland," *Bulletin of the American Geographical Society*, vol. 46, no. 5, 1914, p. 356.

15. The loss of the expedition's theodolite, the crucial device for navigation and positioning, created great distress amongst the men. By building other improvised tools to replace the theodolite, however, the group managed to navigate the ice sheet anyway.

16. Ibid., p. 358.

17. Wegener, *Greenland Journal 1912–1913*, May 13, 1913.

18. Ibid.

19. Ibid., May 26, 1913.

20. "Captain Koch's Crossing of Greenland," p. 359.

21. Wegener, *Greenland Journal 1912–1913*, July 4, 1913.

22. Ibid., July 17, 1913.

23. The published book has never been translated into English. Its German

title is *Durch die weisse Wüste: Die dänische Forschungsreise quer durch Nordgrönland 1912–1913*. It was translated from Danish into German by Else Wegener.

24. J. P. Koch, *Durch die weisse Wüste* (Berlin: Verlag von Julius Springer, 1919), pp. 209–10.

25. Wegener, *Greenland Journal 1912–1913*, June 7, 1913.

26. Greene, *Alfred Wegener: Science, Exploration, and the Theory of Continental Drift*, p. 380.

27. Else Wegener, ed., "Plan and Objects of Alfred Wegener's German Expedition to Greenland," from *Greenland Journey: The Story of Wegener's German Expedition to Greenland in 1930–31 as Told by Members of the Expedition and the Leader's Diary* (London: Blackie & Son Limited, 1939), p. 1.

28. Schwarzbach, *Alfred Wegener: The Father of Continental Drift*. The quoted letter is from chapter 9, "Memories of Alfred Wegener," by Johannes Georgi, p. 156. "My own trouble" likely refers to a diagnosed heart condition that became apparent during the war; Wegener had been beset by spells of dizziness and fainting during moments of battle.

29. Even during his early adventures, Wegener felt burdened by age. In September 1912, while sitting in his tent on the ice sheet and eating polar bear steaks with J. P. Koch, Wegener began daydreaming about his old life in Europe. He wrote in his journal: "One thing is certain: I will not go on another expedition like this voluntarily. I'm too old for this, even if I'm only 31."Alfred Wegener, *Greenland Journal 1912–1913*, September 12, 1912.

30. Meinardus worked as a professor of geography in Göttingen; the seismic technique was developed by Emil Wiechert. The process relied upon using around 150 pounds of explosives on the ice sheet and measuring how the blast waves bounced back from the bedrock.

31. While scientific curiosity was the driving force behind Wegener's Greenland interest, there were other factors. One was his restlessness and love for adventure. Another was his sense of nationalism—that the Greenland mission offered an opportunity to reclaim some of his homeland's waning respect during the postwar period. As historian David Thomas Murphy notes (*German Exploration of the Polar World: A History, 1870–1940* [Lincoln: University of Nebraska Press, 2002]), "the Wegener ex-

pedition set out in part to maintain German scientific prestige and to keep Germany in the front ranks of the exploring nations" (p. 127).

32. The list included his brother Kurt Wegener, who was not in Greenland at the start of the 1930 expedition, as well as several scientists who were to be stationed on the east coast of Greenland, in the third weather station.

33. Wegener wrote a witty and interesting chronicle about this so-called 1929 "reconnaissance" journey, *Mit Motorboot un Schlitten in Grönland*. The book, filled with black and white photographs by Georgi, has never been translated into English.

34. Bringing the heavy motorized sleds up the glacier was particularly difficult—a fact captured in film footage shot by Wegener's team—and required a combination of methods (horses, men, pulleys). Curt Schif, who was in charge of the sleds for the expedition, later wrote that when he first looked at Kamarujuk and considered the difficulty of hauling the machines up the icy incline, it "was enough to make one collapse on the spot." E. Wegener, *Greenland Journey*, p. 61.

35. In summer, when the melting and treacherousness of the glacier path became too much, the men cut a bridle path over the rocky moraine that abutted the glacier. "This was a very laborious task and consumed a great deal of time," Georgi later wrote, "but it was of permanent value, for from August onwards we were able to avoid the glacier itself altogether." Georgi, *Mid-Ice: The Story of the Wegener Expedition to Greenland* (London: Kegan Paul, Trench, Trubner & Co., 1934), p. 25.

36. In 1929, Wegener also had to reckon with the effects of the global stock market crash, which jeopardized some of his expedition's funding and forced him to scurry to find other sources.

37. "Protecting Ponies Worries Explorer," *The New York Times*, April 6, 1930, p. 62.

38. Schwarzbach, *Alfred Wegener: The Father of Continental Drift*, p. 40. Later, Wegener's brother Kurt would observe that the late breakup of the sea ice sealed the fate of the expedition before it even began.

39. Wegener, *Greenland Journal 1930–1931*, August 5, 1930.

40. The question of who was to blame for the lack of orderly deliveries to Eismitte would continue for years afterward, since it was these shortages that made Wegener's late-September dogsled visit necessary. Though

Wegener, as leader of the expedition, should in truth shoulder some of the responsibility, almost certainly Georgi's haste to set up the station and his inadequate oversight of its provisioning shaped the events to come.

41. Wegener, *Greenland Journal 1930–1931*, September 6, 1930.

42. Ibid., August 20, 1930.

43. Freuchen, *Vagrant Viking*, p. 228.

44. Just after Wegener departed for Eismitte on September 21, 1930, he ran into several colleagues and Greenlanders who were returning from Eismitte, where they had brought supplies. They carried a note from Georgi and Sorge, the two scientists now stationed at Eismitte, saying that if they didn't receive more supplies by October 20—they were in particular need of paraffin for their stoves—they would return by foot to the west station. The note is mentioned in the following chapter. Wegener wrote in his journal, "They would not get through but would be frozen to death on the way." In a debate that later ensued, the writing and effect of this note would prove a fundamental point of contention.

45. Curiously, Wegener disliked the idea of heroics. In a letter in the summer of 1930 to his brother Kurt, he wrote: "In short, life here [in Greenland] has its dark side. A person wouldn't be able to put up with most of it, if he didn't know that after a certain number of months are marked off on the calendar, he can go back home and live as he sees fit. And then, thank goodness, the obligation to be a hero ends, too." From Schwarzbach, *Alfred Wegener: The Father of Continental Drift*, p. 40.

8. Digging

1. Georgi, *Mid-Ice*, p. 41.

2. Georgi, *Mid-Ice*, p. 52.

3. E. Wegener, ed., *Greenland Journal 1930–1931*, p. 94.

4. Ibid., p. 189.

5. Georgi, *Mid-Ice*, "Sunday, September 21, 1930," p. 86.

6. E. Wegener, ed., *Greenland Journey*, p. 94.

7. The fateful letter, signed by both Georgi and Sorge, among other things, asserts urgently that the men were in "a new and for us dangerous situation." It demanded the delivery of seventeen large cans of petroleum, more food and gear, and a variety of scientific instruments. "If these

necessities are not here, or their arrival definitely announced, by October 20, we shall start on that day with hand-sledges. But we hope that you will arrive any day with motor-sledges, and that all difficulties will thus be removed." Reprinted in Georgi, *Mid-Ice*, pp. 82–83.

8. E. Wegener, ed., *Greenland Journey*, p. 180.

9. Georgi, *Mid-Ice*, p.118.

10. The sharpness of acuity is why many scientists working in Greenland during the summers hire the Inuit to help them identify research subjects for tagging—bears, narwhal, seals, walrus, and the like.

11. E. Wegener, ed., *Greenland Journey*, p. 113. The letter was written by Wegener to Karl Weiken, a geodesist on his team, at mile 38½ on September 28, 1930; it was brought back to the west coast camp and delivered to Weiken by some of the returning Greenlanders. Wegener continued on his journey to Eismitte.

12. Wegener also offered to remain at Eismitte through the winter with Loewe, if Sorge and Georgi preferred to return to the west coast. The two scientists declined his offer.

13. The list of explorers would include Ludvig Mylius-Erichsen; Ejnar Mikkelsen; and the Swiss meteorologist Alfred de Quervain. It also seems apt to address here the author's Western bias: Before Nansen's arrival, the Inuit had already explored and memorized large areas of their own coastlines, fjords, islands, glaciers, and waterways. European explorers could be seen as merely repeating that exercise. This is undoubtedly true for Greenland's west and northwest coasts, but most parts of the island's northern and eastern coasts were unknown even to the Inuit. The ice cap was moreover viewed by the Inuit as an area of little interest and significant peril.

14. "Explorer Risks Life to Trace Icy Greenland," *New-York Tribune*, October 21, 1923, p. A2. The news story chronicles the harrowing geographical expedition, between 1920 and 1922, of Lauge Koch, the Danish geologist who a few years before had accompanied Rasmussen on the Second Thule Expedition.

15. Kirwan, *A History of Polar Exploration*, p. 321.

16. In 1930, in addition to Wegener's, there were two other expeditions measuring Greenland's weather and aviation potential. One was situated on the west coast and was led by William Hobbs from the University of

Michigan. Another—the British Arctic Air Route Expedition—was located on the east coast and was led by Gino Watkins. The British expedition just barely averted tragedy. One member of the group, August Courtauld, began manning an inland weather station alone on December 6, 1930. Courtauld eventually ran out of fuel and was blocked from exiting his hut as snowdrifts piled up. He was sitting in the pitch dark and cold and had resigned himself to death when he was rescued by colleagues on May 5, 1931.

17. Georgi, *Mid-Ice*, p. 102.

18. Most likely the dropping ceiling was not immediately dangerous but merely represented the usual deformation and densification of firn as snow piles on. Even in modern-day subglacial research—at Camp Century and, more recently, at EGRIP camp, both discussed later in this book—scientists encounter the same phenomenon of lowering ceilings and deformation.

19. E. Wegener, ed., *Greenland Journey*, p. 194.

20. "Fritz Loewe, 1895–1974," *Journal of Glaciology* vol. 14, no. 70 (1975). Also, Michael Piggott, "On First Looking into Loewe's Papers," *University of Melbourne Collections*, issue 3, December 2008.

21. Georgi, *Mid-Ice*, pp. 153–54.

22. Ibid. p. 122.

23. Some of Sorge's research was conducted during the work on a 1933 film called *S.O.S. Iceberg*. Rasmussen and Sorge worked as consultants. The movie, filmed largely in Umanak on Greenland's west coast, was directed by Arnold Fanck. It starred Leni Riefenstahl, who later went on to become a film director favored by Adolf Hitler. Sorge noted that Riefenstahl carried *Mein Kampf* with her everywhere, and constantly praised the chancellor's vision. Ernst Sorge, *With Plane, Boat and Camera in Greenland: An Account of the Universal Dr. Fanck Greenland Expedition* (London: Hurst & Blackett, 1935).

24. Fritz Loewe, "The Scientific Exploration of Greenland from the Norsemen to the Present," public lecture, May 26, 1969, Ohio State University. Institue of Polar Studies, Report No. 35, Byrd Polar Research Center, Ohio State University.

25. Loewe also lectured as a visiting professor at Ohio State University be-

tween 1961 and 1973. After the war, he and Georgi struck up a warm correspondence that included lively debates about the controversy in Germany over Wegener's death and the scientific data they collected. Loewe's career after Eismitte was far more rewarding and successful; he visited Georgi several times in Germany during the 1950s and 1960s. He also traveled to Antarctica, Pakistan, and Greenland (three times) to conduct research. See Jutta Voss, "Johannes Georgi und Fritz Loewe, Zwei Polarforscherschicksale nach 'Eismitte' Aus ihrem Briefwechsel 1929–1971, sowie die gesammelten Schriftenverzeichnisse von J. Georgi und F. Loewe," *Polarforschung* 62 (1994), pp. 151–61.

26. Georgi, *Mid-Ice*, p. 159.

27. "Greenlanders Desert Germans on Ice Cap, Fearing Demons in Terrific Snowstorm," *The New York Times*, October 27, 1930. The article was licensed around the world by Akademia, a syndication service in Heidelberg, which provided it to the *Times*.

28. "German Greenland Expedition: A Feeling of Anxiety," *The Times* (London), Nov. 26, 1930, p. 13. Also see "Germany Fears for Greenland Expedition," *The New York Times*, Nov. 25, 1930, p. 1.

29. E. Wegener, ed., *Greenland Journey*.

30. Ibid., p. 153.

31. Ibid., p. 168–69.

32. A physical during World War I "revealed a pronounced 'click' in his heart sound, which, together with the palpitations, indicated a prolapsed mitral valve." Greene, *Alfred Wegener: Science, Exploration, and the Theory of Continental Drift*, p. 341.

33. E. Wegener, ed., *Greenland Journey*.

34. "Tribute to Wegener by Fellow-Explorer," *The New York Times*, May 10, 1931.

35. Georgi, *Mid-Ice*, p. 220.

36. It seems most convincing to conclude that Wegener's death was caused by several factors and a complex and chaotic situation. Georgi arguably should have done a better job in supplying Eismitte in the early deliveries, so that Wegener would not have to make an emergency run. And Georgi should never have written his note about his intentions to walk back to the west coast station from Eismitte in mid-October. But Georgi

did not cause the cold spring weather that led to the interminable delays in getting materials on the ice sheet. And ultimately, as expedition leader, Wegener was responsible for all decisions and all personnel. Eventually, the arguments over who was at fault for the disaster, though simmering quietly at first, exploded into the public sphere in Germany in the years after, with Kurt Wegener, Alfred's brother, publicly blaming Georgi for much of the tragedy, and with Georgi making a strenuous effort to rebuild his reputation.

37. Georgi, *Mid-Ice*, p. 220.

38. The ice was probably about a thousand feet thicker than Sorge's estimation.

39. E. Wegener, *Greenland Journey*, p. 235.

40. While estimates have long varied—from 20 to 24 feet—regarding how high seas will rise in the event of all Greenland's ice melting, an exhaustive recent study by academic researchers, funded by NASA, put the potential in 2017 at about 24.3 feet. See M. Morlighem et al., "BedMachine v3: Complete Bed Topography and Ocean Bathymetry Mapping of Greenland from Multi-Beam Echo Sounding Combined with Mass Conservation," *Geophysical Research Letters* vol. 44 (September 2017).

9. Machine Age

1. The coordinates of Eismitte were 70°53.8′ N by 40°41′ W; the coordinates of Station Centrale were 70°54′ N by 40°42′ W.

2. Paul-Émile Victor, "Wegener," from *Polarforschung* 40 (Alfred Wegener Institute for Polar and Marine Research & German Society of Polar Research, 1970), pp. 2–3. Translated from the French. A slightly different translation appears in Martin Swarzbach's biography of Wegener, on page 35.

3. Peary also spent the last years of his life helping set up air postal routes that later became the basis for the U.S. airmail system. His claim to have reached the North Pole in 1909 is likely incorrect, a topic discussed in great detail in recent Arctic literature, most persuasively by the writer and explorer Wally Herbert in *The Noose of Laurels* (New York: Atheneum, 1989).

4. Some of the collected letters (in Danish and Norwegian) were compiled

by Eigil Knuth in *Fridtjof Nansen og Knud Rasmussen* (Copenhagen: Gyldendal, 1948).

5. Fridtjof Nansen, *Adventure and Other Papers* (London: The Hogarth Press, 1927). The cited text is from a speech Nansen gave at St. Andrews University in Scotland on November 3, 1926.

6. "Knud Rasmussen, Explorer, Is Dead," *The New York Times*, December 22, 1933, p. 21.

7. Freuchen, *I Sailed with Rasmussen*, p. 224. In later years, Rasmussen and Freuchen had grown apart and rarely communicated.

8. Freuchen's salt-and-pepper beard is preserved in a box at the Arktisk Institut in Copenhagen, where it was shown to the author. He was married twice after the death of Navarana, with his third wife, Dagmar, surviving him.

9. Some scenes from Freuchen's game show appearance are archived on YouTube at goo.gl/nCJ7qq. See also "Explorer is 5th to Win $64,000," *The New York Times*, July 4, 1956, p. 37.

10. Some of the work was conducted under Rasmussen's sponsorship, and some under the supervision of the geologist Lauge Koch, one of Rasmussen's colleagues from the Second Thule Expedition. Rasmussen and Koch continued to work in Greenland, but in different regions; the two men—at that point mortal enemies, following their disagreement about the tragic deaths and aftermath of Second Thule—never spoke again. During the 1930s, both Rasmussen and Koch were separately involved in aerial surveys, done in open-cockpit planes flying at about thirteen thousand feet; the goal was to photograph the glaciers and shoreline of Greenland. The images were then filed away and forgotten. But the pictures—ten thousand or more—would prove to have scientific worth decades later, when the negatives were discovered in an abandoned warehouse outside of Copenhagen. The images offered not just a detailed record of the ice but a detailed record of the ice at a specific *moment* in time. In other words, the photos allowed researchers to precisely compare how much glaciers had receded from the edges of Greenland between Rasmussen's era and today's. See Quirin Schiermeier, "180,000 Forgotten Photos Reveal the Future of Greenland's Ice," *Nature* vol. 535 (July 27, 2016). Also see *The Greenland Ice Sheet: 80 Years of Climate Change Seen from the Air* (Copenhagen: Natural History Museum of

Denmark, 2014). The primary current researchers for the photo-mapping project are Kurt Kjaer and Anders Bjørk; the latter was interviewed by the author in Irvine, California, in 2016.

11. Victor's other Greenland expeditions during the 1930s were impressive feats of anthropology, and also included (a) the yearlong (August 1934–September 1935) immersion on the east coast of Greenland, along with colleagues Gessain and Perez; and (b) a winter stay on the east coast of Greenland (after his June to July 1936 crossing of the ice sheet) that lasted from August 1936 through August 1937. Eigil Knuth, the Dane who accompanied him on his ice sheet crossing, later gained fame from his archaeological digs in Peary Land, on the northwest coast of Greenland, where he identified the remains of early, pre-Inuit Greenlandic settlers. For Victor's descriptions of Rasmussen, see Paul-Émile Victor, *Man and the Conquest of the Poles* (London: Hamish Hamilton, 1964), p. 213.

12. Victor, *My Eskimo Life* (New York: Simon and Schuster, 1939), Appendix: "Preliminary Report of the French Trans-Greenland Expedition, 1936," p. 321.

13. At the end of the 1930s, Victor was doing ethnographic research on the Sami people in northern Norway and in Finland.

14. Victor's Jewish heritage, which he was apparently not yet aware of, could have left him vulnerable to arrest. His parents were originally from eastern Europe and settled in Switzerland (where Victor was born) before moving to the Jura region of western France. Victor's father's full name was Erich Heinrich Victor Steinschneider; he later shortened the family name to Victor.

15. Victor was involved in scouting, which helped him obtain an order from the Ministry of Public Instruction to continue his ethnographic research in Morocco and Martinique. Once in the United States, he traveled extensively, but spent most of his time in New York and Washington, D.C. To earn money, he wrote freelance articles and conducted lectures about his Greenland work (his book on living with the Inuit of east Greenland had been recently published in English). See Daphné Victor and Stéphane Dugast, *Paul-Émile Victor: J'ai toujours vécu demain* (Paris: Robert Laffont, 2015), pp. 278–79.

16. "Here I became familiar with parachuting techniques and the perfor-

mance of aircraft and weasels under northern conditions. I soon realized that by the use of an aircraft and mechanized vehicles it would now be possible to transport the heavy equipment necessary for the work we had contemplated in Greenland." Paul-Émile Victor, "The French Expedition to Greenland, 1948," *Arctic* vol. 2, no. 3.

17. Victor and Dugast, *Paul-Émile Victor: J'ai toujours vécu demain*, p. 294. According to the younger Victor and Dugast, P-É Victor first encountered the weasels when sent to Colorado.

18. Paul-Émile Victor, "Wringing Secrets from Greenland's Icecap," *National Geographic*, January 1956, p. 123.

19. Victor, *My Eskimo Life*, p. 311. Victor also began a serious romantic relationship with Doumidia, a young Inuit woman.

20. John Hanessian, Jr., "Re: Expéditions Polaires Françaises," letter to Mr. Richard Nolte, Institute of Current World Affairs, April 12, 1960. In a detailed assessment of Victor's Greenland work, Janet Martin-Nielsen concludes that his establishment of Station Centrale "marks not a leap from adventure to science, but rather showcases the successful crafting of a modernistic narrative." *Eismitte in the Scientific Imagination* (New York: Palgrave MacMillan, 2013), p. 8. Her insights raise the question as to what extent Victor's promotional techniques could serve climate scientists of the current era. Victor correctly (and uncynically) perceived the benefits of *fusing* the old with the new, of characterizing his forays as grand human adventures that happened to involve science and technology.

21. Victor, "Wringing Secrets from Greenland's Icecap."

22. Hanessian, Jr., "Re: Expéditions Polaires Françaises."

23. To be sure, earlier explorers brought new technologies to Greenland before Wegener and Victor—Nansen's innovative cooking stove and his design for lightweight sleds, the latter of which are still used in Greenland's functioning mid-ice camps (Summit and EGRIP), are two examples. The crucial leap in the 1940s was due to mechanization of transportation (weasels and planes) and portable electronics (radio communications).

24. Fristrup, *The Greenland Ice Cap*, p. 125. Victor wasn't the only scientist who saw the poles as enduring subjects of interest, but he merged that awareness with action and accomplishment long before the Interna-

tional Geophysical Year in 1957 (see chapter 11). His heightened interest in the scientific study of the poles, in short, was prescient.

25. The "great man" theory of history holds that crucial events and discoveries are determined by the unprecedented actions of heroic, trailblazing characters. The theory was articulated by Thomas Carlyle, the British writer, in 1841. Challenges to the theory, which are numerous and varied (and, to this author, more persuasive), hold that the effect of powerful historical figures is contextual and made possible by social and technological conditions. To apply this paradigm to Victor, one can see that his achievements would not have occurred without Wegener's precedent, and moreover without the weasels, airplane drops, and postwar technological and scientific advances. Still, it seems reasonable to conclude that Victor's considerable leadership skills and eagerness to bring machinery and scientists to the ice sheet after the war *accelerated* progress in glaciology by several years.

26. Fristrup, *The Greenland Ice Cap*, p. 131.

27. Victor, "The French Expedition to Greenland, 1948."

28. Victor had considered various places on the west coast to launch his expedition; also, he had asked the local Inuit to survey the coast for him prior to the arrival in 1948 to find a suitable landing. "The results . . . as well as information collected locally, prompted us to visit, first, the point where [Swiss explorer Alfred de] Quervain landed in 1912." They settled on this location. See Paul-Émile Victor, *The French Polar Research Expeditions 1948–1951* (New York: The French Embassy Press and Information Division, 1950), p. 6.

29. Victor, "Wringing Secrets from Greenland's Icecap," p. 122.

30. Victor, *The French Polar Research Expeditions 1948–1951*, p. 11.

31. In some accounts, Victor also describes a "Camp IIIa" that was slightly farther into the ice than Camp III.

32. The water-filled holes varied from a few inches to a few feet in diameter. The latter had been created by the spread of dark dust, known as cryoconite, which blankets part of the ice sheet and hastens melting by absorbing summer sunlight.

33. Victor, "The French Expedition to Greenland, 1948."

34. Victor, "Wringing Secrets from Greenland's Icecap," p. 123.

35. Victor, "Wegener," *Polarforschung* 40 (1970), pp. 2–3.

36. Victor, "Wringing Secrets from Greenland's Icecap," p. 121.

37. Ibid., p. 141.

38. Ibid. Victor quotes Bouché in his article—but Bouché's description of the moment in his own book, *Groenland: Station Centrale*, is slightly different: "A small wind nips us. We shelter near a pile of crates at the kitchen trailer and watch the vehicles as they move away: six black dots recede, then disappear into the snow" (p. 65).

39. Victor, *The French Polar Research Expeditions 1948–1951*, p. 21. The temperatures in the tunnels and storerooms were much lower than in the main hut—usually between –30 and –40 degrees Fahrenheit.

40. Bouché, *Groenland: Station Centrale*, p. 110.

41. Ibid., p. 106.

42. Ibid., Introduction. By the end of the winter, as Victor explained at the start of Bouché's book, Bouché's papers added up to fifteen hundred pages of notes.

43. Ibid., p. 113.

44. Ibid., p. 109.

45. Victor, "Wringing Secrets from Greenland's Icecap," p. 144.

46. Ibid.

47. Ibid.

48. Martin-Nielsen, *Eismitte in the Scientific Imagination*, p. 52.

49. M. Morlighem et al., "Deeply Incised Submarine Glacial Valleys Beneath the Greenland Ice Sheet," *Nature Geoscience* vol. 7 (May 2014).

50. Victor and Dugast, *Paul-Émile Victor: J'ai toujours vécu demain*. The authors chronicle Victor's struggle for funding during this period.

10. The Americans

1. "Birth of a Base," *Life*, September 22, 1952, p. 132. Thule already had a small weather and radio station before the massive 1951 expansion. The building of the base, moreover, was hastened by a new pact signed in 1951 between the United States and Denmark that replaced the 1941 treaty between the two countries. The new treaty allowed the United States generous leeway to establish and enlarge its defense installations.

2. Albert Wohlstetter, "Economic and Strategic Considerations in Air

Base Location," Rand Corporation, 1951. Some estimates put the actual cost of building Thule significantly higher.

3. "The Big Picture: Operation Blue Jay," twenty-seven-minute video, Department of Defense, National Archives and Records Administration, ARC Identifier 2569497, Department of Defense. See also "Birth of a Base," p. 132. Also, "History of Thule," distributed to author at Thule AB in April 2016.

4. "Birth of a Base," p. 132. When Jean Malaurie spoke with some of the first Americans to come ashore, he reported some thought they might be in northern Canada, or even Siberia.

5. Carl Benson, oral history (Karen Brewster, interviewer), June 22, 2001, p. 5. Byrd Polar and Climate Research Center Archival Program, Ohio State University.

6. Thule is situated on Greenland's rocky coast—and a few feet underneath the dirt and rock a permafrost bed is frozen year-round. This made engineering and construction a challenge. The army's solution was to elevate and thickly insulate the floor of all the Thule structures, so that the heat that each building generated would not seep into the permafrost underneath and therefore destabilize the foundations. The idea was mostly successful.

7. No women were assigned "permanent party" to Thule until 1976. "History of Thule," distributed to author at Thule AB.

8. "Defense of Greenland" was signed on April 9, 1941, between Denmark's Henrik Kauffmann and U.S. Secretary of State Cordell Hull. Denmark had already been invaded by Germany, and in effect Kauffmann was representing a government in exile. Whether he therefore had real power to sign the agreement on behalf of Denmark with the United States—seeing as his government was now beholden to a foreign power—is an intriguing historical question beyond the scope of this book, but is discussed in some depth in Denmark's political history literature. In any event, the U.S. government used the agreement as an official justification for occupation.

9. Andrew H. Brown, "Americans Stand Guard in Greenland," *National Geographic*, October 1946, p. 457. "On the very same day, U.S. Marines, vanguard of a survey party, landed in Greenland."

10. Sondrestrom Fjord is now known as Kangerlussuaq and remains the site

of Greenland's international airport, as described in this book's introduction. Its airport code—SFJ—hearkens back to its Danish name.

11. Colonel Bert Balchen et al., "Greenland Adventure," from *War Below Zero* (Boston: Houghton Mifflin Company, 1944), p. 3. For the most part, the Greenland conflicts—exchanging occasional fire with German ships on Greenland's eastern coast or conducting night watches on desolate shores—were justifiably seen as less pivotal than the fiery, high-stakes conflicts on the Normandy shores and Pacific islands. Still, there were several micro-battles in the Arctic—on the island of Spitsbergen, for instance, and on the east Greenland coast.

12. Greenland was deemed valuable by both the Allied and Axis forces, which was why the Germans repeatedly try to set up weather stations on the island's east coast. The weather historian John Ross pointed out, "The Germans had to know how the weather headed their way would affect [the] Luftwaffe's campaign to subdue the Royal Air Force and pave the way for Operation Sea Lion, the planned invasion of England." See Ross, *The Forecast for D-Day* (Guilford, Conn.: Lyons Press, 2014), p. 51.

13. Balchen et al., "Greenland Adventure," from *War Below Zero*, p. 5. In some respects, the notion that Greenland had strategic value long predated World War II—in fact Robert Peary wrote a treatise in 1916 about why Greenland's location made it an ideal site for a naval base. Long before that, moreover—in 1867—U.S. Secretary of State William H. Seward considered the prospect of purchasing Greenland and Iceland, which struck him as rich in natural resources.

14. Nikolaj Petersen, "SAC at Thule: Greenland in the U.S. Polar Strategy," *Journal of Cold War Studies* vol. 13, no. 2 (Spring 2011), pp. 90–115.

15. W. Dale Nelson, "Wanna Buy Greenland? The United States Once Did," Associated Press, May 2, 1991. The AP story was based upon documents in the U.S. National Archives first published by the Danish newspaper *Jyllands-Posten*. See also Natalia Loukacheva, *Arctic Promise: Legal and Political Autonomy of Greenland and Nunavut* (Toronto: University of Toronto Press, 2007), p. 132.

16. Ronald Doel et al., "Defending the North American Continent," from *Exploring Greenland: Cold War Science and Technology on Ice* (New York: Palgrave Macmillan, 2016).

17. Petersen, "SAC at Thule: Greenland in the U.S. Polar Strategy," pp.

90–115. Petersen documents a chain of events that begins with Air Force Secretary Thomas Finletter advocating for bases closer to Russia. In turn Air Force Colonel George Glover asked Balchen to do the study.

18. Balchen recorded (with drama and flair) some of the details of the Sondrestrom project in "Greenland Adventure," from *War Below Zero*.

19. "The Big Picture: Operation Blue Jay." See also "Birth of a Base," p. 132. It is possible that the meeting between Balchen and Rasmussen occurred in 1926 and that the *Life* account (citing 1927) is inexact. Rasmussen was in New York during the fall of 1926 for a lecture tour.

20. At the time, Balchen was helping to set up the Sondrestrom Air Base—then known as Bluie West Eight, now known as Kangerlussuaq airport—on the west side of Greenland, just above the Arctic Circle.

21. Bernt Balchen, *Come North With Me* (New York: E. P. Dutton & Co., 1958), p. 235. Also see Caroll V. Glines, *Bernt Balchen: Polar Aviator* (Washington, D.C.: Smithsonian Institution Press, 1999), p. 143.

22. Petersen, "SAC at Thule: Greenland in the U.S. Polar Strategy," pp. 90–115. Petersen's research into the Balchen archives yielded a memo from December 20, 1950: "Balchen wrote a study concluding that Thule was the only feasible location for an Arctic base and that it offered the 'most favorable' site for a runway for heavy bombers."

23. Jean Malaurie, *The Last Kings of Thule: With the Polar Eskimos, as They Face Their Destiny*, p. 385.

24. The move north did not avert what Malaurie (*The Last Kings of Thule*) later saw as the "social and moral crisis" brought on by their displacement and by the infiltration of Western culture and technology. When he visited Qaanaaq two decades later, he was alarmed by the poor health and pockets of despair amongst the natives, which included some of his old friends and acquaintances, in addition to some men who had traveled with Rasmussen. "The problems posed by the arrival of an ancient society into the modern world remained unsolved," he wrote (p. 415).

25. At the time, in the press and in military publications, efforts to move the Polar Inuit north to modern housing were almost uniformly described as an improvement. The loss of their hunting culture and the long-term effects of displacement—along with thornier questions of legal authority and indigenous rights—were barely considered. Today, an internal history of Thule distributed by American administrators at

the air base concedes: "Although the Inuit were not thrilled with leaving their home, it is said that the noise and smells from the planes and ships had frightened away many of the polar bears, musk ox, seals, narwhals and fish that had previously called the area home and were essential to the Inuit's cultural survival."

26. "As early as 1947 . . . it was pointed out that current Corps of Engineers' equipment, design, construction, and transportation showed inadequacies for arctic operations, and that Engineer laboratory study undertaken in the attempt to correct the inadequacies made evident the necessity of basic, fundamental research into the properties of snow and ice." Richard F. Flint, "Snow, Ice and Permafrost in Military Operations," SIPRE Report 15, September 1953.

27. No doubt Russia's capability was a legacy of its geography and expansionist history in Siberia. Alaska wouldn't become an American state until 1959. Siberia had essentially been part of Russia for several centuries.

28. Victor was a member of the army's Arctic, Desert, and Tropic Information Center (ADTIC), a precursor to CRREL and research divisions discussed later in this chapter that worked in extreme climates.

29. Matthias Heymann et al., "Exploring Greenland: Science and Technology in Cold War Settings," *Scientia Canadensis* 33, no. 2 (2010), pp. 11–42. This paper, along with other research done under the auspices of Aarhus University and Florida State University, is the definitive source on the relationship between Denmark, the U.S. military, and the science conducted in Greenland during the Cold War. See also the collection *Exploring Greenland: Cold War Science and Technology on Ice*, Doel et al., eds.; and Janet Martin-Nielsen, *Eismitte in the Scientific Imagination: Knowledge and Politics at the Center of Greenland*.

30. Flint, "Snow, Ice and Permafrost in Military Operations." Importantly, the goal for the military—and for SIPRE—was both conceptual *and* practical: to gather a more expansive scientific understanding of frozen regions and of ice, and to come up with safe, practical applications for Arctic construction and transportation.

31. The most important drift station in the Arctic Ocean was about twenty square miles and was known as T-3; it was used mainly from 1952 to 1954.

32. SIPRE was founded in 1949 and moved to Wilmette from St. Paul, Minnesota, in 1951. Another foundational organization within the army's engineering corps was ACFEL (Arctic Construction and Frost Effects Laboratory). ACFEL had been formed in 1953 in Boston, the result of combining the corps' existing Frost Effects Laboratory (established 1944) and the corps' Permafrost Division (est. 1945). In 1961, SIPRE and ACFEL were combined into CRREL, the Cold Regions Research and Engineering Laboratory, and relocated to Hanover, N.H. "Both the Frost Effects Laboratory and the Permafrost division . . . had active roles to play in the construction of the airfield at Thule, Greenland." See Edmund A. Wright, *CRREL's First 25 Years: 1961–1986* (Hanover, N.H.: CRREL, 1986).

33. During various author interviews for this book, Carl Benson made the point about how Bader's scientific focus contrasted with his colleagues'; but the insight was made first by author John D. Cox in his explication of Bader's history and of SIPRE. (*Climate Crash: Abrupt Climate Change and What It Means for Our Future* [Washington, D.C.: Joseph Henry Press, 2005]).

34. See, for instance, Henri Bader, *The Physics and Mechanics of Snow as a Material* (Hanover, N.H.: CRREL, July 1962).

35. Edgar May, "They Shiver in August," *Chicago Daily Tribune*, August 21, 1955. Years later, *The Wall Street Journal* noted that Bader's book *Snow and Its Metamorphism* "became to snow what Masters and Johnson later became to sex." Oscar Suris, "There's No Business Like Snow Business in a Period of Thaw," *The Wall Street Journal*, April 23, 1993.

36. Wesley Price, "Every Man Got Out Alive!" *The Saturday Evening Post*, February 9, 1952.

37. Glaciologist Tad Pfeffer would later describe these scientists as "observationalists." See W. Tad Pfeffer, *The Opening of a New Landscape: Columbia Glacier at Mid-Retreat* (Washington, D.C.: AGU Books, 2007), p. 28. Pfeffer's book offers a richer and more complex history of early American glaciology than I've offered here; he also explains how the discipline in the United States evolved separately from European studies in the field. In my conversations with him (Boulder, Colo., May 27, 2014) Pfeffer noted that Robert Sharp at Caltech, who influenced many of the early glaciologists, was himself influenced by a wealthy "gentleman" sci-

entist named Walter Wood, who first exposed him to the glaciers of Alaska.

38. Carl Benson, author interview, August 14, 2015. Charles Bentley, another American glaciologist of the era, quipped that his field was so small that any academic conference to exchange ideas would have to be international, since there weren't enough ice researchers in the United States ranks to actually schedule a formal meeting. See Charles Bentley, oral history, American Institute of Physics, August 6, 2008. Another example is a recollection from Bob Thomas, a glaciologist described in chapter 14: "Glaciology was essentially a dilettante kind of a hobby for people on holiday in the Alps, and they'd see a glacier and they'd go and put a boulder on it and realize it was moving." Robert H. Thomas, oral history (Will Thomas, interviewer) on January 27, 2009 (College Park, Md.: Niels Bohr Library & Archives, American Institute of Physics), aip.org/history-programs/niels-bohr-library/oral-histories/33878).

39. Carl Benson, author interviews, July 11, 2014, August 14, 2015.

40. Regrettably, it took years before glaciology was opened to women scientists, as explained later in this book. Carl Benson recalls that in SIPRE's early days women only worked in clerical or secretarial positions. Even the USGS, which employed a number of female geologists, limited the Arctic areas where they could do research.

41. A number of young researchers—Carl Benson, for instance—were headed to Caltech, where geologist Bob Sharp, who had become interested in the Alaska glaciers, was becoming a mentor to a young generation of glaciologists.

42. Wilford F. Weeks, Polar Oral History Project (Ohio State University, April 17, 2000), pp. 55–56.

43. Price, "Every Man Got Out Alive!"

44. Tony Gow, author interview, Hanover, N.H.: August 6, 2014.

45. Benson recalls that Thule had no proper radar protection at the time. The Americans, in other words, were so intent on getting buildings at Thule up as fast as possible they had left it temporarily vulnerable to surveillance and attack.

46. Carl Benson, author interview, July 11, 2014.

47. Ibid., September 14, 2015.

48. Carl Benson, Polar Oral History Project (Ohio State University, June 22,

2001), p. 6. In his 2001 oral history Benson estimated Bader's time at Site 2 to have been two weeks. In a 2017 author interview—after checking his old logbooks—Benson calculated it was actually thirty-two days.

49. Carl Benson, author interview, July 11, 2014.

50. One of Alfred Wegener's goals during the 1929–30 period was to understand whether Greenland's glaciers were advancing or receding; if it was the latter, it might suggest that Greenland's ice had similarities to the ice sheets that once covered North America and Europe during the Ice Age, tens of thousands of years before the modern era. As the world warmed, those ice sheets had shattered and melted, sometimes with great rapidity. Wegener intended to mark the edges of Greenland's ice cap and make measurements over the course of many years to mark its changes, but his death (and the war) apparently ended that project. Wegener made at least one year of measurements by inserting stakes in the edge of the ice sheet, which in retrospect seems curious but inconclusive, especially because glaciers ebb and flow with natural climate variability, making long-term data sets essential to drawing any conclusions; and Greenland underwent a warm period in the 1930s and 1940s. Nevertheless, before Wegener embarked for Eismitte, one journalist noted: "The melting down of the ice eight feet in a year is taken as first proof of Dr. Wegener's theory that Greenland's ice cap is receding and disappearing, exactly as the glaciers disappeared in North America and Europe in the ice age." *The New York Times*, June 24, 1930.

51. Carl Benson, Polar Oral History Project (Ohio State University, June 22, 2001), p. 14.

52. Carl Benson, author interview, July 11, 2014.

53. Carl Benson, "Stratigraphic Studies in the Snow and Firn of the Greenland Ice Sheet," SIPRE Report 70, July 1962 (Reprinted August 1996).

54. Carl Benson, author interview, July 11, 2014.

55. Chris Polashenski and Zoe Courville, author interview, July 23, 2014. See Polashenski et al., "Observations of Pronounced Greenland Ice Sheet Firn Warming and Implications for Runoff Production," *Geophysical Research Letters* vol. 4 (June 2014). Also see Ned Rozell, "Tracks Across Greenland Ice, 60 Years Apart," Geophysical Institute, University of Alaska–Fairbanks, July 24, 2014.

11. Drilling

1. The depth can depend on location. At Site 2, for instance, the firn turns to ice at 223 feet below the surface.

2. Chester Langway, author interview, Harwich, Mass., May 15, 2013.

3. John Imbrie and Katherine Palmer Imbrie, *Ice Ages: Solving the Mystery* (Cambridge, Mass.: Harvard University Press, 1979), p. 124.

4. Hans Pettersson, "Exploring the Bed of the Ocean," *Nature*, September 17, 1949, p. 468.

5. Raymond S. Bradley, *Paleoclimatology: Reconstructing Climates of the Quaternary*, 3rd ed. (Amherst, Mass.: Elsevier, 2015), p. 453. I'm indebted in this section to Bradley's book for its explanations of paleoclimate science.

6. Andrew Ellicott Douglass, "The Secret of the Southwest Solved by Talkative Tree Rings: Horizons of American History Are Carried Back to A.D. 700 and a Calendar for 1,200 Years Established by National Geographic Expeditions," *National Geographic*, Dec. 1929, pp. 737–70. Douglass's work actually began by following a hypothesis that the growth rates of old trees might help him understand cycles in sunspot activity; when he discerned the value for climatic changes, he changed tack.

7. In more recent times, tree rings have been discovered that go back nearly 14,000 years.

8. Henri Bader, "United States Polar Ice and Snow Studies in the International Geophysical Year," Proceedings of the Symposium at the Opening of the International Geophysical Year, Washington, D.C., June 28–29, 1957.

9. E. Wegener, ed., *Greenland Journey*, p. 147.

10. Carl S. Benson, author interview, July 11, 2014. Bader's paper on bubbles is mostly mathematical in nature and does not delve into the paleoclimate possibilities. See Bader, "The Significance of Air Bubbles in Glacier Ice," *Journal of Glaciology* vol. 1, no. 8 (1950). Perhaps the most crucial advance in utilizing bubbles trapped in ice came slightly later, from the Swedish American scientist Per Scholander. Through experiments in the 1950s, Scholander—who sometimes worked with the Danish scientist Willi Dansgaard—concluded that ice is nearly impermeable to gasses. "This made him wonder whether the air trapped deep in the

Greenland glaciers contains a permanent record of the composition of the atmosphere in earlier periods." See Knut Schmidt-Nielsen, "Per Scholander, 1905–1980: A Biographical Memoir," National Academy of Sciences, Washington, D.C., 1987.

11. Chester C. Langway, Jr. (epigraph), "The History of Early Polar Ice Cores" (Hanover, N.H.: CRREL, January 2008). It seems possible that Bader's early vision of an ice core library was based on a collection of marine sediment cores that was being amassed, beginning in the early 1950s, at Columbia University's Lamont-Doherty laboratory.

12. Epstein was a student of Harold Urey, the Nobel Laureate who is generally credited as the originator of the idea that the isotopic test for oxygen would be useful as a "paleothermometer." In a 1985 oral history (interviewed by Carol Bugé, California Institute of Technology Archives; oralhistories.library.caltech.edu) Epstein notes: "I found that if you sample an ice sequence in a core from an ice cap, you find that there's a nice seasonal variation corresponding to an $^{18}O/^{16}O$ ratio in the ice, with the summer snow being different from the winter snow. And you can then do a lot of stratigraphy and determine how rapidly the snow accumulates . . . We started this whole work." It's arguable, however, that Willi Dansgaard (chapter 12) came up with the idea and spectrometer techniques simultaneously.

13. Carl Benson, oral history (Karen Brewster, interviewer), June 22, 2001, p. 11, Byrd Research Center, Ohio State University. See also Carl Benson, "Stratigraphic Studies in the Snow and Firn of the Greenland Ice Sheet," p. 62. As Benson writes: "Snow deposited in winter has less ^{18}O than that deposited in summer." He also noted Epstein discerned an altitudinal effect on isotopes: "Snow deposited from a single storm shows lower $^{18}O/^{16}O$ ratios at higher altitudes."

14. The geochemistry analysis for ^{16}O and ^{18}O was useful for marine cores, too, following from the idea that plankton skeletons would absorb different amounts of the heavy isotope based on the temperature of the seawater during the period in which they lived. For a deeper discussion of the early isotopic tests and origins, see Imbrie and Imbrie, *Ice Ages: Solving the Mystery*, and Cox, *Climate Crash: Abrupt Climate Change and What It Means for Our Future*.

15. Adrian Howkins, "Science, Environment, and Sovereignty: The Inter-

national Geophysical Year in the Antarctic Peninsula Region," from *Globalizing Polar Science: Reconsidering the International Polar and Geophysical Years* (New York: Palgrave Macmillan, 2011), p. 256.

16. Langway, "The History of Early Polar Ice Cores," p. 9.

17. Fristrup, *The Greenland Ice Cap*, p. 157.

18. "Peter Freuchen Is Dead; Arctic Explorer, Author," *New York Herald Tribune*, September 4, 1957, p. 18. Apparently, Bernt Balchen (who helped chose Thule as a site for the air base) and Donald MacMillan (who overlapped with Freuchen in Greenland years before) were also involved in the television program, a CBS special called *High Adventure with Lowell Thomas*.

19. Henri Bader et al., "Excavations and Installations at SIPRE Test Site, Site 2, Greenland," SIPRE Report 20, April 1955. The pit was dug on a 15-degree angle to make excavation easier, and a rope ladder was installed for access. At the bottom of the 100-foot pit, "a 58-ft core drill hole was sunk by hand drilling from the bottom of the pit to a total depth of 155 ft, measured vertically."

20. Malcolm Mellor, "Undersnow Structures: N-34 Radar Station, Greenland," CRREL Technical Report 132, August 1964.

21. Charles Michael Daugherty, *City Under the Ice: The Story of Camp Century* (New York: The Macmillan Company, 1963), p. 84. Guenther Frankenstein, a SIPRE engineer, recalls, "We had one guy kill himself because there was nothing there. Another guy just went berserk." Guenther "Ginne" Frankenstein, author interview, Hanover, N.H.: October 18, 2015.

22. Chester Langway, author interview.

23. One challenge of drilling into ice is to efficiently remove chips and shavings that get produced as you go deeper; the SIPRE group tried to solve this problem by blowing cold compressed air into the hole as the core was being extracted.

24. Chester Langway, author interview.

25. Chester C. Langway, "A 400 Meter Deep Ice Core in Greenland," *Journal of Glaciology*, March 1958, pp. 216–17.

26. Ibid., pp. 216–17.

27. The drilling at Byrd Station went to 309 meters (1,300 feet); the drilling on the Ross Ice Shelf, at Little America V, encountered water at a depth

of 255 meters (837 feet). See Charles R. Bentley, Bruce R. Koci, "Drilling to the Beds of the Greenland and Antarctic Ice Sheets: A Review," *Annals of Glaciology* 47 (2007).

28. Langway, "The History of Early Polar Ice Cores," pp. 17, 37. He notes that "the U.S. National Academy of Sciences/Committee on Polar Research adjudged the results [of the deep drillings] significant and meriting their highest recommendation for SIPRE to develop a post-IGY deeper ice coring system capable of reaching bedrock depths."

29. The trenches were actually "undercut," meaning that the top of the trench, where it was covered by the metal arch, was narrow (usually eight feet) and that its walls were stepped back and grew progressively wider toward the bottom.

30. Bader and his scientists recognized that by altering the size and temperature of the snow grains, the Peter miller altered the snow's properties. And thus, in the same way that Paul-Émile Victor utilized the weasel to make early travel possible *over* the ice sheet, Bader used the Peter snow miller to make life practicable *under* the ice sheet. Elmer F. Clark, "Camp Century Evolution of Concept and History of Design Construction and Performance," CRREL Technical Report 174, October 1965. The reason the Peter snow hardened to such mechanical strength was that the milling action of the Peter machine "produced a high density (0.55 g/cm^3) granular material with a broad range of grain sizes."

31. The camp at the edge of the ice sheet where Nye stopped—Camp Tuto—was substantial in size (125 buildings and room for 500 people) and served as a station for those coming on or off the ice sheet. Tuto stood for "Thule Take-Off"; and apart from the ramp to the ice sheet, the camp also had a large ice tunnel that U.S. military engineers bored from the edge of the ice sheet with mining equipment for observation. Tuto is now decommissioned and in ruins; the ice sheet ramp collapsed in 2010.

32. Century took its name from being approximately 100 miles from Thule, though in actuality it was 148. The location was chosen because it was effectively the closest point to Thule above the "dry-snow" line, meaning it was not affected by summertime melt, which would have been

problematic for core drilling or—more germane to army officials—a nuclear installation under the ice. The geopolitical intrigue behind Century forms a fascinating backstory that is beyond the scope of this book. As detailed by the Danish academics Henry Nielsen and Kristian H. Nielsen ("Camp Century—Cold War City Under the Ice," *Exploring Greenland: Cold War Science and Technology on Ice*, Ronald E. Doel et al., eds.), the camp construction "had been initiated without Danish government permission." Indeed, their analysis shows the United States seemed well aware of Denmark's deep reservations about the camp's establishment, especially the installation of the nuclear reactor. One reason the Iceworm plan was later shelved, moreover, was due to the prospect of a Danish refusal to allow nuclear weapons in Greenland.

33. A heavy swing could involve anywhere from five to eight tractors, massive loads of diesel fuel, and several dozen sleds and wanigans. The army also conducted "light swings" involving more nimble "polecat" tractors that could travel at about ten miles per hour, thus drastically cutting the travel time from Tuto (see endnote 31, above) to Century.

34. John Frederick Nye, oral history (Paul Merchant, interviewer), The British Library, 2010, sounds.bl.uk.

35. Daugherty, *City Under the Ice*, p. 111.

36. *The City Under Ice*, informational film by the U.S. Army War Office, Research and Development Programs, Report No. 6. Accessible on YouTube. The Cronkite documentary, which is not to be confused with the army's propaganda film, was produced for the program *The Twentieth Century*. It is accessible at cbsnews.com/news/60-minutes-overtime -cronkite-visits-city-under-the-ice.

37. Cassi Feldman, "Cronkite Visits 'City Under the Ice,'" cbsnews.com, January 31, 2016.

38. Austin Kovacs, author interview, Hanover, N.H.: October 19, 2015. In some publications, the army maintained that it was doing rail tests to investigate the potential of creating a more efficient under-ice transport system from Thule to Camp Century.

39. Wayne Tobiasson, author interview, Hanover, N.H., June 2, 2014.

40. Erik D. Weiss, "Cold War Under the Ice: The Army's Bid for a Long-Range Nuclear Role, 1959–1963," *Journal of Cold War Studies* vol. 3,

no. 3 (Fall 2001), pp. 31–58. Weiss notes that some of the reasons be-hind the Iceworm proposal relate to the army's "battle for funds" with the navy and air force. What was obvious, in any case, was that the Iceworm system would have been barely functional, based on the prob-lems of maintaining the integrity of trenches and rooms under the con-stantly moving and changing ice sheet. U.S. military planners estimated the initial investment for Iceworm at $2.37 billion, or nearly $20 billion adjusted for inflation.

41. Chester Langway, author interview.

42. Walter Sullivan, "Army to Drill a Mile-Deep Hole in Greenland Ice," *The New York Times*, August 17, 1963. Bader was interviewed by Sulli-van during a UCLA conference on IGY results; at that point he had left SIPRE to become a professor at the University of Miami, while retaining an advisory role to the U.S. military. In the meantime, SIPRE had merged with another organization to become CRREL—the Cold Re-gions Research and Engineering Laboratory. CRREL was now based in Hanover, New Hampshire.

12. Jesus Ice

1. Some years later, the serious effects of trichloroethylene (TCE) gave rise to health concerns. It is no longer used in ice core drilling. The U.S. Environmental Protection Agency notes that it has been ruled carcino-genic "by all routes of exposure."

2. In later years, scientists discerned that Greenland's ice sheet contained a "brittle ice zone" located between two thousand and four thousand feet down. Ice from this depth required the utmost care when brought up to the surface air; much like a fragile Egyptian relic, it could tolerate only the most delicate touch. In fact, such cores resembled these relics in age. In the brittle zone, the ice comprised snowflakes that fell be-tween four thousand and eight thousand five hundred years ago.

3. Herbert Ueda, author interview, Hanover, N.H., June 2, 2014.

4. Herbert Ueda, oral history (Brian Shoemaker, interviewer), October 23, 2002, Byrd Polar Research Center, Ohio State University.

5. Before Camp Century, Ueda and Hansen tried drilling with the thermal rig (in the summers of 1959 and 1960) at Camp Tuto, at the edge of the

ice sheet, about fourteen miles from Thule Air Base. They began drilling test holes at Century in the summer of 1960.

6. Herbert T. Ueda and Donald E. Garfield, "Drilling Through the Green-land Ice Sheet," CRREL Special Report 126, November 1968.

7. Herbert Ueda, oral history (Brian Shoemaker, interviewer), October 23, 2002, Byrd Polar Research Center, Ohio State University.

8. Some Army Corps engineers have a different recollection of how they procured the drilling rig and recall that it might have been in a storage barn and that Arutunoff himself oversaw the sale.

9. Anthony Gow, author interview, Hanover, N.H.: August 6, 2014.

10. The slow and steady movement of the ice sheet would have eventually wreaked havoc, too, but in the limited number of years Camp Century was open it proved to be much less of a factor than the compression and heat-induced changes.

11. The men also had to master their fear about whether the nuclear reactor would melt down or explode—something Chet Langway in particular worried about. And not every soldier could mentally cope with the iso-lation and under-ice living conditions. Occasionally, someone would try to walk back from Camp Century to Thule (or in one instance, take a tractor). They would be apprehended and moved stateside.

12. Henri Bader, "Scope, Problems, and Potential Value of Deep Core Drill-ing in Ice Sheets" (Hanover, N.H.: CRREL, December 1962).

13. Herbert Ueda, author interview, Hanover, N.H., June 2, 2014.

14. The drilling at Byrd Station was successful—Ueda and Don Garfield reached the bottom in January 1968. However, the drill was lost the fol-lowing year when it became stuck at the bottom of the hole.

15. In late May 1969, Austin Kovacs of the Army Corps of Engineers visited Camp Century to photographically document the effects of the ice sheet three years after the camp had been abandoned. Roving from trench to trench—sometimes a difficult task, as the passages were severely com-pressed in some locations—he captured the camp in the middle stages of collapse, its ceiling arches coming down and its walls closing in. See Austin Kovacs, "Camp Century Revisited: A Pictorial View–June 1969," CRREL Special Report 150, July 1970.

16. The ice cores from Camp Century and other early drilling experiments

have followed a long and complicated journey. First the cores went to New Hampshire; later they went to a commercial freezer in Buffalo, New York, where Langway took a job after CRREL as a geology professor; finally they went to the National Ice Core Laboratory in Denver, where they remain today. Some of the Camp Century cores were shipped to Copenhagen for research, where they are also kept in a commercial freezer near the university.

17. Willi Dansgaard, *Frozen Annals: Greenland Ice Sheet Research* (Copenhagen: Niels Bohr Institute, 2005), p. 50. Dansgaard's first visit to Camp Century in 1964 was to study traces of silicon isotopes. His career before Camp Century involved a variety of projects investigating Greenland's glaciers, including work with Paul-Émile Victor on the EGIG Expedition (Expédition Glaciologique International au Groenland). EGIG involved a European consortium that, beginning in 1959, set out to measure the mass balance of the ice sheet and extract shallow ice cores. When he visited Camp Century in 1964, as Chet Langway notes, Dansgaard imagined "something secret" was happening in the drilling trench. Langway explains: "That wasn't the case, but they wouldn't let anyone in because [drilling engineer] Lyle Hansen got tired of everyone bothering him. He told a major in charge, don't let *anyone* come in. So Dansgaard never got to see it, and I never met him then."

18. J. P. Steffensen, author interview, August 1, 2017.

19. Dansgaard was following a path similar to Caltech's Sam Epstein, who in the mid-1950s helped Carl Benson determine the summer and winter layers in his snow pits when Benson drove around the Greenland ice sheet. They are generally credited with coming up with simultaneous insights on the applications of stable isotope analysis in ice.

20. Estimating the age of each ice core sample was difficult, and Dansgaard based it on what was known of average accumulation, flow, and compaction patterns in that region (and depth) of the ice sheet.

21. Dansgaard, *Frozen Annals: Greenland Ice Sheet Research*, p. 55.

22. Chester Langway, author interview, May 15, 2013. Radiocarbon is only effective as a dating mechanism back to about forty thousand years.

23. Willi Dansgaard et al., "One Thousand Centuries of Climatic Record from Camp Century on the Greenland Ice Sheet," *Science* vol. 166 (October 17, 1969), p. 377.

24. Cox, *Climate Crash: Abrupt Climate Change and What It Means for Our Future*, p. 55.

25. For instance, the country's deepening involvement in wars in Southeast Asia lessened the fear of a cataclysmic battle over Greenland. Janet Martin-Nielsen, the Danish science historian who has written extensively about Camp Century, notes that "the practical realities of operating and fighting in a tropical monsoon climate led to a declining commitment to Arctic research; the focus on Vietnam drew US military strategists' eyes away from the top of the world." See Janet Martin-Nielsen, "'The Deepest and Most Rewarding Hole Ever Drilled': Ice Cores and the Cold War in Greenland," *Annals of Science* vol. 70 (September 2012).

26. The funding streams and number of participants for GISP is quite long. In a summation of the scope and achievements of the project (C. C. Langway, Jr., H. Oeschger, W. Dansgaard, "The Greenland Ice Sheet Program in Perspective," *Greenland Ice Core: Geophysics, Geochemistry, and the Environment*, American Geophysical Union, 1985) Langway acknowledges assistance from a number of academic institutions—SUNY Buffalo, University of Copenhagen, University of Bern, Technical University of Denmark, University of Nebraska–Lincoln, and a number of military divisions in Denmark and the United States, including the army's Cold Regions Research Engineering Laboratory (CRREL). In addition to the NSF, the project received support from the Danish Commission for Scientific Research in Greenland, the Swiss National Science Foundation, the Carlsberg Foundation, and the Danish Natural Science Research Council.

27. Chester Langway, author interview, May 15, 2013.

28. N. S. Gundestrup, S. J. Johnsen, and N. Reeh, "ISTUK: A Deep Ice Core Drill System," from *Ice Drilling Technology*, CRREL Special Report 83–34, December 1984.

29. The Dye stations apparently took their name from Cape Dyer on Baffin Island.

30. Dansgaard, *Frozen Annals*, p. 83.

31. Ibid., p. 86.

32. J. P. Steffensen, author interview, August 1, 2017. All Steffensen quotes in this section are from this conversation.

33. Walter Sullivan, "Ancient Ice Yielding Secrets of Climate," *The New York Times*, August 9, 1981.

34. The machine looked hand-built by mad scientists but was not particularly large: About the size of a dishwasher, it contained an imbroglio of snaking tubes and chambers and was flanked on each side by two lab benches holdings horizontal rows of laboratory vials containing ice samples.

35. Jouzel agreed to let White use his machine in Paris to measure Dansgaard's Greenland ice, but on one condition: He had to do it on the weekends. Apparently, Jouzel's boss did not get along with Willi Dansgaard, so the Danish lab work had to be done discreetly.

36. James White, author interview, Boulder, Colo., May 29, 2014. Some of the rapid climate changes evidenced in the Greenland core were subsequently named (by Wally Broecker) Dansgaard-Oeschger events, or more simply, D-O events.

37. The coldest temperature ever recorded on earth occurred at Vostok on July 21, 1983: –128.6 degrees Fahrenheit. The station was almost certainly the coldest and most remote manned climate camp on earth—an outpost, as the climate historian Spencer Weart noted, where work "was fueled by the typically Russian combination of cigarettes, vodka, and stubborn persistence." (Spencer R. Weart, *The Discovery of Global Warming* [Cambridge, Mass.: Harvard University Press, 2008], p. 126.) Willi Dansgaard once visited the station for a few hours to retrieve a sample of ice; the combination of high altitude, celebratory vodka shots, and sprinting to the runway to make his plane before it departed left him gasping for air. Of course, if he didn't make the plane, he might be left at Vostok for another year, since the station was rarely serviced by air transport. Dansgaard awoke on the floor of the cabin, airborne, being fed oxygen.

38. White adds: "We now have several ice cores from Greenland and they pretty much tell you the same story—that you've got a degree per year for five years, it plateaus, then you have a degree per year for five years after that." A degree Celsius per year is about a hundred times faster than what the earth has experienced in the past century—roughly a degree of temperature rise over a hundred or so years. As it happens, an

impact crater from a large meteorite was discovered under the ice sheet in 2018. Until further research is done, its age (and effect) remain unknown.

13. Deeper

1. James White, author interview, Boulder, Colo., May 29, 2014.
2. More fundamentally, Joseph Fourier in 1827 recognized that earth's atmosphere controls the balance between the solar radiation that strikes earth and the (infrared) energy that returns to space. Fourier is often credited as the discoverer of the greenhouse effect. Foote's work appears to have preceded Tyndall's by several years, but she has only been recognized for her early contributions in the last decade. Her original paper, "Circumstances Affecting the Heat of the Sun's Rays" (*The American Journal of Science and Arts*, New Haven, G.P. Putnam & Co., 1856, p. 382) presciently noted: "The highest effect of the sun's rays I have found to be in carbonic acid gas"—i.e., CO_2.
3. G. S. Callendar, "The Artificial Production of Carbon Dioxide and Its Influence on Temperature," from *The Global Warming Reader: A Century of Writing About Climate Change*, ed. Bill McKibben (New York: Penguin Books, 2012).
4. See Gilbert N. Plass, "Carbon Dioxide and Climate," *Scientific American*, July 1959. The television show was part of the Bell Telephone Science Series, *The Unchained Goddess*, 1958. The episode was directed by Frank Capra, whose great claim to fame was as the director of the Christmas classic *It's a Wonderful Life*.
5. Waldemar Kaempffert, "Science in Review: Warmer Climate on the Earth May Be Due to More Carbon Dioxide in the Air," *The New York Times*, October 28, 1956, p. 191.
6. The two scientists were working at the Geophysical Fluid Dynamics Laboratory, first in Washington, D.C., and then in Princeton, New Jersey. For Manabe's original papers—"The Balance of Energy" (1967) and "The Birth of the General Circulation Climate Model" (1975)—see David Archer and Raymond Pierrehumbert, ed., *The Warming Papers: The Scientific Foundation for the Climate Change Forecast* (West Sussex, U.K.: Wiley-Blackwell, 2011). For context on the historical develop-

ment of Manabe's models, see Weart, *The Discovery of Global Warming*. For the poll on the most important climate papers, see Roz Pidcock, "The Most Influential Climate Change Papers of All Time," July 6, 2015, carbonbrief.org.

7. Mikhail Budyko, "The Effect of Solar Radiation Variations on the Climate of the Earth," *Tellus* 21 (1969), p. 618. Thanks to a citation in Weart's *The Discovery of Global Warming* for helping locate this study.

8. Oil and gas companies were likewise aware of the perils of man-made warming at an early stage, at least as far back as 1968, when the Stanford Research Institute prepared a long study ("Sources, Abundance, and Fate of Gaseous Atmospheric Pollutants," see www.osti.gov/biblio/6852325) on the effects of atmospheric pollutants for the American Petroleum Institute, the industry's trade and lobbying group. Yet fossil fuel companies often adopted a media policy of professing doubt about the effects of CO_2 even as their own internal studies suggested otherwise. See, for instance, Neela Banerjee et al., "Exxon: The Road Not Taken," *Inside Climate News*, September 16, 2015–December 22, 2015. See also Geoffrey Supran and Naomi Oreskes, "Assessing ExxonMobil's Climate Change Communications (1977–2014)," *Environmental Research Letters* vol. 12, no. 8 (August 23, 2017).

9. Chester C. Langway, "The History of Early Polar Ice Cores," p. 31.

10. Scientists measure concentrations of CO_2 in "parts per million," or ppm, which describes how many molecules of carbon dioxide are present in a sample that contains a million molecules of air. Bubbles in ice cores show that before the industrial revolution began in the 1800s, atmospheric CO_2 levels were about 275 ppm. In 1958, Charles Keeling, a scientist at the Scripps Institution of Oceanography, began measuring CO_2 concentrations at an observatory in Hawaii. By Keeling's account, the CO_2 level in 1958 was 315 ppm. By the year 2000, the CO_2 level had climbed to 371 ppm. Since then, increases have accelerated. In 2018, CO_2 levels reached 409 ppm.

11. The 1977 report concluded that while the dangers of CO_2 emissions were serious and inevitable, the present state of knowledge meant that "it is impossible to forecast what might happen to the Greenland and Antarctic ice caps as a result of a rise of several degrees in global average

air temperature." See "Energy and Climate" (Washington, D.C.: National Academy of Sciences, 1977), pp. 8, 52, 56.

12. See, for instance, Jule G. Charney et al., "Carbon Dioxide and Climate: A Scientific Assessment" (Washington, D.C.: National Academy of Sciences, 1979); and William A. Nierenberg et al., *Changing Climate* (Washington, D.C.: National Academy Press, 1983). The prospect of global warming was also presented to the U.S. Congress in testimony by scientists such as NASA's James Hansen during the early 1980s. For a compelling summary of the political and policy battles during this period, see: "Losing Earth," by Nathaniel Rich, *The New York Times Magazine*, August 1, 2018.

13. National Research Council, *Recommendations for a U.S. Ice Coring Program* (Washington, D.C.: National Academy Press, 1986).

14. The political and logistical discussions that preceded the start of GISP-2 are compelling but somewhat tangential to this book, and to the scientific investigations at the site. Much of the credit for getting the project going during the mid-1980s goes to Broecker, of Columbia University's Lamont-Doherty laboratory, who initiated the Boston meeting and helped bring Mayewski into the project. The drilling plans up to that point, Broecker says (author interview, Lamont-Doherty, November 3, 2014), "were just a big mess." Some of the funding disagreements were related to the fact that officials at the National Science Foundation felt they had in effect funded European research in the Dye-3 coring and declined to do so again. Subsequent delays had to do with the refusal on the part of American science funders to replicate the Istuk drill to speed the project's deployment. Instead, the decision was made to build a new deep-coring drill from scratch, partially to add to the expertise of American drillers but also to design a machine that could extract a larger core that could better serve the research needs of the large array of institutions.

15. Paul Mayewski, author interview, February 6, 2018.

16. Paul Andrew Mayewski and Frank White, *The Ice Chronicles: The Quest to Understand Global Climate Change* (Hanover, N.H.: University Press of New England, 2002), p. 62.

17. The GISP-2 drill yielded a 5.2-inch diameter core. Previous cores had

been slimmer, and the Danish core at nearby GRIP was about 3 inches in diameter. One reason for the larger drill and larger core was that a large number of American scientists had professed an interest in working on ice samples; a 5.2-inch core provides far more ice than a 3-inch core. "The great advantage to the 5-inch drill," Mayewski says, "was the fact that we could have a large archive."

18. In addition to Denmark, Switzerland, and France, "Under the auspices of the European Science Foundation (ESF), five other countries— German, the UK, Belgium, Iceland and Italy—joined to launch GRIP." Jean Jouzel, "A Brief History of Ice Core Science Over the Last 50 Years," *Climate of the Past*, November 6, 2013. For the first time, the ranks of scientists, technicians, and ice drillers on the ice sheet included a number of women.

19. GISP-2 and GRIP are located about eighty miles north of Eismitte and Station Centrale. The author flew to the center of the ice sheet several times with the 109th Airlift Wing in summer 2017.

20. GISP-2 was on the western slope of the ice divide, whereas GRIP was on the divide itself; thus GRIP was slightly higher in elevation.

21. This section derives from author interviews with Richard Alley, James White, Kendrick Taylor, Paul Mayewski, Tony Gow, and Joan Fitzpatrick; it is also informed by Richard B. Alley et al., "Visual-Stratigraphic Dating of the GISP2 Ice Core: Basis, Reproducibility, and Application," *Journal of Geophysical Research* vol. 102, no. C12 (November 30, 1997); by Mayewski and White's *The Ice Chronicles*; by Alley's book on ice cores, *The Two-Mile Time Machine* (Princeton, N.J.: Princeton University Press, 2000); and by Kendrick Taylor's essay on the GISP-2 work: "Rapid Climate Change: New Evidence Shows that Earth's Climate Can Change Dramatically in Only a Decade. Could Greenhouse Gasses Flip that Switch?" *American Scientist* vol. 87 (July–August 1999), p. 320.

22. Joan Fitzpatrick, author interview, Denver, Colo., May 28, 2014.

23. Richard Alley, author interview, University Park, Penn., July 30, 2014.

24. The ideal footwear for the ice sheet was developed by the army in the 1950s; known as "extreme cold barrier vapor boots," the shoes contain thick wool padding in the sole to ward off low temperatures. They are often referred to as "bunny boots" or "Mickey Mouse boots" because of their appearance.

25. Kendrick Taylor, "Rapid Climate Change: New Evidence Shows that Earth's Climate Can Change Dramatically in Only a Decade. Could Greenhouse Gasses Flip that Switch?"

26. Todd Sowers, author interview, Copenhagen, November 6, 2017.

27. Herman Zimmerman, author interview, March 2, 2018.

28. Anthony J. Gow, oral history (Brian Shoemaker, interviewer), October 26, 2002, Byrd Polar Research Center, Ohio State University.

29. Mayewski and White, *The Ice Chronicles*, p. 82.

30. Claus Hammer, Paul A. Mayewski, David Peel, Minze Stuiver, "Preface," *Journal of Geophysical Research* vol. 102, no. C12 (November 20, 1997), pp. 26,315–26,316. The only significant shortcoming in the cores was that ice below one hundred thousand years old ("Eemian ice" that dated back to another warm period in earth's history) was folded and not readable. "At the bottom of the ice sheet," Kendrick Taylor explains, "the data sets don't match up between GISP-2 and GRIP. They match up for 98 percent of the core, but not for the bottom 2 percent." Once the initial disappointment about the old ice sunk in, some of the scientists—especially Dorthe Dahl-Jensen, of the Danish group—resolved to go back to the ice sheet on future expeditions to look for better Eemian ice. Indeed, it defined much of their later careers.

31. For more on abrupt climate change, see, for instance, *Abrupt Impacts of Climate Change: Anticipating Surprises*, National Research Council, National Academies Press, Washington, D.C., 2013.

32. Carbon dioxide played a major part in climate changes as well. Around the time of GISP-2, ice cores drilled in Antarctica showed that going back hundreds of thousands of years, whenever earth's temperature rose, so did levels of CO_2; whenever earth's temperature fell, so did levels of CO_2. The correlation did not mean CO_2 *caused* the rise and fall in temperatures. But it seemed to indicate that increasing concentrations of the gas could not only act as a "forcing" agent that made the climate warmer but could also act as a "feedback" agent that made an already warming climate much hotter. To Alley, it seemed clear that many things affected the earth's climate, especially the Milankovitch cycles that changed the way sunlight hit the planet over the course of thousands of years. There are lots of knobs that control the climate, he would note. But while some are "fine tuning knobs," CO_2, he was convinced,

was the big knob. See Richard Alley, "The Biggest Control Knob: Carbon Dioxide in Earth's Climate History," American Geophysical Union conference, 2009, San Francisco.

33. Alley, *The Two-Mile Time Machine*.

34. Richard Alley, whatweknow.aaas.org/richard-alley.

14. Sensing

1. "Nansen, Here, Tells Arctic Flight Plan: Veteran Explorer to Use Graf Zeppelin for Polar Trip Next Year," *The New York Times*, January 23, 1929. Nansen wanted to enlist American support for building a mast for the dirigible in Nome, Alaska. The aging explorer's intensity and bearing is illuminated in eloquent letters he wrote to Brenda Ueland, an American who had come to interview him upon his 1929 arrival in the United States. A passionate epistolary episode followed. See Eric Utne, ed., *Brenda My Darling: The Love Letters of Fridtjof Nansen to Brenda Ueland* (Minneapolis: Utne Institute, 2011).

2. For instance, the Italian explorer Umberto Nobile built and piloted the *Norge*, which was lofted by hydrogen and was used by the Norwegian explorer Roald Amundsen in 1926 to fly over the North Pole, going from Svalbard to Alaska. In 1928, another airship, the *Italia*, also piloted by Nobile, overflew the pole but later crashed on sea ice near Svalbard. Nobile survived but several of his crewmembers did not. The idea of aerial photography and remote sensing dates back much further than Nobile or World War II, however—to the use of airborne cameras attached to balloons, first used in Paris in the late 1850s, and (somewhat later) to cameras attached to pigeons.

3. "Nansen, Here, Tells Arctic Flight Plan: Veteran Explorer to Use Graf Zeppelin for Polar Trip Next Year," *The New York Times*.

4. "Arctic Flight of the Graf Zeppelin," *Science* vol. 70, no. 1826 (December 27, 1929).

5. "Meteorological Observations with *Nimbus I*," *International Geophysics Bulletin* no. 91, January 1965.

6. Bader et al., *Polar Research: A Survey* (Washington, D.C.: National Academy of Sciences, 1970). Henri Bader contributed large portions of this report.

7. "Symposium on Remote Sensing in the Polar Regions," March 6–8,

1968, Arctic Institute of North America. The minutes note: "The big uncertainty in the appraisal of the mass balance of the icecaps is the determination of the loss by icebergs from Antarctica and from Greenland, the difficulty of determining net accumulation over large areas, and the difficulty of measuring melting under the floating ice shelves. For those kinds of problems remote sensing appears to be the only practical solution" (p. 39).

8. *The Terrestrial Environment: Solid-Earth and Ocean Physics* (Washington, D.C.: National Aeronautics and Space Administration, April 1970).

9. Many of the flights were conducted between 1968 and 1974. See Langway et al., "The Greenland Ice Sheet Program in Perspective."

10. Richard Finsterwalder, "Expédition Glaciologique Internationale au Groenland, 1957–60," *Journal of Glaciology* vol. 3, no. 26 (1959), p. 542. The expedition was more commonly known as EGIG.

11. Fristrup's *The Greenland Ice Cap* credits Fritz Loewe with attempting the first serious mass balance calculation in 1936, and made note of Bader's opinion on Greenland's growing mass balance and his methods of calculation.

12. Fristrup, *The Greenland Ice Cap*, p. 282.

13. R. H. Thomas et al., "Satellite Remote Sensing for Ice Sheet Research," NASA Technical Memorandum 86233, 1985. Also of note was the assessment of two British glaciologists who, in writing a 1987 historical assessment about studies of the ice sheet, concluded that "estimates of the balance between [ice] input and [ice] loss are of little value." The margins for error were just too large. Gordon de Quetteville Robin and Charles Swithinbank, "Fifty Years of Progress in Understanding Ice Sheets," *Journal of Glaciology* special issue (1987), p. 33. The authors noted at the time: "Estimates for Greenland lie within the range of 515 ± 115 km^3a^{-1} accumulation, 325 ± 105 km^3a^{-1} for melting, and 240 ± 80 km^3a^{-1} for iceberg discharge."

14. Dorthe Dahl-Jensen, author interview, Copenhagen, November 6, 2017.

15. Joan Fitzpatrick, author interview, Denver, March 28, 2014.

16. Robert H. Thomas, oral history (Will Thomas, interviewer).

17. William Krabill, author interview, March 27, 2018. Krabill notes that precise laser elevation measurements also require that technicians use

GPS to remove any vertical changs in aircraft position *during* the laser data collection.

18. Robert H. Thomas, oral history (Will Thomas, interviewer).

19. John Sonntag, author interview, NASA Goddard Space Center, Greenbelt, Md., February 6, 2018.

20. W. Krabill et al., "Rapid Thinning of Parts of the Southern Greenland Ice Sheet," *Science* vol. 283 (March 5, 1999). W. Krabill et al., "Greenland Ice Sheet: High-Elevation Balance and Peripheral Thinning," *Science* vol. 289 (July 21, 2000). Because of the complexities of converting the volume loss to mass loss—in Greenland, snow compaction makes it difficult to assess the precise mass balance decline—the loss was calculated in km³. The GRACE satellite would later solve that problem. To contextualize the remote sensing data, PARCA also included teams of field scientists working on the ground, digging snow pits to measure accumulation rates and assessing the dynamics of glaciers.

21. John Sonntag, author interview.

22. Jay Zwally, author interview, NASA Goddard Space Center, Greenbelt, Md., February 6, 2018.

23. "ICESat: Ice, Cloud and land Elevation Satellite," NASA informational brochure, undated (circa 2001), 26 pages.

24. "The Terrestrial Environment: Solid-Earth and Ocean Physics."

25. Mike Watkins, author interview, Jet Propulsion Laboratory, Pasadena, Calif., February 13, 2017.

26. Jay Famiglietti, author interview, Jet Propulsion Laboratory, Pasadena, Calif., February 13, 2017.

27. Mike Watkins, author interview, Jet Propulsion Laboratory, Pasadena, Calif., February 13, 2017.

28. Mike Gross, author interview, Jet Propulsion Laboratory, Pasadena, Calif., February 13, 2017.

29. Isabella Velicogna, author interview, University of California Irvine, May 11, 2015.

30. Isabella Velicogna and John Wahr, "Measurements of Time-Variable Gravity Show Mass Loss in Antarctica," *Science*, March 24, 2006; Isabella Velicogna and John Wahr, "Acceleration of Greenland Ice Mass Loss in Spring 2004," *Nature*, September 21, 2006.

31. Berrien Moore, author interview, April 30, 2017.

32. I. Velicogna, "Increasing Rates of Ice Mass Loss from the Greenland and Antarctic Ice Sheets Revealed by GRACE," *Geophysical Research Letters* vol. 36 (October 2009).

33. Ben Iannotta, "Keep Watching the Ice," *Air & Space Magazine*, September 2006.

34. Political denials regarding the evidence of climate change seem even more absurd in light of sea level change data. A reliable tidal gauge record showing a steady upward progression goes back to about 1870; satellite readings go back to the mid-1990s. There is close agreement. See https://sealevel.nasa.gov/news/108/new-study-finds-sea-level-rise -accelerating. "The rate of sea level rise in the satellite era has risen from about 0.1 inch (2.5 millimeters) per year in the 1990s to about 0.13 inches (3.4 millimeters) per year [in 2018]."

35. Alexandra Witze, "Losing Greenland: Is the Arctic's Biggest Ice Sheet in Irreversible Meltdown? And Would We Know It If It Were?" *Nature* vol. 452 (April 17, 2008).

36. Crain oversees Greenland and Antarctic research at the National Science Foundation.

37. Tom Wagner, author interview, Washington, D.C., March 17, 2015.

38. Eric Rignot and Robert H. Thomas, "Mass Balance of Polar Ice Sheets," *Science* vol. 297 (August 30, 2002).

39. There are several events in the relatively recent geological record—known as Meltwater Pulse 1a, 1b, 1c, and so forth—that mark the disintegration of northern ice sheets into the ocean. Meltwater Pulse 1a, for instance, probably occurred about 14,500 years ago, pushed a flotilla of massive icebergs into northern Atlantic waters, and raised sea levels by about 10 meters, or 33 feet, over the course of about 300 years. The ice from Antarctica might have contributed to the sea level rise as well.

40. Eric Rignot, author interview, University of California Irvine, May 11, 2015.

15. A Key

1. Justin Gillis, "As Glaciers Melt, Science Seeks Data on Rising Seas," *The New York Times*, November 13, 2010, p. A1.

2. As it turned out, ICESat's replacement—ICESat-2—wasn't launched into space until September 15, 2018.

3. John Sonntag, author interview, NASA Goddard Space Center, Greenbelt, Md., February 6, 2018.

4. NASA compiles a chart of GRACE measurements of Greenland and Antarctica at climate.nasa.gov. Some years the Greenland total might be less than 286 billion tons. But at other times—a worrisome indicator of the ice sheet's future—the total can go far higher. In 2012, for instance, the year the surface of the ice sheet melted for a few days, the island lost 562 billion tons.

5. John Sonntag, author interview, NASA Goddard Space Center, Greenbelt, Md., February 6, 2018.

6. Joughin et al., "Brief Communication: Further Summer Speedup of Jakobshavn Isbrae," *The Cryosphere* vol. 8 (2014), pp. 209–14.

7. Amanda Schupak, "Greenland Glacier Loses Chunk the Size of Manhattan," cbsnews.com, August 24, 2015.

8. John Sonntag, author interview, in the air over Jakobshavn, Greenland, April 9, 2015.

9. Anker Weidick and Ole Bennike, "Quaternary Glaciation History and Glaciology of Jakobshavn Isbrae and the Disko Bugt region, West Greenland: A Review," Geological Survey of Denmark and Greenland, Bulletin 14, 2007.

10. Alfred Wegener, *Greenland Journal*, mid-to-late August 1913. Accessed digitally at www.environmentandsociety.org/exhibitions/wegener-diaries/original-document. Wegener perceived the retreat by consulting a record of Jakobshavn's calving front from 1902, the last time before his visit that someone had measured it.

11. All David Holland quotes in this chapter, except where noted: Author interviews, Jakobshavn calving front, June 6–12, 2016.

12. A few years later, in the summer of 2018, the Hollands had a similar experience at Helheim Glacier, on the east coast, where Denise captured on video the calving of a four-mile-long tabular iceberg.

13. In 2008, Holland and Bob Thomas, along with a few other colleagues, published a long academic paper that connected the waters in the Jakobshavn Fjord with the acceleration of the glacier. They concluded that a doubling in the speed of the glacier was "triggered" by the arrival of warm water coming over from Iceland. And that warmer water, in turn, was traceable to atmospheric changes in the polar region. Their ulti-

mate point was that if scientists wanted to build better computer models to predict future sea levels, they needed to account for how water temperatures appeared to be a control knob that regulated the movement of any adjacent glaciers. That meant trying to figure out, more generally, how regional ocean and atmosphere circulation brings warm water "to the periphery of the ice sheets." See David M. Holland et al., "Acceleration of Jakobshavn Isbrae Triggered by Warm Subsurface Ocean Waters," *Nature Geoscience* vol. 1 (October 2008), p. 659.

14. See, for example, Pollard, DeConto, and Alley, "Potential Ice Sheet Retreat Driven by Hydrofracturing and Ice Cliff Failure," *Earth and Planetary Science Letters* vol. 412 (February 5, 2015).

15. David Holland, lecture, "Climate Days," Ilulissat, Greenland, June 2, 2015.

16. See, for instance, Jeff Goodell, "The Doomsday Glacier," *Rolling Stone*, May 9, 2017.

17. "West Antarctica Glacier Loss Appears Unstoppable," NASA Jet Propulsion Laboratory, press release, May 12, 2014. The same day, the magazine *Mother Jones* ran a headline on a story by the climate journalist Chris Mooney that read, "This Is What a Holy Shit Moment for Global Warming Looks Like."

16. Meltwater Season

1. See, for instance, Coral Davenport, Josh Haner, Larry Buchanan, and Derek Watkins, "Greenland Is Melting Away," *The New York Times*, October 27, 2015; and Elizabeth Kolbert, "Greenland Is Melting," *The New Yorker*, October 24, 2016. A number of television documentaries—on HBO's *Vice* or on CNN, for instance—carried the same title.

2. National Oceanic and Atmospheric Administration, Arctic Report Card, 2017, arctic.noaa.gov/Report-Card.

3. Arctic Encounter Symposium, Seattle, Wash., April 19, 2018.

4. At a Washington, D.C., conference on the science of the northern regions Julie Brigham-Grette, a glacial geologist, explained: "It isn't just that the Arctic is changing—everything changes. It's the rate of change that is so astounding." Julie Brigham-Grette, presentation, "Arctic Matters," National Academy of Sciences, Washington, D.C.: January 14, 2016.

5. Author interview, Paul Mayewski, March 6, 2018.

6. Chris Mooney, "The Ocean's Circulation Hasn't Been This Sluggish in 1,000 Years. That's Bad News," *The Washington Post*, April 11, 2018.

7. It is possible that sea levels near Greenland will actually go down in the near term, due to the reduced gravitational pull of the diminishing ice sheet on the surrounding water. It is also possible, however, that diminished Antarctic land ice will counterbalance this change. See, for instance, Elizabeth Gudrais, "The Gravity of Glacial Melt," *Harvard* magazine, May–June 2010.

8. Sofus Alataq, author interview, Thule, Greenland, April 26, 2016. See also Jon Gertner, "An Astronaut Finds Himself in Greenland," *The New Yorker*, May 13, 2016.

9. Toku Oskima, author interview, Ilulissat, Greenland, June 3, 2015.

10. This balance—half lost through calving, and half lost through melt—has varied during different years, and many glaciologists expect that as the climate continues to warm, ice lost through meltwater runoff will exceed iceberg losses.

11. Koni Steffen, author interview, Ilulissat, Greenland, June 3, 2015. Swiss Camp was originally situated on the ice sheet's "equilibrium line," where the snow accumulation during winter was balanced by the melt in the summer. In addition to the changes wrought by a warming climate—the eastward and upward movement of the melt zone—the ice under the camp has also been moving west, at about 30 centimeters a day, due to the constant flow of the ice sheet. In all, the camp has moved between 2.5 and 3 kilometers toward the west coast since its establishment in 1990.

12. Greenland's subsurface plumbing is similarly mysterious. In warmer months, water from rivers and streams crosscuts the ice sheet's surface and runs into moulins near the edges, like water swirling and then dropping down a drain. The holes or cracks usually go down to the base of the ice sheet, where the water apparently forms channels between the rock and the ice sheet and tries to find its way out. Sometimes it makes glaciers accelerate, especially in the summer months. But where and how the water gets off the island is not completely understood. Steffen worked with a team that dropped ninety yellow rubber ducks into a moulin, with contact information on the ducks in case they were found.

Only two ducks were recovered—the following year—thanks to a fisherman working in a nearby bay, meaning that somewhere, trapped deep in the Greenland ice sheet, or floating in the waters nearby, there are eighty-eight more rubber ducks. See Jon Gertner, "The Secrets in the Ice," *The New York Times Magazine*, November 15, 2015. The duck experiment began the start of a multiyear study attempting to examine how melting water on Greenland's surface moves over and through the ice sheet and eventually is transported to the sea. Lawrence Wright of UCLA, one of the main investigators of the study, noted in 2015 that his research suggests that rather than a big block of ice, "the ice sheet is porous, like swiss cheese." Coral Davenport, Josh Haner, Larry Buchanan, Derek Watkins, "Greenland Is Melting Away," *The New York Times*, October 17, 2015.

13. A number of scientists are studying the dark zones of the western ice sheet. "The big question is, in a warming world where more of the ice sheet is melting, will more of the ice sheet be colonized by algal blooms, and then get darker? That's a self-perpetuating cycle." Joe Cook and Andrew Tedstone, author interview, Kangerlussuaq, Greenland, July 30, 2017.

14. The ice sheet albedo can also be altered by an increase (or decrease) in bright, fresh snow that falls during a particular season, as well as by melt events on the surface that alter the snow crystals' structural size and reflectivity. See, for instance, Marco Tedesco et al., "The Darkening of the Greenland Ice Sheet: Trends, Drivers, and Projections (1981–2100)," *The Cryosphere*, March 3, 2016.

15. Luke D. Trusel et al., "Nonlinear Rise in Greenland Runoff in Response to Post-industrial Arctic Warming," *Nature*, December 5, 2018.

16. Koni Steffen, *Greenland Is Melting*, short documentary by Jason van Bruggen, youtube.com/watch?v=HKp7OaCQrOM.

17. Koni Steffen, author interview.

18. Tad Pfeffer, "Glaciology Needs to Come Out of the Ivory Tower," *Earth* vol. 57, no. 11, p. 8.

19. Eric Rignot, author interview, Irvine, California, May 11, 2015.

20. Ibid.

21. There remains the possibility that Greenland's ice will be more vulnerable in a warming world. A 2016 study that Alley worked on focused on

bedrock samples extracted at the bottom of the GISP-2 ice core (Joerg Schaefer et al., "Greenland Was Nearly Ice-Free for Extended Periods During the Pleistocene," *Nature* vol. 540, December 8, 2016). The study concluded that Greenland lost the vast majority of its ice sheet during certain warm periods in the past.

22. Jon Gertner, "Is It O.K. to Tinker with the Environment to Fight Climate Change?" *The New York Times Magazine*, April 18, 2017.

23. John C. Moore et al., "Geoengineer Polar Glaciers to Slow Sea-Level Rise," *Nature* vol. 55 (March 14, 2018). The proposal was provocative and included a plan for geoengineering Thwaites Glacier, too. It elicited pushback about its feasibility and outcome. Twila Moon of the National Snow and Ice Data Center responded to *Nature*: "The authors' suggestions (building ocean-bottom sills, installing pinning-point islands and removing subglacial water) might briefly slow [glacial] outflow. However, these strategies could easily cause ice build-up that would overwhelm structural impediments, and further accelerate ice loss."

24. Kevin Anderson, author interview, March 20, 2017.

Epilogue

1. William Colgan et al., "The Abandoned Ice Sheet Base at Camp Century, Greenland, in a Warming Climate," *Geophysical Research Letters* vol. 43 (August 15, 2016).

2. Levermann's group perceived in their projections a close linear relation between the warming and sea-level commitment—rises of 2.3 meters for every *additional* degree (C) of warming.

3. Levermann et al., "The Multimillenial Sea-Level Commitment of Global Warming," Proceedings of the National Academy of Sciences (PNAS), August 20, 2013.

4. Clark et al., "Consequences of Twenty-First-Century Policy for Multi-Millenial Climate and Sea-Level Change," *Nature Climate Change* vol. 6 (February 8, 2016). By "intervention," Clark suggests the possible future deployment of carbon removal technologies.

5. Kerry Emanuel to Miles O'Brien, *PBS NewsHour*, January 10, 2018.

6. Remote sensing efforts of the cryosphere now extend well beyond NASA's work; valuable data is also being collected by the satellites of ESA (European Space Agency) and other countries.

7. Katherine Brooks, "Artist Melts 100 Tons of Arctic Ice to Remind You that Climate Change Is Real," *The Huffington Post*, October 29, 2014.

8. Maria Gallucci, "Dramatic Venice Sculpture Comes with a Big Climate Change Warning," *Mashable*, May 15, 2017. Amongst other striking art projects were Peggy Weil's *88 Cores*, a video chronicle of the ice cores drilled at GISP-2, and Justin Brice Guariglia's large-scale prints that include images taken from Operation IceBridge flights over Greenland.

SOURCES

From the start, this project was conceived as a story about ice and the process of discovery—a book that would trace the scientific exploration of Greenland and explain how this research, aided by technological tools, led to our current moment of environmental awareness. I conducted my reporting between 2012 and 2018, and made six visits to Greenland and seven to Copenhagen. During the course of my research I also made trips to Iceland, where Nansen and Wegener launched their expeditions, and to Munich, to see the satellites for NASA's GRACE mission. In the United States, I visited scientists and engineers in California, Washington, Colorado, Illinois, New Hampshire, Maine, New York, New Jersey, Pennsylvania, Massachusetts, Maryland, and Washington, D.C. I spoke with many more by phone. I gained tremendous insights from scientists quoted in the text, yet I learned just as much from those who are not. In sum, I talked with many more researchers than could plausibly fit into this book's narrative. A list comprising most of these interviews, along with some of the oral histories I used, follows at the end of this section.

Greenland's ice sheet may strike some readers as a novel subject. But it seems necessary to point out that many journalists have been reporting on this topic for years, and in varying respects their articles have served as a catalyst to me. Early on, I was aware of the fragility and importance of the polar regions, thanks in part to Elizabeth Kolbert's pioneering *New Yorker* stories on ice core drilling ("Ice Memory," January 7, 2002) and climate change ("The Climate of Man," April 25, 2005). I also followed Alexandra Witze's Greenland stories in *Nature* and Richard Kerr's in *Science*, as well as the climate-science reports—by Andy Revkin, Justin Gillis, and others—in *The New York Times*. Around 2007, I began writing about climate change at *The New York Times Magazine*. But I only began to think seriously of writing a book about the Arctic in June 2010, when I passed a newsstand in New York City with a terrific *National Geographic* cover story, by Tim Folger, entitled

"Greenland: Ground Zero for Global Warming." A few years later, in the summer of 2012, when I pored over newspaper and Web accounts of how the entire surface of Greenland had melted, it seemed clear to me that this faraway island could be a way to write about the urgency of global warming as well as the applications of technology, my two abiding interests at the time.

I didn't quite realize how much work lay ahead.

The existing body of scholarship on Greenland—on its ice, culture, politics, geography, geology, and climate—is considerable, and comprises the work of countless academics, historians, journalists, and memoirists. In this book's selected bibliography I've listed some of the many hundreds of articles and books I've drawn upon. A few deserve special mention here, especially for readers interested in delving deeper into areas that are only summarized in my treatment. Not only will they encounter a range of scientific and historical analysis that has informed my narrative; they'll enjoy some terrific writing as well.

The best place to begin is Fridtjof Nansen's gripping memoir of his Greenland traverse, *The First Crossing of Greenland*, which was published in two volumes and is available online in both an unexpurgated original version and a more condensed format. Robert Peary's tales of his Greenland years, *Northward Over the Great Ice*, as well as his wife Jo's memoir of this period, *My Arctic Journal*, are likewise crucial in explaining this early era of ice sheet exploration. For a more impartial perspective on the Arctic in the late nineteeth and early twentieth centuries, I benefited in particular from Pierre Berton's *The Arctic Grail*; Wally Herbert's *The Noose of Laurels*; and Lyle Dick's superb *Muskox Land*. Roland Huntford's comprehensive biography of Nansen was also of great value. And Michael Robinson's incisive book, *The Coldest Crucible*, illuminated the cultural, economic, and political goals behind many Arctic expeditions of the 1800s.

One of the pleasures in researching this book was the opportunity to read Knud Rasmussen and Peter Freuchen's stories about their travels, especially Rasmussen's *People of the Polar North* and Freuchen's early chronicle of his wayfaring life, *Arctic Adventure*. Both men have been the subject of biographies that remain untranslated from the Danish. A recent English-language biography of Rasmussen, however—*White Eskimo*, by Stephen Bown—recounts the explorer's life and times in vivid detail, and is necessary reading for those wishing to learn more about the Dane's ethnographic work. Similar language barriers surround the work of Alfred Wegener and Paul-Émile Victor; many of their significant writings, as well as several biographies, remain unavailable to English readers. Fortunately, Mott Greene's monumental recent treatment of Wegener's life and ideas (*Alfred Wegener: Science, Exploration, and the Theory of Continental Drift*) fills in almost all the gaps regarding the German scientist. In addition, the collection of articles written by Wegener's scientific team (*Greenland Journal*, edited by Wegener's widow, Else) presents first-person accounts that were essential to me as I tried to recreate the events of the tragic winter of 1930. Those who do not speak French but remain curious about Victor's life, meanwhile, will have to hunt through dusty bookstores or online listings for *My Eskimo Life*, an out-of-print volume in English that recounts his early travels in Greenland.

The evolution of scientific enquiry regarding the ice sheet is the subject of many impressive books, both old and new. The most comprehensive is Børge Fristrup's massive *The Greenland Ice-Cap*, which in text and photographs traces the chronology of exploration and research on Greenland's ice from the 1500s up to the early 1960s. Also valuable is Janet Martin-Nielsen's more recent *Eismitte in the Scientific Imagination*, which recounts the research goals of Wegener, Victor, and Carl Benson and offers keen insights into the politics and policy behind their forays in the Arctic. The Cold War relationships between science and the military are further delineated in *Exploring Greenland*, which includes some of the exhaustive Greenland research done by academic teams at Aarhus University in Denmark and Florida State University. The Cold War links to ice core science—along with the early discoveries of abrupt climate change—are also explicated in John D. Cox's deep and detailed *Climate Crash*. Meanwhile, for a larger perspective on how the science of ice cores and melting ice sheets became enmeshed in the emerging understanding of climate change, I relied on Spencer Weart's *The Discovery of Global Warming*, which remains the definitive text on the subject.

Finally, this book has been enriched by a number of first-person accounts by scientists who worked on the front lines of glaciology. Among the best are Willi Dansgaard's *Frozen Annals*, Richard Alley's *The Two-Mile Time Machine*, and Paul Mayewski's *The Ice Chronicles*. Chet Langway's booklet *The History of Early Polar Ice Cores* was also indispensible. I was fortunate to encounter many excellent first-person accounts by nonscientists, too—especially the work of Gretel Ehrlich (*This Cold Heaven*), Sarah Wheeler (*The Magnetic North*), and Barry Lopez (*Arctic Dreams*). Their dazzling impressions of the landscape and culture of Greenland offered an equally vital way to understand the stakes, and the potentially catastrophic losses, of a melting Arctic.

Interviews

Waleed Abdalati
Richard Alley
Kevin Anderson
Robin Bell
Carl Benson
Anders Bjørk
Jason Box
Wally Broecker
Joe Cook
Zoe Courville
Dorthe Dahl-Jensen
Ian Fenty
Joan Fitzpatrick
Rick Forster
Guenther Frankenstein
Anthony Gow
Geoff Hargreaves
David Holland
Ian Joughin
Rob Kopp
Austin Kovacs
William Krabill
Kristin Laidre
Chet Langway
Anders Levermann
Paul Mayewski
Mathieu Morlighem

Tom Neumann
Tad Pfeffer
Chris Polashenski
David Pollard
Asa Rennermalm
Eric Rignot
Alex Robinson
Piers Sellers
Malene Simon
John Sonntag
Koni Steffen
J. P. Steffenson
Byron Tapley
Kendrick Taylor
Marco Tedesco
Andrew Tedstone
Wayne Tobiasson
Herb Ueda
Isabella Velicogna
Tom Wagner
Mike Watkins
James White
Josh Willis
Kevin Wood
Herman Zimmerman
Jay Zwally

Oral Histories

Carl Benson (Byrd Polar Research Center/Ohio State University)
Charles Bentley (American Institutes of Physics)
Samuel Epstein (Oral History Project, California Institute of
Technology)
Anthony Gow (Byrd)
Berrien Moore (NASA, Johnson Space Center Oral History Project)

John Nye (British Library/Oral History of British Science)
Byron Tapley (NASA)
Robert H. "Bob" Thomas (American Institute of Physics)
Herb Ueda (Byrd)
James Van Allen (Byrd)
Wilford Weeks (Byrd)

SELECTED BIBLIOGRAPHY

"Birth of a Base." *Life*, September 1952.

"Captain Koch's Crossing of Greenland." *Bulletin of the American Geographical Society* 46, no. 5 (1914).

"Symposium on Remote Sensing in the Polar Regions." Arctic Institute of North America. March 6–8, 1968.

Albarella, Umberto, et al., eds. *The Oxford Handbook of Zooarchaeology*. New York: Oxford University Press, 2017.

Alley, Richard B. *The Two-Mile Time Machine: Ice Cores, Abrupt Climate Change, and Our Future*. Princeton, N.J.: Princeton University Press, 2000.

———. "The Biggest Control Knob: Carbon Dioxide in Earth's Climate History." American Geophysical Union Conference, San Francisco, 2009. youtube.com/watch?v=RffPSrRpq_g.

———. *Earth: The Operators' Manual*. New York: W. W. Norton & Company, 2011.

Alley, Richard B., et al. "Visual Stratigraphic Dating of the GISP2 Ice Core: Basis, Reproducibility, and Application." *Journal of Geophysical Research* 102, no. C12 (November 1997).

Anderson, Alun. *After the Ice: Life, Death, and Geopolitics in the New Arctic*. New York: HarperCollins Publishers, 2009.

Archer, David. *The Long Thaw: How Humans Are Changing the Next 100,000 Years of Earth's Climate*. Princeton, N.J.: Princeton University Press, 2009.

Archer, David, and Raymond Pierrehumbert, eds. *The Warming Papers: The Scientific Foundation for the Climate Change Forecast*. West Sussex, U.K.: Wiley-Blackwell, 2011.

Astrup, Eivind. *With Peary Near the Pole*. Philadelphia: J. B. Lippincott Company, 1899.

Bader, Henri. "United States Polar Ice and Snow Studies in the International Geophysical Year." Proceedings of the Symposium at the Opening of the International Geophysical Year, Washington, D.C., June 28–29, 1957.

———. "The Physics and Mechanics of Snow as a Material." Hanover, N.H.: CRREL, July 1962.

———. "Scope, Problems, and Potential Value of Deep Core Drilling in Ice Sheets." Hanover, N.H.: CRREL, December 1962.

Bader, Henri, et al. "Excavations and Installations at SIPRE Test Site, Site 2, Greenland." SIPRE Report 20 (April 1955).

Bader, Henri, et al. Polar Research: A Survey. Washington, D.C.: National Academy of Sciences, 1970.

Bain, J. Arthur. Life and Explorations of Fridtjof Nansen. London: Walter Scott, Ltd., 1897.

Balchen, Bernt. Come North with Me: An Autobiography. New York: E. P. Dutton & Co., 1958.

Balchen, Bernt, et al. War Below Zero: The Battle for Greenland. Boston: Houghton Mifflin Company, 1944.

Banerjee, Neela, et al. "Exxon: The Road Not Taken." Inside Climate News, September 16–December 22, 2015.

Belanger, Dian Olson. Deep Freeze: The United States, the International Geophysical Year, and the Origins of Antarctica's Age of Science. Boulder: University Press of Colorado, 2006.

Benson, Carl. "Stratigraphic Studies in the Snow and Firn of the Greenland Ice Sheet." SIPRE Report 70 (July 1962).

Bentley, Charles R., and Bruce R. Koci. "Drilling to the Beds of the Greenland and Antarctic Ice Sheets: A Review." Annals of Glaciology 47 (2007).

Berton, Pierre. The Arctic Grail: The Quest for the North West Passage and the North Pole, 1818–1909. New York: Viking Penguin, 1988.

Beukel, Erik, et al. Phasing Out the Colonial Status of Greenland, 1945–54: A Historical Study. Copenhagen: Museum Tusculanum Press, 2010.

Birket-Smith, Kaj. "Knud Rasmussen." Journal de la Société des Américanistes 25, no. 2 (1933).

———. Knud Rasmussen's Saga. Copenhagen: Chr. Erichsens Forlag, 1936.

———. The Eskimos. New York: E. P. Dutton and Company, 1936.

Bloom, Lisa. *Gender on Ice: American Ideologies of Polar Expeditions*. Minneapolis: University of Minnesota Press, 1993.

Bobé, Louis. *Hans Egede: Colonizer and Missionary of Greenland*. Copenhagen: Rosenkilde and Bagger, 1952.

Born, Erik W., et al. *Polar Bears in Northwest Greenland: An Interview Survey About the Catch and the Climate*. Copenhagen: Museum Tusculanum Press, 2011.

Bouché, Michel. *Groenland: Station Centrale*. Paris: Bernard Grasset Editeur, 1952.

Bown, Stephen R. *The Last Viking: The Life of Roald Amundsen*. Boston: Da Capo Press, 2012.

———. *White Eskimo: Knud Rasmussen's Fearless Journey into the Heart of the Arctic*. Boston: Da Capo Press, 2015.

Bradley, Raymond S. *Paleoclimatology: Reconstructing Climates of the Quaternary*. 3rd ed. Amherst, Mass.: Elsevier, 2015.

Broecker, Wallace S., and Robert Kunzig. *Fixing Climate: What Past Climate Changes Reveal About the Current Threat—and How to Counter It*. New York: Hill and Wang, 2008.

Brogger, W. C., and Nordahl Rolfsen. *Nansen in the Frozen World*. Philadelphia: A. J. Holman & Co., 1897.

Brown, Andrew H. "Americans Stand Guard in Greenland." *National Geographic*, October 1946.

Budyko, Mikhail. "The Effect of Solar Radiation Variations on the Climate of the Earth." *Tellus* 21 (1969).

Byers, Michael. *Who Owns the Arctic? Understanding Sovereignty Disputes in the North*. Vancouver: Douglas & McIntyre, 2009.

Caswell, John Edwards. *Arctic Frontiers: United States Explorations in the Far North*. Norman: University of Oklahoma Press, 1956.

Childs, Craig. *Apocalyptic Planet: Field Guide to the Everending Earth*. New York: Pantheon Books, 2012.

Clark, Elmer F. "Camp Century Evolution of Concept and History of Design Construction and Performance." CRREL Technical Report 174 (October 1965).

Climate Change: Evidence & Causes (2014). Published by the Royal Society and the U.S. National Academy of Sciences.

Colgan, William, et al. "The Abandoned Ice Sheet Base at Camp Century,

Greenland, in a Warming Climate." *Geophysical Research Letters* 43 (August 16, 2016).

Conkling, Philip, et al. *The Fate of Greenland: Lessons from Abrupt Climate Change*. Cambridge, Mass.: The MIT Press, 2011.

Counter, S. Allen. *North Pole Legacy: Black, White, and Eskimo*. Amherst: University of Massachusetts Press, 1991.

Cox, John D. *Climate Crash: Abrupt Climate Change and What It Means for Our Future*. Washington, D.C.: Joseph Henry Press, 2005.

Cronin, Thomas M. *Principles of Paleoclimatology*. New York: Columbia University Press, 1999.

Cuffey, K. M., and W.S.B. Paterson. *The Physics of Glaciers*. 4th ed. Burlington, Mass.: Elsevier, 2010.

Cullen, Heidi. *The Weather of the Future: Heat Waves, Extreme Storms, and Other Scenes from a Climate-Changed Planet*. New York: Harper-Collins, 2010.

Dansgaard, Willi. *Frozen Annals: Greenland Ice Sheet Research*. Copenhagen: Niels Bohr Institute, 2005.

Dansgaard, Willi, et al. "One Thousand Centuries of Climatic Record from Camp Century on the Greenland Ice Sheet." *Science* 166 (October 17, 1969).

Daugherty, Charles Michael. *City Under the Ice: The Story of Camp Century*. New York: The Macmillan Company, 1963.

Diamond, Jared. *Collapse: How Societies Choose to Fail or Succeed*. New York: Viking Penguin, 2005.

Dick, Lyle. *Muskox Land: Ellesmere Island in the Age of Contact*. Calgary: University of Calgary Press, 2001.

Diebitsch-Peary, Josephine, with Robert E. Peary. *My Arctic Journal: A Year Among Ice-Fields and Eskimos*. New York: The Contemporary Publishing Company, 1894.

Dillon, Paddy. *Trekking in Greenland: The Arctic Circle Trail*. Milnthorpe, Cumbria: Cicerone, 2010.

Doel, Ronald E., et al., eds. *Exploring Greenland: Cold War Science and Technology on Ice*. New York: Palgrave Macmillan, 2016.

Douglass, Andrew Ellicott. "The Secret of the Southwest Solved by Talkative Tree Rings: Horizons of American History Are Carried Back to

A.D. 700 and a Calendar for 1,200 Years Established by National Geographic Expeditions." *National Geographic*, December 1929.

Dugmore, Andrew J., et al. "Norse Greenland Settlement: Reflections on Climate Change, Trade, and the Contrasting Fates of Human Settlements in the North Atlantic Islands." *Arctic Anthropology* 44, no. 1 (2007).

Egede, Hans. *A Description of Greenland*. London: T. and J. Allman, 1818.

Ehrlich, Gretel. *This Cold Heaven: Seven Seasons in Greenland*. New York: Vintage Books, 2003.

Emmerson, Charles. *The Future History of the Arctic*. New York: Public Affairs, 2010.

Finsterwalder, Richard. "Expédition Glaciologique Internationale au Groenland, 1957–60." *Journal of Glaciology* 3, no. 26 (1959).

Flannery, Tim. *The Weather Makers: How Man Is Changing the Climate and What It Means for Life on Earth*. New York: Grove Press, 2005.

Flint, Richard F. "Snow, Ice and Permafrost in Military Operations." SIPRE Rept. 15 (September 1953).

Folger, Tim. "Viking Weather." *National Geographic*, June 2010.

———. "Why Did Greenland's Vikings Vanish?" *Smithsonian*, March 2017.

Freuchen, Dagmar, and Peter Freuchen. *Peter Freuchen's Adventures in the Arctic*. New York: Julian Messner, Inc., 1960.

Freuchen, Peter. *Arctic Adventure: My Life in the Frozen North*. New York: Farrar & Rinehart, 1935.

———. *Vagrant Viking: My Life and Adventures*. Translated by Johan Hambro. New York: Julian Messner, 1953.

———. *I Sailed with Rasmussen*. New York: Julian Messner, 1958.

———. *Men of the Frozen North*. Cleveland, Oh.: The World Publishing Company, 1962.

———. *The Peter Freuchen Reader*. New York: Julian Messner, 1965.

Freuchen, Peter, and Finn Salomonsen. *The Arctic Year*. New York: G. P. Putnam's Sons, 1958.

Friesen, T. Max, and Owen K. Mason, eds. *The Oxford Handbook of the Prehistoric Arctic*. New York: Oxford University Press, 2016.

Fristrup, Børge. *The Greenland Ice Cap*. Seattle: University of Washington Press, 1967.

Gad, Finn. *The History of Greenland I: Earliest Times to 1700*. Translated by Ernst Dupont. Montreal: McGill-Queen's University Press, 1971.

Georgi, Johannes. *Mid-Ice: The Story of the Wegener Expedition to Greenland*. Translated by F. H. Lyon. London: Kegan Paul, Trench, Trubner & Co., 1934.

Gilberg, Rolf. "Inughuit, Knud Rasmussen, and Thule: The Work of Knud Rasmussen Among the Polar-Eskimos in North Greenland." *Inuit Studies* 12 (1988).

Glines, Carroll V. *Bernt Balchen: Polar Aviator*. Washington, D.C.: Smithsonian Institution Press, 1999.

Goodell, Jeff. *The Water Will Come: Rising Seas, Sinking Cities, and the Remaking of the Civilized World*. New York: Little, Brown and Company, 2017.

Green, Fitzhugh. *Peary: The Man Who Refused to Fail*. New York and London: G. P. Putnam's Sons, 1923.

Greene, Mott T. *Alfred Wegener: Science, Exploration, and the Theory of Continental Drift*. Baltimore: Johns Hopkins University Press, 2015.

The Greenland Ice Sheet: 80 Years of Climate Change Seen from the Air. Copenhagen: Natural History Museum of Denmark, 2014.

Grotzinger, John, et al. *Understanding Earth*. 5th ed. New York: W. H. Freeman and Company, 2007.

Gundestrup, N. S., et al. "ISTUK: A Deep Ice Core Drill System." CRREL Special Report 83–34 (December 1984).

Hansen, James. *Storms of My Grandchildren*. New York: Bloomsbury, 2009.

Harper, Kenn. *Give Me My Father's Body: The Life of Minik, the New York Eskimo*. South Royalton, Vt.: Steerforth Press, 2000.

Hayes, J. Gordon. *Robert Edwin Peary: A Record of His Explorations, 1886–1909*. London: Grant Richards & Humphrey Toulmin, 1930.

Henderson, Bruce. *True North: Peary, Cook, and the Race to the Pole*. New York: W. W. Norton and Company, 2005.

Henriksen, Niels. *Geological History of Greenland: Four Billion Years of Earth Evolution*. Denmark: Geological Survey of Denmark and Greenland (GEUS), 2008.

Henson, Matthew. *A Negro Explorer at the North Pole*. New York: Frederick A. Stokes Company, 1912.

Herbert, Wally. *The Noose of Laurels: Robert E. Peary and the Race to the North Pole*. New York: Atheneum, 1989.

Heymann, Mattias, et al. "Exploring Greenland: Science and Technology in Cold War Settings." *Scientia Canadensis* 33, no. 2 (2010).

Hobbs, William Herbert. *An Explorer-Scientist's Pilgrimage: The Autobiography of William Herbert Hobbs*. Ann Arbor, Mich.: J. W. Edwards, Inc., 1952.

———. *Peary*. New York: The Macmillan Company, 1936.

Holland, David M., et al. "Acceleration of Jakobshavn Isbrae Triggered by Warm Subsurface Ocean Waters." *Nature Geoscience* 1 (October 2008).

Huntford, Roland. *Nansen*. London: Abacus, 2001.

Imbrie, John, and Katherine Palmer Imbrie. *Ice Ages: Solving the Mystery*. Cambridge, Mass.: Harvard University Press, 1979.

Irmscher, Christoph. *Louis Agassiz: Creator of American Science*. Boston: Houghton Mifflin Harcourt, 2013.

Jensen, Johannes V., et al. *Bogen om Knud: Skrevet af hans Venner*. Copenhagen: Westermann, 1943.

Jones, Gwyn. *The Norse Atlantic Saga: Being the Norse Voyages of Discovery and Settlement to Iceland, Greenland, and North America*. New York: Oxford University Press, 1986.

Joughin, I., et al. "Brief Communication: Further Summer Speedup of Jakobshavn Isbrae." *The Cryosphere* 8 (2014).

Jouzel, Jean. "A Brief History of Ice Core Science Over the Last 50 Years." *Climate of the Past*, November 2013.

Jouzel, Jean, et al. *The White Planet: The Evolution and Future of Our Frozen World*. Translated by Teresa L. Fagan. Princeton, N.J.: Princeton University Press, 2013. First published in 2008 by Odile Jacob.

Kane, Elisha Kent. *Arctic Explorations in the Years 1853, '54, '55*. Hartford, Conn.: American Publishing Company, 1881.

———. *Arctic Explorations: The Second and Last United States Grinnell Expedition in Search of Sir John Franklin*. Hartford, Conn.: American Publishing Company, 1881.

Keely, Robert N., and G. G. Davis. *In Arctic Seas: The Voyage of the "Kite" with the Peary Expedition*. Philadelphia: The Thompson Publishing Co., 1883.

Kent, Rockwell. *Greenland Journal*. New York: Ivan Obolensky, Inc., 1962.

Kintisch, Eli. "Why Did Greenland's Vikings Disappear?" *Science* 354 (November 2016).

Kirwan, L. P. *A History of Polar Exploration*. New York: W. W. Norton & Company, 1959.

Kleivan, Inge, and Ernest S. Burch, Jr. "The Work of Knud Rasmussen Among the Inuit: An Introduction." *Inuit Studies* 12 (1988).

Knight, Clayton, and Robert C. Durham. *Hitch Your Wagon: The Story of Bernt Balchen*. Drexel Hill, Penn.: Bell Publishing Company, 1950.

Knuth, Eigil. *Fridtjof Nansen og Knud Rasmussen*. Copenhagen: Gyldendal, 1948.

Koch, J. P. *Durch die weisse Wüste*. Berlin: Verlag von Julius Springer, 1919.

Kolbert, Elizabeth, and Francis Spufford, eds. *The Ends of the Earth: An Anthology of the Finest Writing on the Arctic and Antarctic*. New York: Bloomsbury USA, 2007.

Kolbert, Elizabeth. *Field Notes from a Catastrophe: Man, Nature, and Climate Change*. New York: Bloomsbury USA, 2006.

Kovacs, Austin. "Camp Century Revisited: A Pictorial View—June 1969." CRREL Special Report 150 (July 1970).

Kpomassie, Tété-Michel. *An African in Greenland*. Translated by James Kirkup. New York: Horace Brace Jovanovich, 1983. First published in 1981 by Flammarion.

Krabill, W., et al. "Rapid Thinning of Parts of the Southern Greenland Ice Sheet." *Science* 283 (March 1999).

Krabill, W., et al. "Greenland Ice Sheet: High-Elevation Balance and Peripheral Thinning." *Science* 289 (July 2000).

Langway, Chester C. "A 400 Meter Deep Ice Core in Greenland." *Journal of Glaciology* 3 (March 1958).

———. "The History of Early Polar Ice Cores." Hanover, N.H.: CRREL, January 2008.

Langway, Chester C., et al., eds. *Greenland Ice Core: Geophysics, Geochemistry, and the Environment*. Washington, D.C.: American Geophysical Union, 1985.

Lansing, Alfred. *Endurance: Shackleton's Incredible Voyage*. New York: Basic Books, 2007.

Launius, Roger D., et al., eds. *Globalizing Polar Science: Reconsidering the In-*

ternational Polar and Geophysical Years. New York: Palgrave Macmillan, 2010.

Loewe, Fritz. "The Scientific Exploration of Greenland from the Norsemen to the Present." Public Lecture, Ohio State University, May 26, 1969.

Loomis, Chauncey C. Weird and Tragic Shores: The Story of Charles Francis Hall, Explorer. New York: Alfred A. Knopf, 1971.

Lopez, Barry. Arctic Dreams: Imagination and Desire in a Northern Landscape. New York: Charles Scribner's Sons, 1986.

Loukacheva, Natalia. Arctic Promise: Legal and Political Autonomy of Greenland and Nunavut. Toronto: University of Toronto Press, 2007.

Lurie, Edward. Louis Agassiz: A Life in Science. Baltimore: The Johns Hopkins University Press, 1988. First published in 1960 by the University of Chicago Press.

MacMillan, Donald. Four Years in the White North. New York: The Medici Society of America, 1925.

Malaurie, Jean. The Last Kings of Thule: With the Polar Eskimos, as They Face Their Destiny. Translated by Adrienne Foulke. New York: E. P. Dutton, Inc., 1982. First published in 1976 in France.

———. Ultima Thule: Explorers and Natives in the Polar North. Translated by Willard Wood and Anthony Roberts. New York: W. W. Norton & Company, 2003.

Martin-Nielsen, Janet. "'The Deepest and Most Rewarding Hole Ever Drilled': Ice Cores and the Cold War in Greenland." Annals of Science, September 2012.

———. Eismitte in the Scientific Imagination: Knowledge and Politics at the Center of Greenland. New York: Palgrave Macmillan, 2013.

Maxtone-Graham, John. Safe Return Doubtful: The Heroic Age of Polar Exploration. New York: Charles Scribner's Sons, 1988.

Mayewski, Paul Andrew, and Frank White. The Ice Chronicles: The Quest to Understand Global Climate Change. Hanover, N.H.: University Press of New England, 2002.

McCannon, John. A History of the Arctic. London: Reaktion Books, 2012.

McCoy, Roger M. Ending in Ice: The Revolutionary Idea and Tragic Expedition of Alfred Wegener. New York: Oxford University Press, 2006.

McCullough, David. The Path Between the Seas. New York: Simon & Schuster, 1977.

McGhee, Robert. *The Last Imaginary Place: A Human History of the Arctic World*. New York: Oxford University Press, 2005.

McKibben, Bill. *Eaarth: Making a Life on a Tough New Planet*. New York: St. Martin's Griffin, 2010.

McKibben, Bill, ed. *The Global Warming Reader: A Century of Writing About Climate Change*. New York: Penguin Books, 2012. First published in 2011 by OR Books.

Mellor, Malcolm. "Undersnow Structures: N-34 Radar Station, Greenland." CRREL Technical Report 132 (August 1964).

Menocal, Aniceto Garcia. *Report of the U.S. Nicaragua Surveying Party* (1885). Washington, D.C.: Government Printing Office, 1886.

Mikkelsen, Ejnar. *Lost in the Arctic*. New York: George H. Doran Company, 1913.

Mirsky, Jeannette. *To the Arctic! The Story of Northern Exploration from Earliest Times to the Present*. New York: Alfred A. Knopf, 1948.

Moore, John C., et al. "Geoengineer Polar Glaciers to Slow Sea-Level Rise." *Nature* 555 (March 14, 2018).

Morlighem, M., et al. "Deeply Incised Submarine Glacial Valleys Beneath the Greenland Ice Sheet." *Nature Geoscience* 7 (May 2014).

Morlighem, M., et al. "BedMachine v3: Complete Bed Topography and Ocean Bathymetry Mapping of Greenland from Multibeam Echo Sounding Combined with Mass Conservation." *Geophysical Research Letters* 44 (September 2017).

Murphy, David Thomas. *German Exploration of the Polar World: A History, 1870–1940*. Lincoln: University of Nebraska Press, 2002.

Mylius-Erichsen, Ludvig. *Grønland*. Copenhagen: Gyldendal, 1906.

Nansen, Fridtjof. *Farthest North: The Epic Adventure of a Visionary Explorer*. New York: Skyhorse Publishing, Inc., 2008. First published in 1897 by Harper & Brothers.

———. *The First Crossing of Greenland*. Translated by Hubert Majendie Gepp. 2 vols. London: Longmans, Green, and Co., 1890–1919.

———. *In Northern Mists: Arctic Exploration in Early Times*. London: William Heinemann, 1911.

National Research Council. *Energy and Climate: Studies in Geophysics*. Washington, D.C.: The National Academies Press, 1977.

―――. *Recommendations for a U.S. Ice Coring Program*. Washington, D.C.: The National Academies Press, 1986.

―――. *Seasonal to Decadal Predictions of Arctic Sea Ice: Challenges and Strategies*. Washington, D.C.: The National Academies Press, 2012.

―――. *Abrupt Impacts of Climate Change: Anticipating Surprises*. Washington, D.C.: The National Academies Press, 2013.

―――. *The Arctic in the Anthropocene: Emerging Research Questions*. Washington, D.C.: The National Academies Press, 2014.

―――. *Arctic Matters: The Global Connection to Changes in the Arctic*. Washington, D.C.: The National Academies Press, 2015.

Nickerson, Sheila. *Harnessed to the Pole: Sledge Dogs in Service to American Explorers of the Arctic, 1853–1909*. Fairbanks: University of Alaska Press, 2014.

Nordhaus, William. *The Climate Casino: Risk, Uncertainty, and Economics for a Warming World*. New Haven, Conn.: Yale University Press, 2013.

Oceanography 29, no. 4 (December 2016). Special Issue on Ocean-Ice Interaction. Published by the Oceanography Society.

Officer, Charles, and Jake Page. *A Fabulous Kingdom: The Exploration of the Arctic*. 2nd ed. New York: Oxford University Press, 2012.

Oreskes, Naomi. *The Rejection of Continental Drift. Theory and Method in American Earth Science*. New York: Oxford University Press, 1999.

Peary, Robert. *Northward Over the "Great Ice": A Narrative of Life and Work Along the Shores and Upon the Interior Ice-Cap of Northern Greenland in the Years 1886 and 1891–1897*. 2 vols. London: Methuen & Co., 1898.

―――. *Secrets of Polar Travel*. New York: The Century Co., 1917.

Péroz, Francis. *L'exploration polaire française: une épopée humaine*. Pontarlier, France: Éditions du Belvédère, 2015.

Petersen, Nikolaj. "SAC at Thule: Greenland in the U.S. Polar Strategy." *Journal of Cold War Studies* 13, no. 2 (Spring 2011).

Petterson, Hans. "Exploring the Bed of the Ocean." *Nature*, September 1949.

Pfeffer, W. Tad. "Glaciology Needs to Come Out of the Ivory Tower." *Earth* 57, no. 11 (2012).

Pfeffer, W. Tad. *The Opening of a New Landscape: Columbia Glacier at Mid-Retreat*. Washington, D.C.: AGU Books, 2007.

Piggott, Michael. "On First Looking into Loewe's Papers." *University of Melbourne Collections*, no. 3 (December 2008).

Plass, Gilbert N. "Carbon Dioxide and Climate." *Scientific American*, July 1959.

Polashenski, Chris, et al. "Observations of Pronounced Greenland Ice Sheet Firn Warming and Implications for Runoff Production." *Geophysical Research Letters* 41 (June 2014).

Pollack, Henry. *A World Without Ice*. New York: Penguin Group, 2010.

Poulsen, Hans Holt, et al. *Nomination of Aasivissuit-Nipisat, Inuit Hunting Ground Between Ice and Sea*. Sisimiut: Qeqqata Municipality, 2017.

Randers, Jorgen. *2052: A Global Forecast for the Next Forty Years*. White River Junction, Vt.: Chelsea Green Publishing, 2012.

Rasmussen, Knud. *The People of the Polar North: A Record*. London: Kegan Paul, Trench, Trübner & Co., 1908.

———. "The Second Thule Expedition to Northern Greenland, 1916–1918." *Geographical Review* 8, no. 2 (August 1919).

———. *Greenland by the Polar Sea: The Story of the Thule Expedition from Melville Bay to Cape Morris Jesup*. Translated by Asta and Rowland Kenney. London: William Heinemann, 1921.

———. *Across Arctic America: Narrative of the Fifth Thule Expedition*. New York: Greenwood Press, 1969.

Reynolds, E. E. *Nansen*. London: Geoffrey Bles, 1932.

Rich, Edwin Gile. *Hans the Eskimo: His Story of Arctic Adventure with Kane, Hayes, and Hall*. Boston and New York: Houghton Mifflin Company, 1934.

Riffenburgh, Beau. *The Myth of the Explorer: The Press, Sensationalism, and Geographical Discovery*. London: Belhaven Press, 1993.

Rignot, Eric, and Robert H. Thomas. "Mass Balance of Polar Ice Sheets." *Science* 297 (August 30, 2002).

Rink, Hinrich. *Danish Greenland: Its People and Its Products*. London: Henry S. King & Co., 1877.

Robeson, G. M., ed. *Narrative of the North Polar Expedition, U.S. Ship Polaris, Captain Charles Francis Hall Commanding*. Washington, D.C.: GPO, 1876.

Robin, Gordon de Quetteville, and Charles Swithinbank. "Fifty Years of Progress in Understanding Ice Sheets." *Journal of Glaciology*, special issue (1987).

Robinson, Bradley, and Matthew Henson. *Dark Companion: The Story of Matthew Henson*. Robert M. McBride & Company, 1947.

Robinson, Michael F. *The Coldest Crucible: Arctic Exploration and American Culture*. Chicago: University of Chicago Press, 2006.

Ross, John. *The Forecast for D-Day: And the Weatherman Behind Ike's Greatest Gamble*. Guilford, Conn.: Lyons Press, 2014.

Rozell, Ned. "Tracks Across Greenland Ice, 60 Years Apart." Geophysical Institute, University of Alaska–Fairbanks, July 2014.

Sagan, Scott D. *The Limits of Safety: Organizations, Accidents, and Nuclear Weapons*. Princeton, N.J.: Princeton University Press, 1993.

Schiermeier, Quirin. "180,000 Forgotten Photos Reveal the Future of Greenland's Ice." *Nature* 535 (July 27, 2016).

Scholander, P. F. *Enjoying a Life in Science: The Autobiography of P. F. Scholander*. Fairbanks: University of Alaska Press, 1990.

Schwarzbach, Martin. *Alfred Wegener: The Father of Continental Drift*. Madison, Wis.: Science Tech, Inc., 1986. First published in 1980 by Wissenschaftliche Verlagsgesellschaft.

Scoresby, William. *An Account of the Arctic Regions with a History and Description of the Northern Whale-Fishery*. Edinburgh: Archibald Constable and Co., 1820.

Sejersen, Frank. *Rethinking Greenland and the Arctic in the Era of Climate Change: New Northern Horizons*. New York: Routledge, 2015.

Sides, Hampton. *In the Kingdom of Ice: The Grand and Terrible Polar Voyage of the USS* Jeannette. New York: Anchor Books, 2015.

Smith, Laurence C. *The World in 2050: Four Forces Shaping Civilization's Northern Future*. New York: Dutton, 2010.

Sörensen, Jon. *The Saga of Fridtjof Nansen*. Translated by J.B.C. Watkins. New York: W. W. Norton & Company, Inc., 1932.

Sorge, Ernst. *With Plane, Boat, & Camera in Greenland: An Account of the Universal Dr. Fanck Greenland Expedition*. London: Hurst & Blackett, Ltd., 1935.

Stafford, Marie Peary. *Discoverer of the North Pole: The Story of Robert E. Peary*. New York: William Morrow & Company, 1959.

Stefansson, Vilhjalmur. *Greenland*. New York: Doubleday, Doran & Company, Inc., 1942.

———. *The Fat of the Land*. New York: The Macmillan Company, 1960.

Supran, Geoffrey, and Naomi Oreskes. "Assessing ExxonMobil's Climate Change Communications (1977–2014)." *Environmental Research Letters* 12, no. 8 (August 2017).

Sweet, William. *Kicking the Carbon Habit: Global Warming and the Case for Renewable and Nuclear Energy*. New York: Columbia University Press, 2006.

Taylor, Kendrick. "Rapid Climate Change: New Evidence Shows that Earth's Climate Can Change Dramatically in Only a Decade. Could Greenhouse Gases Flip that Switch?" *American Scientist* 87 (July–August 1999).

The Terrestrial Environment: Solid-Earth and Ocean Physics. Washington, D.C.: National Aeronautics and Space Administration, April 1970.

Thomas, R. H., et al. "Satellite Remote Sensing for Ice Sheet Research." NASA Technical Memorandum 86233 (1985).

Ueda, Herbert T., and Donald E. Garfield. "Drilling Through the Greenland Ice Sheet." CRREL Special Report 126 (November 1968).

Utne, Eric, ed. *Brenda My Darling: The Love Letters of Fridtjof Nansen to Brenda Ueland*. Minneapolis: UTNE Institute, 2011.

Velicogna, Isabella. "Increasing Rates of Ice Mass Loss from the Greenland and Antarctic Ice Sheets Revealed by GRACE." *Geophysical Research Letters* 36 (October 2009).

Velicogna, Isabella, and John Wahr. "Acceleration of Greenland Ice Mass Loss in Spring 2004." *Nature* 443 (September 21, 2006).

———. "Measurements of Time-Variable Gravity Show Mass Loss in Antarctica." *Science* 311 (March 24, 2006).

Victor, Daphné, and Stéphane Dugast. *Paul-Émile Victor: J'ai toujours vécu demain*. Paris: Robert Laffont, 2015.

Victor, Paul-Émile. *My Eskimo Life*. Translated by Jocelyn Godefroi. New York: Simon and Schuster, 1939.

———. "The French Expedition to Greenland, 1948." *Arctic* 2, no. 3 (December 1949).

———. *The French Polar Research Expeditions 1948–1951*. New York: The French Embassy Press and Information Division, 1950.

————. "Wringing Secrets from Greenland's Icecap." *National Geographic*, January 1956.

————. *Man and the Conquest of the Poles*. Translated by Scott Sullivan. London: Hamish Hamilton, 1964.

————. "Wegener." *Polarforschung* 40 (1970).

The Vinland Sagas. Translated by Keneva Kunz. London: Penguin Classics, 2008.

Voss, Jutta. "Johannes Georgi und Fritz Loewe, Zwei Polarforscherschicksale nach 'Eismitte' aus ihrem Briefweschel 1929–1971, sowie die gesammelten Schriftenverzeichnisse von J. Georgi und F. Loewe." *Polarforschung* 62 (1994).

Wager, Walter. *Camp Century: City Under the Ice*. Philadelphia: Chilton Books, 1962.

Walker, Gabrielle. *Antarctica: An Intimate Portrait of a Mysterious Continent*. New York: Houghton Mifflin Harcourt, 2013.

Ward, Peter D. *The Flooded Earth: Our Future in a World Without Ice Caps*. New York: Basic Books, 2010.

Weart, Spencer R. *The Discovery of Global Warming*. 2nd ed. Cambridge, Mass.: Harvard University Press, 2008.

Weems, John Edward. *Peary: The Explorer and the Man*. London: Eyre & Spottiswoode, 1967.

Wegener, Alfred. *Greenland Journal 1906–1908*. environmentandsociety.org/exhibitions/wegener-diaries/expedition1.

————. *Greenland Journal, 1912–1913*. environmentandsociety.org/exhibitions/wegener-diaries/expedition2.

————. *Mit Motorboot und Schlitten in Grönland*. Bielefeld and Leipzig: Verlag von Velhagen & Klasing, 1930.

Wegener, Else, ed. *Greenland Journey: The Story of Wegener's German Expedition to Greenland in 1930–31 as Told by Members of the Expedition and the Leader's Diary*. London: Blackie & Son Limited, 1939.

Weidick, Anker, and Ole Bennike. "Quaternary Glaciation History and Glaciology of Jakobshavn Isbrae and the Disko Bugt Region, West Greenland: A Review." Geological Survey of Denmark and Greenland, Bulletin 14 (2007).

Weiss, Erik D. "Cold War Under the Ice: The Army's Bid for a Long-Range

Nuclear Role, 1959–1963." *Journal of Cold War Studies* 3, no. 3 (Fall 2001).

Wheeler, Sara. *The Magnetic North: Notes from the Arctic Circle*. New York: Farrar, Straus and Giroux, 2009.

Williams, David B. *Cairns: Messengers in Stone*. Seattle: Mountaineers Books, 2012.

Witze, Alexandra. "Losing Greenland: Is the Arctic's Biggest Ice Sheet in Irreversible Meltdown? And Would We Know If It Were?" *Nature* 452 (April 17, 2008).

Woodward, Jamie. *The Ice Age: A Very Short Introduction*. Oxford, U.K.: Oxford University Press, 2014.

Wright, Edmund A. *CRREL's First 25 Years: 1961–1986*. Hanover, N.H.: CRREL, 1986.

Zuckoff, Mitchell. *Frozen in Time: An Epic Story of Survival and a Modern Quest for Lost Heroes of World War II*. New York: Harper Perennial, 2014.

INDEX

Page numbers in *italics* refer to illustrations and maps.

ABOUT THE AUTHOR

Jon Gertner is the author of the bestselling book *The Idea Factory: Bell Labs and the Great Age of American Innovation* and a longtime contributor to *The New York Times Magazine*. He lives in New Jersey with his wife and children.

jongertner.net
Twitter: @jongertner

ABOUT THE TYPE

This book was set in Goudy Old Style, a typeface designed by Frederic William Goudy (1865–1947). Goudy began his career as a bookkeeper, but devoted the rest of his life to the pursuit of "recognized quality" in a printing type.

Goudy Old Style was produced in 1914 and was an instant bestseller for the foundry. It has generous curves and smooth, even color. It is regarded as one of Goudy's finest achievements.